Food for the Few

Food for the Few

Neoliberal Globalism and Biotechnology
in Latin America

EDITED BY GERARDO OTERO

University of Texas Press *Austin*

Requests for permission to reproduce material from this work should be sent to:
    Permissions
    University of Texas Press
    P.O. Box 7819
    Austin, TX 78713-7819
    www.utexas.edu/utpress/about/bpermission.html

♾ The paper used in this book meets the minimum requirements of ANSI/NISO
z39.48-1992 (R1997) (Permanence of Paper).

Library of Congress Cataloging-in-Publication Data

    Food for the few : neoliberal globalism and biotechnology in Latin America /
edited by Gerardo Otero. — 1st ed.
        p.   cm.
    Includes bibliographical references and index.
    ISBN 978-0-292-72613-0
    1. Agriculture—Economic aspects—Latin America.   2. Agricultural
biotechnology—Latin America.   3. Farmers—Latin America.   4. Produce
trade—Latin America.   5. Agriculture—Social aspects—Latin America.   I.
Otero, Gerardo.
HD1790.5.F66 2008
338.1098—dc22

                                                            2007043301

In memory of Fred Buttel

Contents

Acknowledgments

Throughout this project, I benefited from a Standard Grant from the Social Sciences and the Humanities Research Council of Canada, which I thankfully acknowledge. Thanks to Yolanda Massieu for suggesting the first part of the book's title. She also contributed by assessing some of the initial chapter proposals for this volume. Manuel Poitras offered invaluable help in the initial stages of the editorial process, including the review of several proposed chapters. Special thanks are due to Gabriela Pechlaner, my doctoral student since 2002. During this time, not only did she make impressive progress through her program of study, completing a wonderful dissertation in May of 2007, but she was also my research assistant and main interlocutor on this project. As a research assistant, she provided such talented help that she became a coauthor of three of the chapters that follow.

My greatest debt is of course to my contributors, for patiently revising their chapters more than once, first at my request, and then sometimes at the request of external reviewers for the University of Texas Press. These anonymous reviewers, particularly the first one, offered excellent suggestions that helped us to strengthen the book. I deeply appreciate their constructive critiques and positive readings.

This book is dedicated to the memory of Fred Buttel, pioneer in the social study of biotechnology's impacts. He and I engaged in a friendly debate over the character of biotechnology in the early 1990s,[1] but the fact remains that his work was greatly influential and inspirational for my own. Furthermore, Fred was one of my most supportive and generous mentors through the middle stages of my professional career. His prolific and crucial contributions to the sociology of agriculture and food, to the sociology of science and technology, and to environmental sociology speak for themselves. I can just hope that our volume is a modest tribute to the life and work of such an exemplary scholar, mentor, and friend.

Gerardo Otero
North Vancouver
January 2007

Note

1. Frederick H. Buttel, "How Epoch Making Are High Technologies? The Case of Biotechnology," *Sociological Forum* 4, 2 (1989): 247–260; "Beyond Deference and Demystification in the Sociology of Science and Technology: A Reply to Otero," *Sociological Forum* 6, 3 (1991): 567–577; Gerardo Otero, "The Coming Revolution of Biotechnology: A Critique of Buttel," *Sociological Forum* 6, 3 (1991): 551–565 (reprinted pp. 47–61 in *The Biotechnology Revolution?* ed. Fransman Martin, Junne Gerd, and Roobeek Annemieke, Oxford, UK, and Cambridge, USA: Blackwell).

Introduction

GERARDO OTERO

Latin America's agriculture has been one of the economic sectors most negatively affected by the neoliberal reform set off in the 1980s. In most countries, a broad program of agricultural liberalization was launched under pressure from the United States and suprastate organizations, such as the International Monetary Fund and the World Bank. Economic liberalization generally included the unilateral lifting of protectionist policies, the opening of agricultural markets by lowering or eliminating tariffs and quotas, the privatization and/or dismantling of government corporations for rural credit, infrastructure, commercialization, and technical assistance, the end or even reversal of land reform policies, and the radicalization or reorientation of food policies focused on the internal market toward an export-based agricultural economy. These extensive reforms had profound, often negative, consequences for the agricultural sectors of Latin American countries and for a high proportion of agricultural producers. Impacts have been compounded by the fact that reforms in Latin America were not accompanied by a corresponding liberalization of agricultural trade and production in advanced capitalist countries, which continue to heavily subsidize and protect their farm sectors with billions of dollars, thus placing Latin American producers at a competitive disadvantage. "Neoliberal globalism" is what we call the ideology driving this set of reforms, both to describe their content and to highlight the fact that such policies can be changed with a different outlook.

The biotechnology revolution of the 1990s, which has inundated the countryside of some countries and supermarkets around the world with transgenic crops and other new products, was superimposed on the reforms brought about under the impetus of neoliberal globalism. From their beginnings at the laboratory stage in the 1980s, agricultural biotechnologies generally and genetic engineering in particular were described as potent tools for sustainable development and for ending famine, food insecurity, and malnutrition. It is well known that such problems are disproportionately concentrated in developing countries, which also happen to have large proportions of their population engaged in agriculture. Because of modern agriculture's technological bias, however—focused mostly on enhancing the productivity of large, specialized, capital-intensive farms—most developing-country peasants and small-family farmers have been rendered "inefficient." Millions have become excluded as producers by the new agricultural policies and technologies. All too often peasants and farmers have been transformed into wage workers for the capitalized farms, and countless have enlarged the ranks of the unemployed. Many of these people have participated in the growing trends toward international migration, separating family members for extended periods of time or permanently from their communities.

Yet, for millennia, peasant farmers have been directly responsible for preserving vast amounts of plant biological diversity. In fact, given the vagaries of nature, developing countries possess the largest plant biological diversity on earth, as well as the largest problems of soil depletion and environmental degradation. Input-intensive, capitalized farmers cannot preserve biological diversity, as modern agriculture has a clear bias toward monocropping based on modern plant varieties. Ironically, plant breeders producing these modern varieties depend on the availability of plant genetic diversity offered and preserved by small peasants. If the latter become extinct, so will the raw materials for further plant breeding. Adding agricultural biotechnology to this scenario, combined with the policies associated with neoliberal globalism, can only exacerbate the trends of social polarization and environmental degradation.

The first wave of studies about socioeconomic and environmental impacts of agricultural biotechnologies in the 1980s and 1990s used a prospective approach, because the products at stake were still in a laboratory stage; only a few medical applications entered the market in the 1980s. By the turn of the twenty-first century, however, a multiplicity of agricultural biotechnologies has been implemented on farms around the world, with the highest concentration in the Americas, North and South.

This book offers the general and specialized public a solid collection of empirically based studies written by social scientists. Represented disciplines among contributors are anthropology, economics, geography, political science, and sociology. Yet each chapter adopts an interdisciplinary perspective about the concrete socioeconomic and environmental impacts of agricultural biotechnologies in Latin America. These studies capture both the central issues that have come about with the application of agricultural biotechnology, and emerging alternatives for a sustainable agriculture.

In the general chapters on Latin America (1–4), the authors provide the theoretical and historical background to locate biotechnology in the context of modern agriculture and neoliberal globalism, the experience of the Green Revolution, the issues associated with global governance of biosafety, and the perils for smaller countries of relying on "absentee expertise" for shaping local legislation. A variety of national experiences (Chaps. 5–10) are then addressed, from widespread adoption of transgenic crops in countries such as Argentina, or somewhat restricted adoption as in Mexico, to the creation of a zone free of genetically modified organisms in Brazil that was ultimately bulldozed by neoliberal policies in 2004. Social-movements and bottom-up perspectives, as well as a research agenda for future research, are offered in the last two chapters.

Rather than being journalistic accounts, or predominantly normative or prospective in orientation, the chapters in this book are based on empirical evidence and emphasize interdisciplinary socioeconomic analysis. Most of our findings are rather somber. Yet there are also signs of emerging alternatives. We thus hope that our modest contribution to understanding such trends and alternatives will help activists and policy makers to transform this somber reality in a socially and environmentally sustainable direction.

Food for the Few

Neoliberal Globalism and the Biotechnology Revolution: Economic and Historical Context

GERARDO OTERO

Agricultural biotechnology is poised to "make deserts bloom," solve the world's food problems, and put an end to hunger. Or is it? Industry proponents and others advocated similar views in regard to the previous agricultural revolution—the Green Revolution of the 1970s. When agricultural biotechnology was still in the laboratory or field-trial stage of its development, in the 1980s and early 1990s, most observers considered that it would have a revolutionary effect on agricultural production and therefore a profound impact on agrarian social structures and the environment. The question remained whether such an impact would be positive or negative for society. Industry advocates and some scholars, such as D. Gale Johnson, editor of *Economic Development and Cultural Change*, continue to argue that the greatest problem with biotechnology is the political forces preventing its faster diffusion, and hence its ability to benefit the "millions of small and poor farmers who could gain if GMO [transgenic crop] varieties were available" (Johnson 2002, 4).

In sharp contrast with this view, critical observers tended to make rather ominous predictions about the impacts of biotechnology: agrarian social structures were to become further polarized, with fewer and larger farmers overwhelmingly dominating the scene while others bankrupted; negative environmental repercussions would overwhelm from such causes as rising use of agrochemicals; and biodiversity losses and increased crop vulnerability—prompted by increased crop homogeneity—raised food security concerns (Buttel, Kenney, and Kloppenburg 1985; Goodman, Sorj, and Wilkinson 1987; Kloppenburg 1988).

Given these polarized stances on biotechnology's potential, it is extremely important to offer an empirically based assessment of its actual impact on agrarian social structures now that several transgenic crops and other genetically modified organisms (GMOs) have been implemented in the fields. Rather than making extrapolations into prospective studies of future impacts or venturing informed opinions, we now have the capability to ascertain the actual impact of biotechnology on the basis of field data.

The empirical results from the chapters in this book provide indications that biotechnology's actual impact may be a far cry from the considerable optimism displayed by industry proponents or by economists like Johnson (2002). More accurately, our findings support the view that, while production and productivity are indeed increasing significantly, their benefits are not necessarily accruing to small farmers or the hungry. In this and the following chapters my contributors and I will confirm that regional and social polarization is already taking place at an accelerated pace with the introduction of agricultural biotechnology. Furthermore, in contrast to Johnson's prediction about its environmental benefits, the vast majority of transgenic crops have emerged from a determined technological paradigm that makes an increased use of agrochemicals not only more likely but also necessary: transgenic crops that are resistant to herbicides account for about 70 percent of the market (see Chap. 2, this volume).

We use the phrase *biotechnology revolution* primarily with regard to its socioeconomic and environmental impacts, but not because it is functioning as a revolutionary technology that will transcend the current petrochemical era of agriculture. In fact, we argue that biotechnology is captured within the same technological paradigm of modern agriculture and the Green Revolution. In analogy with the Kuhnian concept of a "scientific paradigm," Giovanni Dosi defines a "technological paradigm" as a "'model' or 'pattern' of solution of selected technological problems, based on selected principles derived from natural sciences and on selected material technologies" (Dosi 1984, 15). According to Dosi, technological paradigms move along "technological trajectories," or "a pattern of 'normal' problem solving activity" (Dosi 1984, 15). These technological paradigms also have powerful "exclusion effects," such that "the efforts and the technological imagination of engineers and of the organizations they are in are focused in rather precise directions while they are 'blind' with respect to other technological possibilities" (Dosi 1984, 15).

Finally, we argue that the ideological and policy context of neoliberal globalism further exacerbates the social and regional polarizing trends of the biotechnology revolution. Neoliberal globalism is variously known as *Structural Adjustment Programs*, the *Washington Consensus*, the *Wall Street–Treasury Complex*, *Liberal Productivism*, and the *New World Order.*

Mexico's debt crisis of 1982 fundamentally challenged the protectionist, inward-looking, and state-centered development model that had been in place since the 1930s in the larger countries of Latin America. By the mid-1980s, a series of neoliberal reforms were introduced to substantially cut the government deficit and steer the national economies toward an export orientation. Neoliberal policies have involved eliminating most subsidies, dismantling or privatizing state-run firms, allowing the entry of foreign products, promoting foreign capital investment, and deregulating most sectors of the economy, not least the agricultural sector, which may have been the most protected throughout the world. By 2002, after nearly two decades of neoliberal policies, *BusinessWeek*'s correspondent in Mexico reported, "Farmers are getting plowed under" (Smith 2002, 53).

This introductory chapter will begin by situating biotechnology in its broader technological and economic context. The next section outlines the emergence of modern agriculture in the United States since the 1930s, which provides the historical background for the introduction of biotechnology in Latin America. "Modern agriculture," conceived as a technological paradigm, has come to dominate capitalist agriculture throughout the world. The last section briefly describes how Latin America has been affected by this technological paradigm and outlines the organization of the rest of this book. We will see how most of the chapters, based on the actual, empirical implementation of biotechnology in Latin America, support the view that it is becoming a revolutionary technology; revolutionary not in transcending modern agriculture as a technological paradigm, but for its detrimental social and environmental impacts. Are there any alternatives? These will be discussed at least in outline form in most chapters, but particularly in the last two chapters of this volume.

Biotechnology in the Third Technological Revolution

Although the very definition of *biotechnology* has been the subject of controversy, the meaning we intend refers to the "new biotechnologies." We use the definition provided early on by the General Accounting Office (GAO) of the U.S. Congress: "Today, biotechnology is generally considered to be a component of high technology, and the 'new biotechnologies' are those resulting from recently developed, sophisticated research techniques, including plant cell and protoplast culture, plant regeneration, somatic hybridization, embryo transfer, and recombinant DNA methods" (GAO 1986, 10).

While the analysis of biotechnology's socioeconomic impact was still in its early years, one of its most prominent pioneers, Frederick Buttel, switched away from his original position that biotechnology would be an "epoch-making" technology, the way electronics and informatics are (1989a, 1989b). My debate with Fred Buttel's later position can be found elsewhere (Otero 1991, 1995). Two points that Buttel made are particularly relevant here, however: first, Buttel minimized biotechnology's impact on the basis that it was a substitutionist technology (one that replaces an existing crop or product); second, he minimized its impact on the basis that the technology was applicable only to declining sectors (agriculture and manufacture), rather than to leading ones (services), making it a "subsidiary technical form." His main quibble with biotechnology, in short, was that it would not transcend modern agriculture as a technological paradigm.

Even if biotechnology were limited to substitutionism, I argue, it remains capable of facilitating profound changes in productivity, the international division of labor, and even the environment, all of which can generate major social changes. Substitutionism is the ability of new products of technology to substitute for previously existing products. A classic and early example of this was the introduction of high fructose corn syrup in the United States. From 1978 to 1987, this corn-based sweetener produced with new enzymatic techniques replaced 42 percent of the sugar used in the United States. Slightly more than 40 percent of caloric sweeteners continue to come from this source (Otero and Flora, forthcoming). The switch was profoundly damaging for several Caribbean countries and the Philippines, whose foreign exchange largely came from these sugar exports (Ahmed 1988; Otero 1992). Consequently, mere "substitutionism" can have profoundly damaging effects on primary goods–exporting countries, in many of which a majority continues to live off agriculture.

A most basic concern with the debate over biotechnology's revolutionary status lies in the false dilemma presented by asking the question at all. More important, we must ask first whether there has been a "third technological revolution" in effect (Mandel 1978, 1995) since the 1990s or indeed an "information age" of capitalism (Castells 2000a, 2000b, 2004; Pérez 2002). Does this new technological revolution or era—based on electronics, informatics, new materials, and biotechnology—actually represent a new ascendant phase of capitalism? If we can answer this affirmatively, then we could ask what the place of biotechnology is in the information age, as Manuel Castells has attempted. While acknowledging that each technology has its own rhythm of development in the context of these dynamics, we need to keep the question of the "third technological revolution" or the "information age" as a conceptually unified phenomenon, not one made up of several, juxtaposed, revolutions based on each of the new technologies.

Capitalism as a world economy entered a period of profound crisis in the late 1960s and early 1970s (Aglietta 1979; Mandel 1978), as its heavy-industry-based development began to decline (Piore and Sabel 1984). The restructuring of the world economy, which started in the 1980s, was predicated on productivity increases, which in turn depended on new technologies (Hastopoulos, Krugman, and Summers 1988; Young 1988; Thurow 1987). By the early 1990s, most state-socialist societies had been drawn back into capitalism, largely due to their productivity problems and lack of technological innovation. At that point, the conditions were in place for a new "long wave" of capitalist development to set in. Ernest Mandel (1978), building on Kondratiev's studies of capitalism's economic history, proposed that its development took place within "long wave" cycles that lasted around fifty years. Each long-wave cycle has been associated with a major technological revolution. These cycles contain an upswing phase and a recessionary phase that last about twenty-five years each. In his seminal work, *Late Capitalism*, Mandel (1978) anticipated in the 1970s that a "third technological revolution" was brewing after the capitalist crisis that started in the late 1960s and saw the phenomenon of "stagflation" in the 1970s: stagnation or no economic growth combined with inflation—unprecedented in the history of capitalism.

Using Mandel's conceptualization, in my debate with Buttel I predicted that biotechnology would be inserted in the coming "third technological revolution" that would account for a new era of capitalist growth in the 1990s (Otero 1991). By now, it has been amply demonstrated that a new upswing phase of capitalist development started in the early 1990s, with this decade having seen one of the longest expansionary phases of economic growth (see M. J. Mandel 1996, 2004 for data).[1] Therefore, rather than engaging in a partial analysis of whether any given technology is revolutionary or not, the task at hand, for economic sociology at least, is to decode the implications of the "reformation of capitalism" (Sklair 1989, 2002) for the new international division of labor and the new information age (Castells 2000a).

To the extent that new technologies have been at the core of neoliberal economic restructuring since the 1980s, biotechnology has played a major role in transforming agriculture, whether by direct adoption of biotechnology's products or indirectly by its substitutionist impacts.

Moreover, the biotechnology industry is a healthy rather than a declining one, but due to short-term financial problems it has undergone several major episodes of concentration in the hands of the chemical and pharmaceutical giants.

In fact, this emerging market structure does question whether there will be a biotechnology industry per se. Rather, biotechnology has primarily become an enabling technology that has allowed the existing pharmaceutical and petrochemical industries to expand their profitability.[2] This new industry structure has undoubtedly had a major bearing on the ways biotechnology products are disseminated in the world. The contrasts in this regard with the "Green Revolution," which was promoted by public and semipublic institutions, are profound. Finally, biotechnology will cause or facilitate ascending rates of productivity growth. The obvious consequence is that large proportions of the labor force currently in agriculture have or will become redundant. Whether resulting unemployment will be mitigated by economic expansion in other sectors is indeed an important question. For the shape of the new distribution of income and standards of living of people in the midst of neoliberal economic restructuring depends largely on such changes.

Although the bulk of biotechnology research and development is taking place in advanced societies, deployment of its fruits has implications for the world economy as a whole. As Iftikhar Ahmed suggested early on, the "application of biotechnologies to agriculture would automatically affect 60 per cent of the Third World population who depend on agriculture alone for their livelihood" (1989, 553). By the turn of the twenty-first century, this figure may be closer to 50 percent, still making biotechnology's impact potentially quite severe. Given the vast heterogeneity existing among Third World societies, they will be affected differentially by the information age, depending on the profile of their socioeconomic structures. Some have recently become industrialized, precisely on the basis of new technologies; the larger countries have a certain potential to jump on the bandwagon of the technological revolution, and others might be integrated to the world economy simply as producers of cheap labor power, while still others could be marginalized from the main economic trends. Thus, one thing that should be looked at more closely is the new stratification of underdeveloped societies that is bound to emerge. An initial formulation has been provided by Manuel Castells (2000a, 2000b, 2004).

Castells sees the majority of Third World countries as being condemned to economic obsolescence, unemployment, misery, hunger, illness, and violence in their large urban centers. Countries in the latter stratum either would be marginalized from the world economy or would experience a "perverse integration" (Castells 2004) through the production and export of illegal crops. One alternative to be explored in this book is what happens to those countries that adopt foreign-developed biotechnologies, and those whose crops are directly or indirectly affected by new transgenic varieties produced in advanced countries.

In sum, biotechnology potential is best viewed within the context of a "third technological revolution" or the information age, and there is significant evidence to refute calls to diminish its revolutionary potential. Rather than transcending the technological paradigm represented by modern agriculture, however, biotechnology will deepen its effects in social polarization and environmental degradation. Such effects are only exacerbated by the ideological and policy context of neoliberal globalism.

Modern Agriculture in the United States: Social and Environmental Implications

Given that the United States continues to be the leading country in terms of technological innovation in agriculture, new developments emerging in this country directly and indirectly affect agriculture in dependent nations. Hence an analysis of the impact of biotechnology must be set in a world perspective, looking at what has happened in the United States and how this impacts Latin American countries.

Broadly speaking, the "power age" of agriculture began in the United States in the 1860s, with the widespread introduction of mechanical horse-drawn harvesting and threshing equipment, new ploughs and disc harrows, etc. Our focus, however, is on technological developments around the biological components of agriculture, primarily seeds. In this narrower sense, modern agriculture is best dated in the 1930s with the introduction of hybrid corn, and continued with further improved crop varieties, chemical fertilizers, pesticides, insecticides, herbicides, and increasingly sophisticated agricultural machinery. As a package, these technological innovations gave a tremendous boost to productivity, although large numbers of farmers were displaced from their occupations and had to look for other employment (Cochrane 1979; Kloppenburg 1988).

The main problem that agriculture in the United States had to face before the "power age" was a relative scarcity of food in the context of strong population growth. Given this problem, the achievements of modern agriculture were spectacular. The shift from horses to tractors, which started as early as the 1920s, signaled the advent of the power age of agricultural production. With the power age came unprecedented productivity increases, to such an extent that supply outstripped domestic demand, and the exportation of surpluses became the dominant U.S. agricultural strategy (Berlan 1991; Orden, Paarlberg, and Roe 1999).

Prior to the power age, agricultural production had many similarities with peasant production systems. Production was balanced on a careful system of crop rotation to preserve soil fertility, and draft animals provided the fertilizer for subsequent crops. It required minimal inputs, and cash incomes were low. The power age essentially moved production from a technically sophisticated system—requiring careful management to meet the needs of both current production and required inputs for the following year—to a technologically sophisticated system, requiring capital (Berlan 1991, 121). The necessary externally provisioned agricultural inputs have increased exponentially since the advent of the power age: animal fertilizer had to be replaced with artificial fertilizer; tractors were joined by threshers, sprayers, pesticides, hybrid plants . . . . These inputs were characterized by an acute substitution of labor power for capital. Between 1920 and 1970, while "the use of purchased inputs nearly doubled," the "quantity of labor declined by nearly three-fourths" (Knutson, Penn, and Flinchbaugh 1998, 215).

An unprecedented rural-urban migration process accompanied this displacement of agricultural labor power from farming. The robust industrialization process that was taking place in the United States before and after the depression of the 1930s, however, absorbed most of the workers and farmers displaced in agriculture. It is this "release" of labor from agriculture that "contributed materially to the overall development of the economy and the transformation of the United States from an agricultural to an industrial economy" (ibid., 215). Today less than 2 percent of the workforce in the United States is dedicated to farming. Social polarization of the U.S. agrarian social structure also accompanied this transition. Despite an increasing population, the total number of farms declined from its peak of almost 7 million in 1935 to about 2 million in 1997, with the greatest decline occurring from the 1940s to the 1960s (USDA, *Agricultural Fact Book: 2001–2002*, 24). The reduction of farms of specific types is further informative of this trend. From 1964 to 1997, the number of farms (in thousands) that sold hogs and pigs dropped from 803 to 102, while the number of milk producers dropped from 1,134 to 117; wheat from 740 to 244; cotton from 324 to 31. The least affected was the cattle sector, which only dropped from 1,991 to 1,011 (MacDonald and Denbaly n.d., 5).

Along with the decline in the number of farms has come an increase in their average size. The almost 7 million farms in 1935 had an average size of 155 acres. By 1964 there were less than half the farms—3.16 million—averaging 352 acres per farm. By 1997 there were 1.91 million farms, averaging 487 acres per farm (ibid.). Of course, these figures do not represent an across-the-board size increase in remaining farms. Instead, since 1974, both small farms (under 50 acres) and large farms (over 500 acres) have increased in numbers, while mid-range farms have declined. The share of farms with over 500 acres increased from 4 percent in 1935 to 18 percent in 1997 (USDA, *Agricultural Fact Book: 2001–2002*, 25). Farm size can be an ineffective indicator of agricultural productivity, however, and more can be revealed by data on farm sales class.

As recorded by USDA, sales classes of farms are made up of "small family farms," "large" and "very large" family farms, and "nonfamily farms." While farms with sales of $500,000 or more (very large farms) accounted for only 3 percent of farms, they accounted for 52 percent of agricultural production. In the sixteen-year period between 1982 and 1997 (for which constant dollar data was available), only the larger sales class showed consistent increases, with the number of farms with sales $500,000 or more growing the most rapidly (USDA, ERS website, Briefing Room, farm structure). In contrast, "small family farms" (having sales less than $250,000) accounted for a whopping 91 percent of U.S. farms in 1998 (Hoppe and MacDonald 2001, 3). Most of these farms, however, reported income inadequate even to cover expenses (ibid., 4). Fifty-six percent of small farms had sales less than $10,000 in 1999, and accounted for only 2 percent of agricultural production (USDA, ERS website, Briefing Room, farm structure). A vast number of small farms are therefore practically irrelevant for agricultural production, with the majority of them relying on off-farm income (Hoppe and MacDonald 2001, 4). While the "small family farm" class includes retirement and residential/lifestyle farms, it also represents those impoverished farmers who have been rendered "inefficient" by the technological paradigm of modern agriculture, which contains an unequivocal bias in favor of large-scale production. Therefore, only early adopters of technological innovations, who must be well endowed financially and have superior managerial skills, have been able to stay afloat in U.S. agriculture (Cochrane 1979).

The new, capital-intensive structure of agriculture contained a profound change in the agrarian social actors. This change set the stage for the new agro-industrial complex, and the two main sectors that would escort the activity of farming into the modern age: the inputs-producing component, heavily oligopolistic, in which producers have substantial control over *selling* prices of their goods; and the processing and distributing component, which presents an oligopsonistic structure, i.e., one with few buyers of farm products who possess a tight control over their *buying* prices. Distributors of farm produce now include huge supermarket chains, some with transnational operations in many countries, such as Wal-Mart. Concentration of these market participants has long become the norm and is getting deeper. For example, the share of value added by the top 100 food processors has grown steadily from 51 percent in 1967 to 75 percent in 1997 (MacDonald and Denbaly n.d., 5). Even in the context of current struggles over market share, Campbell's Soup controls 70 percent of market share of wet soup, with 90 percent home penetration (Fischer 2000). The most important subsectors in inputs and processing display such high "concentration indexes" that 50 percent of sales or purchases, respectively, are controlled by only four or fewer giant firms. The four largest firms in flour milling increased their processing share from 33 percent in 1977 to 62 percent in 1997, while soybean milling increased from 54 percent to 83 percent for the same time period (MacDonald and Denbaly n.d., 6). Agribusiness concentration is even more prominent in meatpacking: for example, the share of the slaughter of the four largest firms in steer and heifer meatpacking increased from 36 percent in 1980 to 78 percent in 1998 (MacDonald and Denbaly n.d., 6).

Farms themselves, although becoming fewer and larger, still constitute a competitive sector, in the classical economic sense that no single producer or small group of producers controls their selling or buying prices. Hence input producers and agricultural processors and distributors wield much market power.

Five Main Problems of the Current State of American Agriculture and the Rise of Supermarket Power

From a brief survey of work by keen students of U.S. agriculture, we can identify the five main issues facing it today. Most of them are interrelated, but each will be discussed in turn. Finally, the rise of global supermarket chains and private standards promoted by the World Trade Organization (WTO) introduce a new twist in the global agrifood system.

First, farming has ceased to be profitable for most producers. Excepting those family and nonfamily farms that fall into the large and very large category, farming has ceased to be profitable for those squeezed between oligopolistic input markets and oligopsonistic processors and purchasers. The high degree of concentration of both these sectors is exacerbated by the fact that many of the same firms are leaders in different industries and related businesses, forcing farmers to deal with a limited number of large agribusinesses in a range of different contexts (MacDonald and Denbaly n.d., 2). In sum, direct producers confront increasing production costs at the same time that their forward linkages with an oligopsonistic structure make it very hard for them to retain the benefits of their productivity increases. While some of these benefits are transferred to consumers via lower prices, the greatest part of such benefits is actually accrued by the large processing firms (Goldberg et al. 1990). The advent of contract farming, while containing a number of producer benefits (e.g., a reduction in price risks and easier acquisition of debt financing), comes with a direct loss of control over production decisions, and leads to further exploitation by large corporations (Lewontin 2000).

A second major shift resulting from modern agriculture is the emergence of agribusiness as producers, promoters, and disseminators of agricultural technologies. In the beginning of the agricultural revolution of the 1930s, the creators and disseminators of farm technologies were the agricultural universities of the United States (the Land Grant Colleges) and, on a world level, what later became the International Agricultural Research Centers (Kloppenburg 1988). These institutions had a public or semipublic character, with mandates that had a social orientation, such as contributing to the viability of small or medium-sized farms. In contrast to this, the new producers of technology—and this is increasingly the case with biotechnology—are large transnational corporations (TNCs). The mandate of TNCs is to maximize profits, not attend to social priorities. The potential for significant environmental and health impacts of incautiously directed biotechnologies raises the prospect of social repercussions considerably more wide-reaching than agrarian social and regional polarization.

Third, the American model of modern agriculture has resulted in such severe environmental problems that it now calls into question the sustainability of its continued development. While the use of traditional plant breeding techniques produced improved plant varieties, it also created two mutually reinforcing complications: plant genetic diversity declined as local varieties were left behind in favor of improved ones, and the resulting increased homogeneity rendered new varieties more vulnerable to pests. Concerns about the resulting narrowing of the genetic base were already being raised in 1936; however, they became unambiguous during the 1970s corn blight, when the year's harvest was hit by a disease organism that "attacked a cytoplasmic character carried by over 90 percent of American corn varieties" (Kloppenburg 1988, 163).

Further, the intensive use of machinery and petrochemicals has resulted in severe problems of soil degradation, underground water contamination, and the appearance of chemical residues in food, which have been directly linked to human cancer and genetic diseases (National Research Council 1989). Soil erosion is a significant problem in the United States. While erosion rates have declined with a number of conservation efforts (e.g., conservation tillage and the retirement of highly erodible cropland), it is still widespread. The actual onsite impact of erosion differs by region, soil type, and crop; however, production losses are estimated to range from 3 to 31 percent for a 100-year period. The annual economic losses due to erosion in the United States are estimated at $56 million (USDA, ERS website, "Global Resources and Productivity").

Contaminations from the requisites of agricultural production raise equally serious questions about the sustainability of the current high input style of U.S. agricultural production. In the northern Gulf of Mexico, for example, 7,000 square miles have been rendered a hypoxic "dead zone" as a result of high nitrogen loads (which reduce oxygen), killing fish, crabs, and other marine life. The majority of the nitrogen load is believed to be the result of fertilizer and manure runoff from agricultural lands, carried down by the Mississippi River (Ribaudo et al. 2001).

Fourth, the food consumption pattern promoted by U.S. modern agriculture is heavily inclined to producing animal protein, namely meat and dairy products. The social issues associated with this dietary shift from grains and cereals have three aspects: food security, environmental effects, and health impacts. With respect to food security, the dietary emphasis on meat and dairy involves dedicating enormous quantities of grain to produce feed for livestock. This is a food pattern that is rife with inefficiency and that promotes inequality. The irrationality of this form of producing proteins is evident. Despite the fact that malnutrition and starvation remain persistent and severe problems, the amount of global cereal production dedicated to feed has more than doubled, from 273 million metric tons in 1961 to 685 million metric tons in 2001. The cereal supply dedicated to feed in developed countries in 2001 was almost 460 million tons: this is well over 50 percent of the amount of grain consumed for food in developing countries, where 78 percent of the world's population is concentrated (FAOSTAT).

While not as blatant, the environmental costs of the high meat diet are also pervasive. As it can take up to ten times the amount of grain to produce 1 kilogram of meat, more land is needed to produce less food. Water resources are also inefficiently used when nutrition is meat rather than plant based. Further, ranching itself contributes to soil erosion. Disposal of manure and urine produced by livestock animals is another major environmental concern, as they penetrate into drinking and groundwater supplies. Lastly, methane, a greenhouse gas, is produced in the digestive process of cows and other ruminants. These animals are estimated to account for 15 percent of the world's methane production.

With respect to human health, alarm bells have been loud and clear regarding the high-fat, high-cholesterol, high-protein, low-fiber, meat and dairy–based American diet. In 1988, the Surgeon General reported that 70 percent of all U.S. diseases were diet related. Excess weight increases the risk for heart disease, hypertension, diabetes, and cancer; heart disease is the number one killer in the United States. Currently, the United States is facing a growing epidemic of overweight and obesity. In 1999, an estimated 61 percent of U.S. adults and 13 percent of children and adolescents were overweight. The direct and indirect financial cost of these health impacts amounted to $117 billion in 2000 (United States Department of Health and Human Services, 13 December 2001). Ironically, by the start of the twenty-first century, North American diets had added substantial quantities of grain products that further exacerbate the obesity epidemic.

Fifth, the American diet has not only come to emphasize meat and dairy, but has also come to be characterized by the high degree of prepared and processed foods sold in supermarkets. This consumption shift from raw agricultural products to processed foods exacerbates many of the social, health, and environmental concerns already noted, as processors and supermarkets—who are becoming global in reach and production methods—mediate the relationship of consumer to agricultural product. Not only has increased food manufacturing and processing facilitated the growth of the agrifood complex (Friedmann 1995)—as the high profits attached to the added value of processed food are a key factor for capital accumulation—but media-facilitated consumption choices further entrench dominance by a few large corporations and reduce the potential for change.

Finally, there has been an ironic twist to the fifth trend mentioned above that may give a new type of consumer increased power over the food supply. It also modifies the relative power of big processors and big supermarket chains, some of which now operate globally. As Larry Busch and Carmen Bain have observed, the institutions and regulations facilitated by the WTO to promote trade have placed the private-sector food retailers at the center in the transformation of the global agrifood complex (Busch and Bain 2004). One of the ironic consequences of the various WTO agreements that were designed to harmonize standards for the international marketplace, they argue, was the rise of private standards.

The rise of private standards is primarily, but not solely, promoted by the retail sector, as it is the last link to consumers. Thus, enforcement of WTO standards has led to a variety of social contradictions of concern not only to NGOs, but also to many consumers. At the same time, the increasingly global scope of supermarkets has shifted the balance of power from processors to an oligopoly of retailers, who compete for the attention of consumers—many of whom are more health conscious and have higher disposable incomes than consumers of the past. New consumers, at least in the wealthy countries of the North and the rich of the South, base their preferences on quality rather than price. Fresh and store-branded products have absorbed a lot of this consumer interest, rendering retailers more vulnerable to consumer perceptions in case of failure and in case of revelations provided by NGOs over social, environmental, or health inadequacies associated with their products. Consequently, given the tight profit margin, retailers have necessarily become highly conscious of product standards and have often instigated standards in excess of those required by government agencies.

"In short," say Busch and Bain, "the private sector has jumped ahead of the public sector, substituting consumer demand for citizen demand, market accountability for governmental accountability" (2004, 335). While such private re-regulation is responsive to consumer demand, there are significant emerging issues with respect to accountability, transparency, and democratic input, among others, as a result of the rise of private standards and big supermarket power.

U.S. State Support of Biotechnology

In spite of free-trade rhetoric, the U.S. government has worked hard to facilitate the development of its biotechnology industry. While transnational corporations in the agricultural-inputs sector have become the crucial agents in producing and disseminating modern agricultural technologies, this took place in close association with the U.S. government (through the U.S. Department of Agriculture) and the U.S. Land Grant Universities. The latter produced the science with government funding, and firms developed the new inputs for modern agriculture. Similarly, research at International Agricultural Research Centers (IARCs) can be easily monopolized by transnational corporations (TNCs), which can even patent very similar plant varieties to those developed by the IARCs. It is a complex situation in which resources developed or preserved by public or semipublic institutions may be monopolized by large TNCs. This governmental or public support was not limited to economic and research support, however, but occurred through policy and law as well, as will be presently discussed. Although U.S. farmers also participated in this alliance, they never played any determining role as to what technologies were to be produced or developed; they were merely recipients of innovations that responded to the profit-maximizing interests of TNCs.

Therefore, state-facilitated TNC concentration appears to be the norm in biotechnology as developed countries vie for technological hegemony, with the United States the consistent leader. As most developing nations do not have the resources available to create appropriate technologies (Otero 1989a), they will very likely be locked in as technology consumers—"reduced to importing expensive high-tech packages created for developed-nation purposes" (Peritore 1995, 15). China (and India to a lesser extent), among developing countries, may be one of the very few possible contenders in high technology innovation (*Economist* 2005).

Control over genetic resources is another method with which developed states attempt to secure biotechnological dominance (Carlson 2004). While gene banks have become a priority in developed nations, the frequent refusal of these banks to grant germplasm to other nations is promoting a system in which monopoly of the world's genetic resources will become a form of power (Kloppenburg 1988). Ironically, as externally directed agricultural technology compromises the genetic heritage of developing countries, they will become increasingly beholden to developed nations: "Thus, in some sectors, less developed countries' genetic-environmental erosion could actually enhance corporate profitability and control" (Peritore 1995, 17; McAfee, this volume).

Biotechnology adds another feature to the food insecurity of developing nations that production-for-export markets and food import dependency already created. As already discussed, shifts in consumption patterns allowed for the development of the agrifood complex. This complex is now an internationally structured food production complex, where processed foods are assembled from globally sourced components, much like the "world car" (Friedmann 1992, 1995). Technological advances that assist in the isolation of "generic ingredients" increasingly allow for the substitution of agricultural products in profit-determined global sourcing strategies (Friedmann 1991, 67), such as the replacement of Caribbean and Philippine sugars with the corn-derived sweetener already noted. The resulting fluidity of the international division of labor has considerable implications for national economic strategies and trade stability.

Biotechnological research is already under way that might substitute key crops from developing nations, such as cocoa, coffee, vanilla, and rubber (Peritore 1995, 22). The resolution of these would have obvious and severe impacts on the vulnerability of developing nations. Further, it increases the capability of TNCs to source their ingredients in such a way as to externalize environmental costs onto regions outside the view of their target market, and thus maintain consumer support for their products.

Patents and the Monopolization of Genetic Resources

A patent is the existing legal form to protect industrial property. In the United States a patent grants the monopoly over the use of protected inventions—which may be either products or industrial processes—for seventeen years, with some exceptions. The holder of a patent title may grant a license to those willing to pay royalties, although licenses may also be granted without asking payment of royalties, which is rather common practice in many U.S. universities that do not have much marketing experience.

Joan Robinson, the well-known British economist, defined patents as follows:

> A patent is a device to prevent the diffusion of new methods before the original investor has recovered profit adequate to induce the requisite investment. The justification of the patent system is that by slowing down the diffusion of technical progress it ensures that there will be more progress to diffuse . . . Since it is rooted in a contradiction, there can be no such thing as an ideally beneficial patent system, and it is bound to produce negative results in particular instances, impeding progress unnecessarily even if its general effect is favorable on balance. (Cited by Nelkin 1984, 15)

A most defining characteristic of the biotechnology industry is its frenzy for patents. By 1987, the *Economist* touted that "in high technology, patents are symbols of virility. The number of patents filed by a company is a sign of its inventiveness—and its future prosperity" (1987, 82). In the particular case of biotechnology, the rise of patent applications is so frantic that the U.S. Patents and Trademarks Office (PTO) has suffered an enormous backlog in processing them. The total backlog of pending applications, including appeals and amendments, climbed to 6,907 in 1987, up from 5,837 a year earlier (Crawford 1988). By 2003, the total backlog of biotechnology patent applications had reached 500,000, and it is expected to reach 1 million by 2008 (Reichardt 2003). In part, this backlog has doubtless resulted from the supportive context for biotechnological development the U.S. government has attempted to create, such as initiated by the Reagan administration (Otero 1992).

Contrasting views on the favorable balance of the current patent system abound, particularly with reference to biotechnology and the internationalizing of patent agreements. One of the multiple problems posited by the proliferation of patents is the tremendous inequality of conditions among nation-states to produce inventions. Thus, the countries with strong systems of science and technology will have the greatest advantages of monopolizing the commercial use of knowledge. This situation induces an additional mechanism of economic polarization and differentiation among countries, as developed countries have prepared well to protect the industrial property of their firms internationally through promoting homogeneous patent legislation across the globe. In fact, one of the central elements of the Uruguay Round of the General Agreement on Tariffs and Trade (GATT), which lasted from 1987 to 1993, was to homogenize intellectual property rights protection across the globe. The results were instituted in the Trade Related Intellectual Property Rights Agreements (TRIPs) of the World Trade Organization (WTO), the new name for GATT as of 1994 (Buttel 2000).

Vandana Shiva claims that American patent law is guilty of the "myth of ignorance as innovation," whereby innovations unknown in the United States but practiced elsewhere could be patented (Shiva 2000, 502). This system allows U.S. corporations to patent gains made by centuries of indigenous cultivation. Shiva argues that the resulting "biopiracy" is so extensive that it amounts to a form of recolonization. American patent laws, however, would not be so disadvantageous to developing countries if it were not for their globalization through international trade agreements. The WTO's TRIPs are rooted in U.S.-style patent law, and attempts by developing countries to reform the agreements are defeated on the basis that "the WTO cannot be subordinated to other international agreements" (ibid., 507).

While seemingly at the greatest expense to developing nations, the American determination to support its biotechnology industry is not without any repercussions at home. There have been increasing frictions over the impacts of biotechnology patents and increasing TNC dominance on local producers. For example, in 1999 a class action lawsuit was launched on behalf of American soy farmers, "charging that [Monsanto] has not conducted adequate safety testing of engineered crops prior to release and that the company has tried to monopolize the American seed industry" (Halweil 2000, 203). The U.S. government's steadfast resistance to labeling GMOs is also raising some consumers' ire, and even without labeling, consumer concern has in fact been able to harm the biotech industry's profitability (Montague 2000; Halweil 2000). (Kathy McAfee expands on these issues in Chap. 3 below.)

In sum, increasing privatization of knowledge, along with the increasing concentration of the biotechnology industry and the agrifood complex generally, suggests great difficulties for applying the technical progress that it produces at low social costs. With the huge gap that separates the scientific production systems of advanced capitalist countries as compared to developing countries, the commodification and monopolization of knowledge can only have further polarizing effects within and between countries. In fact, most of the evidence indicates that the patterns of diffusion of biotechnology products will increase the presence of TNCs in developing nations, resulting in increased social polarization, the further loss of plant genetic diversity, and other negative impacts. Let us turn to how the biotechnology revolution and neoliberal globalism impact Latin America.

Agricultural Biotechnology in Latin America and Organization of This Book

Although there were heavy human costs in displacement from agriculture to industry, from rural to urban areas in the United States, by the mid-1950s it could be said that most displaced people from agriculture were successfully absorbed in the urban centers. With respect to the question of whether agriculture is now a declining sector, as Buttel (1989a) suggested in the late 1980s, the answer depends on whether one measures decline by production and productivity increases or, as he seemed to suggest implicitly, by the proportion of the labor force absorbed by the sector. Despite a constantly reducing labor force, agriculture's productivity has been historically quite dynamic—indeed surpassing the rest of the U.S. economy during the 1980s in its productivity rate of growth (National Research Council 1989, 33), and producing enough food for an exportable surplus (Orden, Paarlberg, and Roe 1999). On the other hand, if labor force participation were the measure of dynamism, then the tertiary or "services" sector (with over 70 percent of the U.S. labor force) would indeed surpass agriculture. But *capitalist* productivity growth is usually measured not by labor absorption but by how much was produced per hour, per hectare, or per unit of capital. Thus, when talking about the labor productivity associated with a given technology, the more productive it is, the more labor it tends to displace.

Without the robust process of industrialization that the United States experienced after the 1930s, the resulting negative socioeconomic consequences of modern agriculture for Latin American countries have been much graver. Massive migrations to the cities occurred, but little employment could be found. The "tertiary" sector in the cities swelled rapidly, very often outside of the formal economy (Portes, Castells, and Benton 1989), thus absorbing people below their working capacities and diminishing their living standards (Portes and Hoffman 2003). This was certainly a sign of productive "dynamism" in agriculture, but it masked severe under- and unemployment of people who had been displaced from farming (Bartra 2004). Changes of this type since the mid-1980s, once neoliberal globalism was in full force in most Latin American countries, have led population scholars to label the twenty-first century "The Age of Migration" (Castles and Miller 2003). Latin American agriculture therefore became more productively dynamic, but it absorbed fewer people, causing a social decline that quite arguably was "revolutionary" in impact.

By now, this book's discussion has been placed in its larger analytical context of how biotechnology is part of the third technological revolution, and how its effects have been exacerbated by the ideological and policy context of neoliberal globalism. This technology and policy combination proves to have a "revolutionary" impact in the sense described above: Given the technology's effect of increasing the productivity rate of growth and its large-scale bias, small peasant farmers are increasingly rendered inefficient and expelled from agriculture. The withering away of small peasants as cultivators has consequences not only for the social structures, but also for the endangerment of biological diversity, to the extent that peasants have been the "curators," so to speak, of many of the region's crops' biodiversity. Therefore, while Latin American agricultures may be producing a greater share of exports of grains, fruits, and vegetables, their ability to feed their own people has decreased, thus becoming dependent on increasing food imports; or increasing numbers of people simply go hungry (more on this in the chapters below).

Chapter 2 by Gerardo Otero and Gabriela Pechlaner moves on to discuss Latin America within the regional frame of the Americas. It discusses how the technological paradigm of U.S. modern agriculture was eventually transferred to the developing world through the "Green Revolution." It expands the discussion on the U.S. dietary pattern and how it has also been transferred to Latin America in substitution of some of its grains and cereals as the basis of local diets. Such dietary transfer has had major negative effects on Latin America's peasant farmers by displacing their crops from domestic markets; and on the health of consumers, many of who are now also afflicted by an obesity epidemic in the midst of malnutrition. The social and environmental sustainability of these patterns is discussed and questioned.

One of the key mechanisms by which the U.S. government has pushed the agenda of TNCs throughout the world is by homogenizing scientific and legal frameworks to protect intellectual property rights. Chapters 3 and 4 address the specific ways in which such promotion is taking place. Kathy McAfee's Chapter 3 discusses how the U.S. state has exported the bioengineered crops and the myths developed about how they are expected to perform molecular miracles in the solution to world hunger. It points to continuities between transgenic crops and problems of the Green Revolution noted in Chapter 2. McAfee raises the question of whether genetic engineering can open a new path to agricultural development in Latin America. She then offers a survey of the published evidence about the actual, rather poor performance of transgenic crops, the problems of resistance and genetic erosion, and the reasons why transgenic crops do not represent a new technological direction in agriculture, but rather a continuation of the modern agricultural paradigm. McAfee then outlines various key geographies of difference that distinguish agriculture in the United States from farming in most developing countries, pointing to the risks of globalized, deregulated biotechnology and to the limitations of any standardized, technology-centered approach to agricultural productivity. She concludes with an outline of more promising alternatives to improving Latin American agriculture based on bottom-up perspectives that take into account local farmers and agroecological conditions.

In tandem with promoting a "science" approach to setting world standards, pushing neoliberal globalism as the dominant ideology has put considerable pressure on Latin American countries to construct regulatory frameworks that prepare the path for a smooth introduction of transgenic crops. Chapter 4 by Kees Jansen and Esther Roquas explores the interactions evolving between Latin American countries and international organizations to establish such frameworks in the shortest time possible. It raises a series of critical questions about the current trend to harmonize biosafety regulations. Jansen and Roquas argue that the problem of developing biosafety regulatory frameworks cannot be reduced to the relatively simple problem of how to increase local scientific and regulatory capacity. The issue is much broader and requires proper attention to controversies and contrasting views around biotechnology, the heterogeneity of national political cultures and socioeconomic conditions, and the complexities of regulation making and implementation in weak developing states. The making of biosafety regulation is thus basically a sociopolitical issue and not a technocratic one. The latter receives ample attention of international organizations; the former receives hardly any.

Chapters 5–10 are intended to add depth to the broad trends noted in the previous chapters by looking at how they materialize in various case studies and with a focus on more specific issues. Manuel Poitras's Chapter 5 offers the political economy of changes resulting in Mexico's legislation to allow and promote transgenic agriculture, as well as the main agro-industrial profile characterized by the existence of only one major biotechnology company in the country. On the basis of primary and secondary data, he contests the scale-neutrality argument which claims that small-scale peasant farmers will benefit from the new biotechnology. Poitras analyzes state responses to the new agricultural biotechnologies by reviewing the regulation of their use in Mexican agriculture, and the position of public research on this matter. Finally, he presents the private sector's outlook regarding these technologies by looking at the transgenic products that have been commercialized in Mexico by the turn of the twenty-first century, as well as the strategies of the only life science company of Mexican capital, called Pulsar. He concludes that the mode of introduction of genetic engineering technologies into the Mexican countryside in the current context of neoliberal globalism does not favor their use by small peasant farmers.

In Chapter 6, Liz Fitting assesses the direct and indirect but devastating effects of genetic engineering in North American corn production for Mexico's maize producers. Given that Mexico is the biological center of origin for corn, it contains the greatest biodiversity of this particular crop, estimated at over 10,000 local varieties (Nadal 2000). In order to protect this biodiversity, Mexico's government banned the importation of transgenic corn for agricultural production. One would think that in this case no particular effects of biotechnology would be felt. Yet, in the Tehuacán Valley, one of the sites of original maize domestication ten to twelve millennia ago, transgenic corn was found in farmers' fields. Such discovery amplified a controversy about the extent to which transgenic corn imports (for animal feed or to produce high-fructose corn syrup), largely from the United States and Canada, pose a threat to biological diversity in the crop's center of diversity and origin. The introgression of transgenic corn in her Tehuacán Valley case study is situated within the context of what Fitting designates as the "neoliberal corn regime": cuts in subsidies to small-scale producers and the policy gaps in agricultural biotechnology regulation. Social issues explored by Fitting include the insufficient resources to enforce regulation and to educate and support small-scale corn producers, and increasing hardships faced by peasant cultivators, such as labor migration and the regional decline in agriculture under the neoliberal corn regime.

Chapter 7, by Gerardo Otero, Manuel Poitras, and Gabriela Pechlaner, offers a comparative study of state policies and farmers' adoption patterns of agricultural biotechnologies in North America. The main focus is on the regional experience of the La Laguna dairy farmers in North-Central Mexico with the controversial recombinant bovine somatotropin (rBST, a milk productivity-enhancing hormone for dairy cows), within a North American context. The first part offers comparative consideration regarding the regulation of rBST around the world, focusing on debates in the region of the North American Free Trade Agreement (NAFTA). The second section analyzes the economics of rBST use in the La Laguna region of Mexico, raising significant doubts about the profitability of using rBST in this region. The next section presents fieldwork material on the qualitative assessments by capitalist dairy farmers in La Laguna of rBST's yield and profitability performance, why they adopted it, did not adopt it, or stopped using it, as well as Monsanto's marketing strategy in the region. The final section offers a political-economy and cultural interpretation of these qualitative data, as well as a research agenda based on the empirical questions that remain for a better understanding of the larger political economy of agricultural biotechnologies in the Americas.

Argentina was considered one of the world's "food baskets" for the better part of the twentieth century. By the turn of the twenty-first century, however, Argentina's overspecialization in transgenic soy production had resulted in an agricultural and hunger crisis. This tragedy is analyzed in Chapter 8 by Miguel Teubal. He examines the irony that one of the major food producers in the world, which fed its people and routinely had agricultural surpluses for export, is now facing famine among its population. This paradox has resulted from the combination of neoliberal state policies that favor export promotion and the biotechnology revolution around transgenic crops. The result has been an agricultural system that has come to overspecialize in the production of transgenic soybeans for export. Teubal points toward alternative policies for science and technology, based on local universities, and the diversification of food production for the internal market.

Given Brazil's importance as Latin America's leading agriculture producer, Wendy Jepson, Christian Brannstrom, and Renato Stancato de Souza offer in Chapter 9 an analysis of the emblematic case of attempting to keep Rio Grande do Sul, a state south of Brazil, free of genetically modified crops. The Brazilian judicial system has maintained an injunction on commercial planting of GM crops while allowing GM crop experimentation. However, experimentation without commercialization is only a temporary policy. This chapter draws on semistructured interviews, government documents, and other primary sources to assess whether Brazil's so-called "GM-free" status was politically and legally sustainable. First, the authors review the regulatory framework, the main actors, and government and research institutions involved in the biotechnology debate. This review illustrates how federal- and state-level policies are often in conflict over the experimentation, regulation, and commercialization of GM crops. More often, nongovernmental groups' legal and popular protests over current policies further shape the extent to which these technologies are regulated by the judicial system and accepted by the general public. Finally, the authors discuss the potential of unconventional alliances between key farmer producer groups opposed, at present, to GM crops and oppositional NGOs to maintain GM-free commodity markets. From this perspective, they argue that, even if the federal courts allow commercial planting of GM crops, it is not inevitable that GM technology will saturate all production systems in all areas of Brazil. State-federal political conflict, skepticism by key farmer groups, and activism by nongovernmental organizations will shape the geography of GM production in Brazil, probably maintaining significant GM-free areas.

Finally, for Brazil, Shuji Hisano and Simone Altoé explore in Chapter 10 farmers' acceptance of GMOs. They approach this issue from the viewpoint of "globalization versus localization," as well as "industrialization versus sustainability" discourses, because they grasp this issue as a manifestation of the increasing integration of Brazil into the global re-structuring of the agrifood system. Brazil can be divided into two main areas of soybean cultivation, the traditional (based on small family farms) and the *cerrados* (based on large corporate-like farms). Hisano and Altoé focus on the southern part of Brazil, since it is a traditional soybean-producing area. Concerned about the situation of small family farmers, as well as the general reality of farmers in that region, they examine the extension programs and discuss whether public institutions benefit small family farmers and help to transform their reality in the region. They focus on EMATER/RS, an extension institution in the state of Rio Grande do Sul working with a sustainable-agriculture approach based on agro-ecology. This kind of response may work well when dealing with both GMO issues and the predicament of small farmers facing globalization and industrialization.

In Chapter 11, Manuel Poitras examines the social movements that have emerged to contest the current use of genetic engineering, situating them in the context of the restructuring of state-society relations. The central concept developed is that of "technological hegemony," which incorporates the hypothesis that major technologies usually carry a veneer of political neutrality in order to be pervasively adopted, and eventually become part of the dominant "technological hegemony." In many parts of the world, genetic engineering has failed to become hegemonic, and has instead attracted protest and mobilization against its use. The causes for this failure are examined here, with particular reference to Mexico. Drawing from primary as well as secondary sources, social movements contesting the fledging technological hegemony of GMOs in Mexico are then examined. This analysis situates the politics of GMOs in the context of the broader struggles of the Mexican peasant and indigenous social movements, mainly in relation to their emergence from the authoritarian-corporatist structure of the developmentalist years, and to their core political claims to autonomy and self-management.

My concluding Chapter 12, written with Gabriela Pechlaner, draws the main lessons from the Latin American experience with the biotechnology revolution in agriculture. It focuses on what have been the main political-economy determinants of adoption and diffusion of the technology. It sums up its main impacts, who have been the main players in promoting it—corporations, governments, and adopters—and who have been the main challengers. Far from the idyllic images created by its proponents, biotechnology's introduction, which has coincided with the advancement of neoliberal globalism, has wreaked havoc on Latin America's agriculture and its capacity to feed its own people. The chapter ends with a prospective outlook as to the potential alternatives that might be more socially and environmentally sustainable, as well as a proposed agenda for future research.

Notes

1. As far as I know, Michael J. Mandel is no relation to Ernest Mandel. He is the Chief Economics Editor of *BusinessWeek*. Interestingly, though, since the late 1990s M. J. Mandel has been one of the main proponents of the "New Economy" notion to characterize the new expansionary phase of capitalism, based on increased productivity growth from new technologies.

2. Thanks to Martin Kenney for this insight.

References

Aglietta, Michel. 1979. *A Theory of Capitalist Regulation: The U.S. Experience*. London: New Left Books.

Ahmed, Iftikhar. 1988. "The Bio-revolution in Agriculture: Key to Poverty Alleviation in the Third World?" *International Labour Review* (Geneva, ILO). Vol. 127, No. 1.

———. 1989. "Advanced Agricultural Biotechnologies: Some Empirical Findings on Their Social Impact." *International Labour Review* (Geneva, ILO), Vol. 128, No. 5.

Angell, Marcia, et al. 2004. *Buying In or Selling Out? The Commercialization of the American Research University*. Piscataway, NJ: Rutgers University Press; London: Eurospan.

Arndt, Michael, and Adam Aston. 2004. "U.S. Factories: Falling Behind." *Business-Week*. 24 May 2004. Accessed online 2 September 2004. http://www.business week.com/@@6v3Nu4QQJJywghYA/magazine/content/04_21/b3884128 _mz057.htm.

Bartra, Armando. 2004. "Rebellious Cornfields: Toward Food and Labour Self-sufficiency." pp. 18–36 in Otero 2004.

Baumol, William J. 1989. "Is There a U.S. Productivity Crisis?" *Science*, Vol. 243, 3 February, pp. 611–615.

Berlan, J.-P. 1991. "The Historical Roots of the Present Agricultural Crisis." In *Towards a New Political Economy of Agriculture*. Ed. W. H. Friedland, L. Busch, F. H. Buttel, and A. P. Rudy. Boulder: Westview Press. pp. 115–136.

Blumenthal, David, Michael Gluck, Karen Seashore Louis, and David Wise. 1986. "Industrial Support of University Research in Biotechnology." *Science*, Vol. 231, 17 January.

Busch, Lawrence. 2000. *The Eclipse of Morality: Science, State, and Market*. New York: Aldine de Gruyter.

Busch, Lawrence, and Carmen Bain. 2004. "New! Improved? The Transformation of the Global Agrifood System." *Rural Sociology* 69(3):321–346.

Busch, Lawrence, William Lacy, and Jeffrey Burkhardt. 1991. *Plants, Power, and Profit: Social, Economic, and Ethical Consequences of the New Biotechnologies.* Cambridge, Mass.: B. Blackwell.

Buttel, Frederick H. 1989a. "How Epoch Making Are High Technologies? The Case of Biotechnology." *Sociological Forum* 4(2): 247–260.

———. 1989b. "Social Science Research on Biotechnology and Agriculture: A Critique." *Rural Sociologist*, Summer, pp. 5–15.

———. 2000. "The World Trade Organization and the New Politics of GMOs." Paper presented at the Annual Meetings of the Rural Sociological Society. Washington, D.C. 14–17 August.

Buttel, Frederick, Martin Kenney, and Jack Kloppenburg. 1985. "From Green Revolution to Biorevolution: Some Observations on the Changing Technological Bases of Economic Transformation in the Third World." *Economic Development and Cultural Change* 34(1). October.

Carlson, Laura. 2004. "Conservation or Privatization? Biodiversity, the Global Market and the Mesoamerican Biological Corridor." In *Mexico in Transition: Neoliberal Globalism, the State, and Civil Society.* Ed. Gerardo Otero. London: Zed Books; Nova Scotia: Fernwood Publishing, pp. 52–71.

Castells, Manuel. 2000a. *Information Age,* vol. 1: *The Rise of the Network Society.* 2nd Edition. Oxford and Malden, Mass.: Blackwell Publishers.

———. 2000b. *Information Age,* vol. 3: *End of Millennium.* 2nd Edition. Oxford and Malden, Mass.: Blackwell Publishers.

———. 2004. *Information Age,* vol. 2: *The Power of Identity.* 2nd Edition. Oxford and Malden, Mass.: Blackwell Publishers.

Castles, Stephen, and Mark J. Miller. 2003. *The Age of Migration: International Population Movements in the Modern World.* New York and London: Guilford Press.

Cleaver, Harry. 1972. "Contradictions of the Green Revolution." *Monthly Review* 2(1). June.

Cochrane, Willard. 1979. "Theory of the Treadmill," from *The Development of American Agriculture.*

Crawford, Mark. 1987. "Biotechnology's Stock Market Blues." *Science*, Vol. 238, 11 December, pp. 1503–1504.

———. 1988. "Patent Claim Build-up Haunts Biotechnology." *Science*, Vol. 239, 12 February, p. 723.

Dosi, Giovanni. 1984. *Technical Change and Industrial Transformation.* London: Macmillan.

Economist. 1987. "The Pitfalls of Patents." *Economist*, 9 May, p. 82.

Economist. 2005. "A Market for Ideas: A Survey of Patents and Technology." *Economist*, 22 October, p. 58.

FAOSTAT (Food and Agriculture Organization of the United Nations, Statistical Database). Available online: http://apps.fao.org/. Accessed October 2004.

Fischer, J. 2000. "Why Aren't You Eating Soup?" Fool.com. 19 May 2000. http://www.fool.com/dripport/2000/dripport000519.htm. Accessed November 2003.

Friedmann, Harriet. 1991. "Changes in the International Division of Labour: Agri-food Complexes and Export Agriculture." In *Towards a New Political*

Economy of Agriculture. Ed. W. H. Friedland, L. Busch, F. H. Buttel, and A. P. Rudy. Boulder: Westview Press. pp. 65–93.

Friedmann, H. 1992. "Distance and Durability: Shaky Foundations of the World Food Economy." *Third World Quarterly*, Vol. 13(2): 371–383.

———. 1995. "Food Politics: New Dangers, New Possibilities." *Food and Agrarian Orders in the World-Economy*. Ed. P. McMichael. Westport, Conn.: Greenwood Press. pp. 15–34.

GAO (General Accounting Office, U.S. Congress). 1986. "Biotechnology: Agriculture's Regulatory System Needs Clarification." (Report to the Chairman, Committee on Science and Technology, U.S. House of Representatives), GAO/RCED 86–59, March.

Goldberg, Rebecca, Jane Rissler, Hope Shand, and Chuck Hassebrook. 1990. *Biotechnology's Bitter Harvest: Herbicide-Tolerant Crops and the Threat to Sustainable Agriculture, A Report of the Biotechnology Working Group* (U.S.).

Goodman, David, Bernardo Sorj, and J. Wilkinson. 1987. *From Farming to Biotechnology*. Oxford: Basil Blackwell.

Halweil, B. 2000. "Portrait of an Industry in Trouble." In *Environmental Politics Casebook: Genetically Modified Foods*. Ed. N. Miller. Boca Raton, Fla.: Lewis Publishers, 201–204.

Hastopoulos, George N., Paul R. Krugman, and Laurence H. Summers. 1988. "U.S. Competitiveness: Beyond the Trade Deficit." *Science*, Vol. 241, 15 July, pp. 299–307.

Hewitt de Alcántara, Cynthia. 1978. *Modernización de la agricultura mexicana*. Mexico City: Siglo XXI Editores.

Hoppe, R. A., and J. M. MacDonald. 2001. "America's Diverse Family Farms: Assorted Sizes, Types, and Situations." ERS Agriculture Information Bulletin No. 769.

Johnson, O. Gale. 2002. "Biotechnology Issues for Developing Countries." *Economic Development and Cultural Change* 51(1):1–4.

Kenney, Martin F. 1986. *Biotechnology: The University-Industrial Complex*. New Haven: Yale University Press.

Kleinman, Daniel L. 2003. *Impure Cultures: University Biology and the World of Commerce*. Madison: University of Wisconsin Press.

Kleinman, Daniel L., and Jack R. Kloppenburg, Jr. 1988. "Biotechnology and University-Industrial Relations: Policy Issues in Research and the Ownership of Intellectual Property at a Land Grant University." *Policy Studies Journal* 17(1).

Kloppenburg, Jack R., Jr. 1988. *First the Seed: The Political Economy of Plant Biotechnology, 1492–2000*. New York: Cambridge University Press.

Kloppenburg, Jack R., Jr., Daniel L. Kleinman, and Gerardo Otero. 1988. "La Biotecnología en Estados Unidos y el Tercer Mundo." *Revista Mexicana de Sociología*, Year L, No. 1, January–April.

Knutson, R., J. Penn, and B. Flinchbaugh. 1998. *Agricultural and Food Policy*. Fourth Edition. Upper Saddle River, N.J.: Prentice Hall Inc.

Lewontin, R. C. 2000. "The Maturing of Capitalist Agriculture: Farmer as Proletarian." *Hungry for Profit*. Chapter 5, pp. 93–106.

MacDonald, James M., and Mark Denbaly. United States Department of Agriculture [USDA], Economic Research Service [ERS]. Undated. "Concentration in Agribusiness." Available online: www.farmfoundation.org/tampa/macdonald.pdf. Accessed November 2003.

Mandel, Ernest. 1978. *Late Capitalism*. London: New Left Books.

———. 1995. *Long Waves of Capitalist Development: The Marxist Interpretation*. 2nd Revised Edition. London and New York: Verso.

Mandel, Michael J. 1996. *High Risk Society: Peril and Promise of the New Economy*. New York: Times Business.

———. 2004. *Rational Exuberance: Silencing the Enemies of Growth and Why the Future Is Better Than You Think*. New York: HarperBusiness.

Montague, P. 2000. "#695—Biotech In Trouble—Part 1, May 04, 2000." From Environmental Research Foundation Home. Rachel's Environment & Health News. *Environmental Politics Case Book: Genetically Modified Foods*, ed. N. Miller. Boca Raton, Florida: Lewis Publishers. 201–204.

Nadal, Alejandro. 2000. "Corn and NAFTA: An Unhappy Alliance." *Seedling: The Quarterly Newsletter of Genetic Resources Action International* (Barcelona). 17(2): 10–17.

National Research Council. 1989. *Alternative Agriculture*. Washington: National Academy Press.

Nelkin, Dorothy. 1984. *Science as Intellectual Property: Who Controls Research*. New York: Macmillan.

Nowak, Pete. 1989. "Agricultural Biotechnology: A Perspective from the Bottom Up," *Rural Sociologist*, Summer, pp. 17–19.

Orden, David, Robert Paarlberg, and Terry Roe. 1999. *Policy Reform in American Agriculture: Analysis and Prognosis*. Chicago and London: University of Chicago Press.

Otero, Gerardo. 1989a. "Ciencia, nuevas tecnologías y universidades." *Ciencia y Desarrollo* (Mexico City). Vol. XV, No. 87, July–August, pp. 49–59.

———. 1989b. "Commodification of Science: Biotechnology in the United States and Mexico." Paper Presented at the Meetings of the Society for Social Studies of Science, Irvine, California, 15–18 November.

———. 1991. "The Coming Revolution of Biotechnology: A Critique of Buttel." *Sociological Forum* 6(3): 551–565.

———. 1992. "The Differential Impact of Biotechnology: The Mexico–United States Contrast." pp. 117–126 in *Biotechnology: A Hope or a Threat*. Ed. Iftikhar Ahmed. London: Macmillan.

———. 1995. "The Coming Revolution of Biotechnology: A Critique of Buttel." In *The Biotechnology Revolution?* Ed. Fransman Martin, Junne Gerd, and Roobeek Annemieke. Oxford, UK, and Cambridge, USA: Blackwell. (Reprint of Otero 1991.)

Otero, Gerardo, ed. 2004. *Mexico in Transition: Neoliberal Globalism, the State and Civil Society*. London: Zed Books; Nova Scotia: Fernwood Publishing.

Otero, Gerardo, and Cornelia Buttler Flora. Forthcoming. "Sweet Protectionism: State Policy and Employment in the Sugar Industries of the NAFTA Countries." In Juan Rivera and Scott Whiteford, eds., *NAFTA, Agriculture and Small Peasant Producers*. University of Scranton Press.

Pérez, Carlota. 2002. *Technological Revolutions and Financial Capital: The Dynamics of Bubbles and Golden Ages*. Cheltenham, UK; Northampton, Mass.: E. Elgar.

Peritore, N. 1995. "Biotechnology: Political Economy and Environmental Impacts." *Biotechnology in Latin America: Politics, Impacts, and Risks*, ed. N. P. Peritore and A. K. Peritore. Wilmington, Delaware: Scholarly Resources, Inc., pp. 1–36.

Peterson, Thane, and Larry Armstrong. 1990. "Roche's Big Buy May Set Off a Shopping Frenzy." *BusinessWeek*, 19 February, No. 3146, p. 38.

Piore, Michael J., and Charles F. Sabel. 1984. *The Second Industrial Divide: Possibilities for Prosperity*. New York: Basic Books, Inc.

Portes, Alejandro, Manuel Castells, and Lauren A. Benton. 1989. *The Informal Economy: Studies in Advanced and Less Developed Countries*. Baltimore, Md.: Johns Hopkins University.

Portes, Alejandro, and Kelly Hoffman. 2003. "Las estructuras de clase en América Latina: Composición y cambios durante la época neoliberal." Serie Políticas Sociales #68. Naciones Unidas: CEPAL, División de Desarrollo Social, Santiago de Chile. (An English version is available.)

Reichardt, D. 2003. "Biotech Overwhelms U.S. Patent Office: Complicated Science, Process Create Patent Backlog." *Atlanta Business Chronicle*. Atlanta, Ga. Website: http://www.bizjournals.com/atlanta. Accessed November 2003.

Ribaudo, M., R. Heimlich, R. Claassent, and M. Peters. 2001. "Least-Cost Management of Nonpoint Source Pollution: Source Reduction versus Interception Strategies for Controlling Nitrogen Loss in the Mississippi Basin." *Ecological Economics*. Excerpted as "'Dead Zone' in the Gulf: Addressing Agriculture's Contribution." Amber Waves. USDA. ERS.

Shiva, V. 2000. "North-South Conflicts in International Property Rights." *Peace Review* 12(4): 501–508.

Sklair, Leslie. 1989. *Assembling for Development: The Maquila Industry in Mexico and the United States*. Boston: Unwin Hyman.

———. 2002. *Globalisation: Capitalism and Its Alternatives*. Oxford: Oxford University Press.

Smith, Geri. 2002. "Farmers are Getting Plowed Under: With Tariffs Disappearing, U.S. Exports to Mexico May Soar." *BusinessWeek*, 18 November.

Thurow, Lester C. 1987. "A Weakness in Process Technology." *Science*, Vol. 238, 18 December.

United States Department of Agriculture (USDA), Economic Research Service (ERS). 2001. "Farm and Rural Communities: Hired Farm Labor—Comparing the U.S. & Mexico." *Agricultural Outlook*. Shields, D., Economics Ed., ERS, Market and Trade Economics Division. January–February 2001.

United States Department of Agriculture (USDA), Economic Research Service (ERS). 2006. "Global Resources and Productivity." Chapter 3.5 in Agricultural Resources and Environmental Indicators. http:www.ers.usda.gov/publications/arei/eib16/Chapter3/3.5/. Accessed 11 November 2007.

United States Department of Agriculture (USDA), Office of Communications. 2003. *Agricultural Fact Book: 2001–2002*. March 2003. http://www.usda.gov.ca.

Wade, Nicholas. 1974. "Green Revolution: A Just Technology Often Unjust in Use." *Science*, Vols. 186 and 187, 20 and 27 December.

Young, John A. 1988. "Technology and Competitiveness: A Key to the Economic Future of the United States." *Science*, Vol. 241, 15 July.

CHAPTER 2

Latin American Agriculture, Food, and Biotechnology: Temperate Dietary Pattern Adoption and Unsustainability

GERARDO OTERO AND GABRIELA PECHLANER

The main proposition addressed in this chapter is that, with the globalization of capitalism, national agricultures in Latin America have increasingly conformed to temperate-climate food consumption and production patterns. Because the Green Revolution has been effectively transferred, at least to the regions with irrigated agriculture, Latin America has become technologically dependent. Adopting dietary patterns of temperate countries and technological dependency entails undesirable social and environmental implications. Socially, farm structures tend to become quite polarized, with fewer and larger farmers surviving, and the rest rendered bankrupt or productively redundant. Environmentally, modern technologies have taken agriculture to an unsustainable point: soil erosion, land and water contamination, and decreased genetic diversity are just a few of the problems that bring into question the sustainability of this production model. Furthermore, the diet based on meat and dairy products has become dangerous to people's health, for it is clearly associated with increased incidence of heart disease and various cancers.

On the positive side, it should be said that these problems, which first emerged in the United States (see Chap. 1, this volume), have prompted the attempt to explore alternative agricultural practices. A number of studies have found that alternative agricultural practices, which enhance the biological interactions of the environment and keep chemical inputs at a minimum, are not only friendly to the environment but can also be economically profitable (Altieri 2001; National Research Council 1989). This "movement" toward an alternative agriculture is still a minority trend in the United States, but there is some indication that it is growing. Certified organic agriculture, for example, more than doubled between 1992 and 1997, and shows continued strong market signals (Greene 2001, 19).

Nonetheless, by 2001, still only 0.3 percent of total farmland was certified organic, although there are important variations. For example, while top American field crops continue to have low rates of organic certification (e.g., corn, 0.1 percent, soy, 0.2 percent, and wheat, 0.3 percent), fruits, vegetables, and herbs show strong organic trends (e.g., apples, 3 percent, and lettuce, 5 percent) (USDA, ERS, "Data," 2002).

The question is whether alternative agriculture may become established to any significant degree in the current socioeconomic and institutional context. This context includes, on one hand, that agricultural biotechnology and genetic engineering have been added to the mix of "modern agriculture." On the other hand, the biotechnology revolution has converged with neoliberal globalism and its attendant policies of market orientation, deregulation, privatization, withdrawal of state subsidies, and so on. This institutional, ideological, and policy context represents a big contrast to that in which the Green Revolution was introduced: the milieu of a nationally oriented development model, focused on internal markets, protectionism, public agricultural research, state subsidies for farm production, and so forth.

The technological paradigm of modern agriculture involves a specific package of inputs made up of hybrid and other high-yielding plant varieties; mechanization; agrochemical fertilizers and pesticides; and irrigation. The "Green Revolution" is the name adopted by this technological package when it was exported to developing countries. While the Green Revolution technically began in Mexico in 1943, with a program promoting high-yielding wheat varieties (Hewitt de Alcántara 1978), its origin and initial development were located in the agriculture of the United States, dating from the 1930s (Kloppenburg 1988). This exported package then became the "technological paradigm" for modern agriculture throughout the twentieth century (see Chap. 1, this volume).

After its initial success with wheat in Mexico, the technological package quickly spread to Asia and other parts of the developing world, spurring a "revolution" of increased agricultural productivity. Globally, the crops most affected by the modern agricultural paradigm, in terms of hybrid and improved plant varieties, were corn, rice, and wheat—the most important food crops in the world. The rest of the technological package was extended to a large number of crops through massive applications of chemicals based on hydrocarbons. This is particularly the case in the production of fruits and vegetables (Murray and Hoppin 1990; Thrupp 1991).

The technological package of the Green Revolution was not applied across the board in developing countries, however. As a complete package it was adopted mostly in irrigated agricultural areas in Latin America, the Near East, and North Africa; while in Asia, a selective adoption was made which was much less mechanization-intense. In contrast, few regions in Sub-Saharan Africa have adopted the Green Revolution technologies (Evenson and Gollin 2003; ILO 1988). One of the consequences of this differential adoption of technological innovation is that vast regions of the developing world have become largely marginalized from the world economy. If we take "irrigated area" as a rough indicator of those regions that were suitable to adopt Green Revolution technologies, the proportion of arable lands that were irrigated in developing countries was only 26 percent by 2001 (FAOSTAT). Although some of the best rain-fed regions have adopted modern technologies, the largest proportion of rain-fed cultivated area in developing countries has remained without them (Paarlberg 1988). Therefore, some 74 percent of arable lands in developing countries are excluded from the modern agricultural paradigm, and thus from the global economy. Differential adoption (and impacts) did not only occur between world regions, of course, but also within regions, as will be discussed in more detail later.

There could be an optimistic side to such regional differential adoption. Those "marginalized" regions of the world might have the possibility of moving directly beyond the technological paradigm of modern agriculture into one that maximizes the biological synergies of ecosystems. This could increase production and expand employment opportunities in such a way that the economic and environmental sustainability of agriculture are preserved. The current economic and institutional trends lead more to pessimism than to optimism, however. Trends of the last two decades of the twentieth century indicate that rather than contributing to produce an "alternative agriculture," biotechnology and genetic engineering are controlled by the same private economic actors that ultimately became central to the "technological paradigm" of modern agriculture. Furthermore, the agricultural-inputs industry, on one hand, and the agricultural distribution and processing industries, on the other, have both become greatly concentrated in just a few economic agents. The consequences of this will be illustrated in this chapter.

We argue that new products resulting from biotechnology reinforce and deepen the structural changes initiated by the Green Revolution and that, given the different institutional and economic policy context in which biotechnology emerged, it is having even greater socially and regionally polarizing effects. Moreover, we maintain that the major obstacle to overcome in order to move in the direction of an alternative agriculture, both in the United States and in Latin American countries, is the current market structure of input producers and agricultural distributors and processors in the agrifood complex. This market structure is made up of a set of oligopolistic transnational corporations (TNCs), which are directing the trajectory of biotechnology research in ways that are socially and environmentally problematic (Heffernan and Hendrickson 2002; Hendrickson and Heffernan 2005). Most notably on the inputs side, rather than developing products to make farmers less dependent on the use of agrochemicals, TNCs have been advancing research and development that further entrench and extend the petrochemical era of agriculture.

The most blatant example of this entrenchment can be found in the genetic traits selected for promotion: the largest proportion of transgenic or genetically modified (GM) crops launched since 1995 are programmed to be herbicide resistant, allowing for seed and herbicide to be sold together, as the central components of a new technological package. The increasing vertical integration between input suppliers on one hand and output processors on the other promotes this course and creates a power imbalance that further decreases the potential for democratic change.

This chapter is divided into three main sections. The first and largest section identifies the social and environmental problems that have emerged with the transfer of modern agriculture to Latin American countries, some with aggravated consequences to those that emerged in the United States (Chap. 1, this volume). The second section addresses the temperate dietary pattern adoption that has taken place in the region, and notes its associated consequences. The third section places the emergence of the biotechnological revolution in the context of these previous sections—adding the centrality of transnational corporations or TNCs—and discusses its impact on the social and environmental issues first raised by the Green Revolution. Two differences of biotechnology are highlighted—environmental particularities and the private corporate dynamics of TNCs—and the implications of such differences are outlined. In the conclusion we argue for the democratization of agricultural research agendas, so that new technologies can be developed that are friendly to small producers and to the environmental sustainability of agriculture.

The Green Revolution in Latin America

In this section we argue that the impacts of modern agriculture in Latin America largely parallel those outlined for the United States in Chapter 1 of this volume, except that they have been even more socially polarizing and detrimental to the environment. Two major differences should be noted from the outset between the United States and the Latin American experiences with modern agriculture. First, whereas the United States underwent a robust industrialization process that was able to absorb the rural masses displaced from agriculture, Latin American countries have had to struggle with growing unemployment and underemployment in cities unable to effectively absorb the incoming rural masses (Portes 1989; Portes, Castells, and Benton 1989; Eckstein 1990; Gilbert 1998; Castles and Miller 2003). Further, while this urbanization process took some time in the United States, the transition away from subsistence and small agriculture in some regions of Latin America has occurred at a significantly faster pace. The highly inequitable distribution of land from at least Spanish colonization onward made things worse in Latin America, also with the failure of agrarian reforms and the subsequent crumbling of agrarian structures (Thiesenhusen 1989, 1995; Bryceson, Kay, and Mooij 2000; Petras and Veltmeyer 2005).

Second, although some important regional differences exist in U.S. agriculture, based mostly on the social organization of production, the modern technological paradigm was adopted across the board. In contrast to this, only irrigated regions of Latin America were capable of profitably adopting such a technological model. Due to these two differences, the problems of social and environmental issues have been much graver in Latin America than in the United States.

The Green Revolution

The Green Revolution was first launched as a means to forestall red revolutions in the developing countries through increasing food production. In the early Green Revolution years, concerns were high about a population growth that was projected to outpace food production capabilities. With respect to increasing food production, the Green Revolution was phenomenally successful. Despite subsequent losses from soil degradation (as will be discussed), global cereal yields have consistently increased since the 1960s. From 1966 to 1990, global wheat production increased over 75 percent, while rice production doubled (Davies 2003, 125). At the national level, this sometimes had dramatic results. For example, with a 77 percent area adoption of high-yield varieties (HYVs), Indonesia witnessed a 276 percent increase in production (Davies 2003, 125). For developing countries more generally, the average cereal yields increased 2.3 percent annually; 2.1 percent per year in Latin America (Weibe 2003, 10). It is estimated that 50 percent of the improved yields were a result of genetic improvements and 50 percent were due to the increased use of conventional inputs (Byerlee et al. 2000, cited in Weibe 2003, 10)—the combined technological package of modern agriculture.

Consequently, in contrast to Malthusian fears that population growth would outpace food production, post–Green Revolution production increases have not only met, but even outpaced, global population growth. While global population increased 110 percent from 1950 to 1990, global cereal production increased 174 percent for the same time period (Dyson 1996, 101). While this increased production is by no means evenly distributed over the globe, the very fact of it indicates the potential for new technologies to subvert a demographically induced food crisis, and fuels arguments for further technological increases, such as those brought about by biotechnology. For this reason, it is important to directly address what was the role of these technologies with respect to hunger alleviation.

While the Green Revolution was indisputably successful in increasing food production in vulnerable regions, it is decidedly less certain what its actual impact on hunger has been. Critics of the Green Revolution have argued that its increased food production has not necessarily translated into less hunger. Not to overstate the issue, but the number of food-insecure people in the developing world has declined overall. For example, the number of food insecure has declined from 960 million in 1969–1971 to 791 million in 1995–1997 (Pinstrup-Andersen and Pandya-Lorch 2000, 3). There is a great degree of variation between countries, however, with some improving and others declining in their proportion of food secure. Lappé, Collins, and Rosset argue that while total food availability increased 11 percent per person and while the number of hungry people fell 16 percent over the key Green Revolution decades (between 1970 and 1990), removing China from the analysis actually results in an 11 percent increase in the number of hungry people in the rest of the world (Lappé, Collins, and Rosset 1998, 61). In South America, for example, while there was an 8 percent increase in per capita food supplies, the number of hungry people actually increased by 19 percent (ibid.). Such contrasting results are reflective of the different social structures and their respective impacts on social inequality.

Therefore, questions about inequality, distribution of benefits and losses, and persistent hunger have followed hard on the heels of Green Revolution successes with higher yields. Consequently, much attention has been paid to investigating the "inevitability" of the new technologies and their impact on poverty alleviation: do high-yield ([HY] or "modern variety" [MV]) seeds necessarily benefit the poor, as it seems reasonable to assume? Early proponents of the Green Revolution claimed that the technology knew no distinction between large and small farmers—that it was scale-neutral. Lipton (1989) argued that high-yield varieties do not transform power structures and relations of production because "MV [Modern Variety] technology is too 'seriable,' 'separable' and 'single unit' for that."[1] Already by the late 1980s, however, numerous studies had raised the possibility that such optimism may not be warranted. Poor people can be affected by the new technology as small farmers, as laborers, and as net purchasers of food (Das 2002). Therefore, the impact of the technology on all three aspects would need to be determined in order to make any assumptions about poverty alleviation.

Most significantly, despite their theoretical "scale-neutrality," Green Revolution technologies have ultimately been found to disproportionately disadvantage small producers. The best yields require the application of the petroleum-based package of fertilizers and pesticides, which increases the capital required for profitable farming. Decreased genetic diversity and increased intensity of modern production heighten disease and pest problems, which affect even non–Green Revolution farmers, and further increase the incentive for chemical inputs. At the same time, "credit, tenural and marketing arrangements have tended to favour the adoption of the new technologies by larger rather than smaller farmers" (Conway and Barbier 1990, 22). Small farmers cannot gain volume discounts, or "hold out" for the best crop price; they pay substantially higher interest rates to local moneylenders; and "government subsidized credit overwhelmingly benefits the big farmers" (Lappé, Collins, and Rosset 1998, 67–68). Those who do not adopt the technologies are further disadvantaged by the decreased price for their products, while adopters can offset price declines with increased yields. Thus small peasants are rendered "inefficient" by the introduction of new technologies: because they cannot afford to adopt them, they are condemned to bankruptcy and/or living in poverty and marginality. This trend has become particularly exacerbated since the onset of neoliberal reform in the 1980s (Rubio 2001, 2004; von Bertrab 2004).

Numerous other factors, such as labor and migration, can further influence the impact of Green Revolution technologies. Overall, however, the findings have strongly critiqued assumptions of the technology as necessarily being "pro-poor." A 1995 report that reviewed "every research report published on the Green Revolution over a thirty-year period" was quite revealing. Out of the 300 reports reviewed, 80 percent were found to conclude that inequality increased as a result of the Green Revolution (Boletim da Commissão Pastoral da Terra-CPT, cited in Lappé et al. 1998, 65). As a direct result of these experiences with the Green Revolution, "economists have learned that the social and economic environments people live in have more to do with who benefits from new technology than the specific characteristics of the technology itself" (Arends-Kuenning and Makundi 2000).

In sum, agricultural-productivity-enhancing technologies are not necessarily to the inevitable benefit of the poor and hungry. Similarly, from this analysis of the Green Revolution we may conclude that poverty is not a "thing" but a "socio-spatial process" (Das 2002, 25). Consequently, "whether the GR or any technology will enhance or mitigate poverty is a contingent matter" (Das 2002, 19).

In the end, the new production system introduced into developing countries has had a polarizing effect. Food production has increased, and while the total number of hungry people is expected to continue to decline in many places (excluding sub-Saharan Africa), the nutritional gap—the "gap between available food and food needed to meet the minimum daily caloric intake requirements estimated by FAO" (Weibe 2003, 43)—is predicted to increase for many regions. In 2000, the nutritional gap in Latin America and the Caribbean was 0.7 million metric tons. This is predicted to rise to 0.9 million metric tons by 2010 (Shapouri and Rosen 2000, cited in Weibe 2003, 44). Nonetheless, the Green Revolution was not so much a plot as the result of extending the capitalist and U.S. agricultural models to the Third World.

Modern versus Traditional Production and Labor: Social and Regional Polarization

Having outlined the general social and regional polarization trends linked to the introduction of the new technologies in developing countries, we will now take a moment to look more specifically at the Latin American experience. We will first discuss these trends as a result of the new technologies, and then we will address environmental problems.

Mexico was the first Third World country which imported Green Revolution technologies for its agricultural development, and shared in their expansion and testing. The Green Revolution was more polarizing in Mexico and many other developing countries than in the United States because most of the productive units in Mexico were of subsistence, peasant type, with few possibilities of making capital investments in the technological package. Latin American government policies have tended to favor landowners and agricultural companies. Given the heavily mechanized character of their operations, they tend to employ few people and to orient their production toward exports, rather than the national markets. In contrast, subsistence farmers who do produce for local and regional markets have received little or no government support; their main role has been to keep urban food prices down (Dyson 1996, 191).

Consequently, only a tiny minority of rural producers—agricultural entrepreneurs—were to become the basis for the increased production and productivity of commercial and export crops (Sanderson 1986; Mares 1987; Otero 1999). Governmental attitude has focused on agricultural intensification of "high potential agricultural areas" (Pichón and Uquillas 1997, 489), and therefore, the polarization that occurred within communities and between producers was replicated on a larger scale between regions.

> In Latin America, Green Revolution technologies were only introduced in the humid pampas of South America, the irrigated areas of Mexico, the Caribbean islands, the Pacific coast, and some tropical areas. During the period between 1961 and 1990, 71 percent of the increased production in the LAC region resulted from increased yields obtained mainly from these areas, while the remaining 29 percent was obtained from area expansion. (Pichón and Uquillas 1997, 489)

Rice provides a good example of the polarization dynamic that is set in motion by the American agricultural package. A summary report on the economic impact of improved varieties of rice and beans by the Centro Internacional de Agricultura Tropical (CIAT) indicates that rice has been integrated into Latin American diets to the point where it has now surpassed wheat, maize, cassava, and potatoes as a source of calories, with per capita rice consumption increasing from 10 kilograms in the 1920s to 30 kilograms by the 1990s (Schoonhoven and Pachico 1998, 2). This transition to greater rice consumption has in part been the result of changing diets due to urbanization, as will be discussed. Therefore, improvement in yields of such an important food crop could have important implications for hunger alleviation. Modernization of rice varieties and agricultural practices has in fact caused a much celebrated doubling of production yields with only moderate increases in the area planted (ibid., 6)—a rather significant feature in the context of increasing population pressures.

The benefits of increased rice yields, however, have not been evenly distributed. Productivity gains of irrigated areas have benefited the producers in these areas, and they have also benefited consumers (with a 50 percent decline in the price of rice, in real terms, in the last thirty years). Without the benefits of adopting the technology and its consequential increase in productivity, however, the drop in rice prices comes at a substantial loss to those farmers outside the irrigated areas. The transition from subsistence agriculture to commercial agriculture, therefore, necessarily implies the semi-proletarianization of a large proportion of "traditional producers." We say *semi*-proletarianization because the inability of the rest of the macro economy to absorb bankrupt peasants in jobs with livable wages condemns many of them to continue trying to scrape a living from agriculture. By themselves, however, neither land nor agricultural wages are sufficient sources of income for the vast majority of rural dwellers (for figures on Mexico, see Otero 1999, Chap. 4).

A comparative study by the USDA-ERS on U.S. and Mexican hired farm labor is quite illustrative of the agricultural dynamics in Mexico. The report notes that the gap between "modern" and "traditional" farms in Mexico has widened "due to large differentials in organization, technology, and financing" (USDA, ERS, "Farm," 2001, 16). The "modern sector" of Mexico's agriculture has benefited from freer trade in North America, "while offering seasonal employment to farmworkers from the traditional agricultural sector" ("Farm," 16). Consequently, this "sizable pool" of available laborers acts as a comparative advantage for the Mexican modern agricultural sector as it seeks out new export markets (ibid., 14).

Viewed from the perspective of the traditional farmer, however, these dynamics are rather less rosy, as the "plentiful supply" of laborers means more competition for limited employment. In fact, 1.4 million of the 2.3 million hired farmworkers are migrants, chasing a series of seasonal and part-time jobs (ibid., 15). This employment varies greatly by region. It has decreased at an average 7.6 percent annually in the central states between 1996 and 1999, largely as a result of the land and labor absorption of urbanization (ibid.). At the same time, agricultural employment is growing in the southern states, which have "relatively high levels of poverty and a larger indigenous population" (ibid.). Further, in real terms, agricultural wages in Mexico have been decreasing 4.3 percent annually on average between 1989 and 2000. Given the context, there is nonetheless "little evidence of a single commodity or activity in Mexico's agriculture facing difficulties obtaining hired labor" (ibid.). As a result, large proportions of Mexicans migrate from rural areas in search of employment, often as far as the United States, where they are drawn by the huge wage differential between the two countries ($3.60 in Mexico versus $66.32 in the United States for eight hours' labor in 2000). In 1998, Mexican laborers in the United States (57 percent of whom were undocumented) made up 78 percent of all U.S. farmworkers (ibid., 14).

In sum, even these brief statistics clearly illustrate the hardship involved in the proletarianization or semi-proletarianization of traditional laborers that results from the introduction of modern agriculture. Mexican migration to the United States had reached such proportions by the twenty-first century that some analysts talk about Mexico's loss of "labor sovereignty" (Bartra 2004). Related to this phenomenon, dollar remittances from Mexican migrants in the United States have reached all-time highs. From being the fourth source of foreign revenue in the early 1990s, by the turn of the century they became the second source, after oil, beating manufacturing exports and tourism (Delgado Wise 2004).

Environmental Issues:
Contamination, Soil Degradation, and the Loss of Genetic Diversity

The pollution and health problems associated with the intensive use of agrochemicals are often more acute in Latin America, due to the greater laxity of legislation and its insufficient enforcement. It is widely known that many chemicals that have been banned or restricted in the United States continue to make their way into Third World markets (Restrepo 1988; Murray and Hoppin 1990). Further, farmers with little formal education from these countries are much more likely to exceed the necessary amount of pesticides and herbicides. The problems of Third World adoption of modern agriculture, however, are not limited to regulatory laxity or procedural lack of knowledge. The biggest environmental issues of U.S.-style intensive agricultural practices—soil degradation and erosion and chemical contamination from modern agricultural inputs—are both replicated and exacerbated in developing countries. To these impacts, we must add the exacerbated health effects on humans. The World Health Organization estimates 25 million cases of pesticide poisonings occur each year the world over. The tragedy for developing countries is that while 80 percent of pesticides are used in developed countries, 99 percent of pesticide poisonings occur in developing countries (United States Embassy Website, Tokyo, Japan, 2002).

A key argument behind Green Revolution technologies is that, due to their land-saving capabilities, the use of these high-input, high-yield varieties is actually an environmentally sound way of meeting the food demands of a growing population. For example, estimates are that it would have required double the planted area to achieve the same increases in rice yields that were attained through the irrigation and variety improvements in Latin America (Schoonhoven and Pachico 1998, 7). On the other hand, there are also reports that even on the productive lands, returns from the high-yield varieties are beginning to decline (World Development Report [WDR] 2003; Conway and Barbier 1990). The Global Assessment of Soil Degradation (GLASOD) estimated that since 1945, 38 percent of the world's cropland has been degraded to some extent. Estimates for Latin America are that degradation has affected 51 percent of cropland (Weibe 2003, 15).

Soil erosion is "substantially greater in tropical developing countries, where soils, rainfall, and agricultural practices are more conducive to erosion" than in the United States (WDR 1992, 56). Crop yields are declining as a result of soil degradation, but this is less of a problem in the aggregate than for specific countries and crops: most notable for developing counties are reductions in yields for maize (WDR 1992, 56). The deterioration of agricultural soils is particularly significant for dryland areas, such as can be found in large rural pockets of Mexico (WDR 2003, 61). Again, erosion rates differ by crop, region, and soil type, but weighted average annual erosion rate estimates for Latin America range from 8.7 tons per hectare for potatoes to 15.0 tons per hectare for maize (Weibe 2003, 31).

Soil degradation has a variety of causes. Scholars of high technologies in agriculture agree that the erodibility of marginal lands is the reason why intensive technologies like those of the Green Revolution and biotechnology should only be used on less erodible lands. The 2003 World Development Report found that although government and the private sector had been emphasizing the development of lands with commercial potential—on the assumption that as these were developed, less productive lands would be abandoned—these assumptions had been flawed. Not only does much of the rural population of developing countries remain on fragile lands, but it is estimated that the population on fragile lands has actually doubled since 1950 (WDR 2003, 59), as increased population pressures are forcing people into lands that are of marginal agricultural quality. Soil degradation does not only result from direct erosion, but from chemical inputs, such as fertilizer. Between 1961 and 1998, global fertilizer consumption increased 4.1 percent per year (Weibe 2003, 11). While the growth in fertilizer consumption is declining slightly, specifically in developed countries, this consumption is still projected to increase 0.9 percent per year to 2030, with associated onsite and offsite contamination impacts (Weibe 2003, 11).

Another aspect of the technological package of modern agriculture is irrigation. Agriculture accounts for 70 percent of water withdrawals globally, and 90 percent in developing countries. Globally, irrigated cropland has increased an average 1.9 percent annually since 1961—six times the growth in cropland area (Weibe 2003, 12). Irrigation, however, is not only about water withdrawals. The extensive irrigation practices of modern agriculture have created soil degradation problems; most notably salinization (elevated salt concentrations). Irrigated lands in developing countries increased from 100 million hectares in 1961 to over 200 million hectares in 2001. Already by 1992, 24 percent of all irrigated lands suffered from salinization due to poor irrigation practices (WDR 1992, 57).

Most significant of all of the environmental problems brought about by the Green Revolution in Latin American countries is the loss of plant genetic diversity. Due to the vagaries of nature, the regions of greatest plant genetic diversity are located mostly in the South, in today's developing countries—in the so-called "Vavilov centers of diversity." When homogeneous plant varieties, improved with plant breeding techniques, are introduced on a massive scale in these centers of genetic diversity, the local traditional cultivars are abandoned. Such cultivars represent plants that had been domesticated by generations of peasants, and when they are replaced by improved varieties, they simply become extinct (Fowler and Mooney 1990). The genetic diversity contained in these traditional cultivars, paradoxically, has been the very raw material for plant breeders to introduce a number of desirable features in the improved varieties. If the latter replace the former, plant breeders (and nature) will be left with insufficient raw material to continue their work.

The main cause of extinction is the replacement of traditional, local varieties by commercial, improved ones, by ten times over any other cause. Habitat destruction is the next leading cause of genetic erosion. In Europe, it is expected that 75 percent of varieties currently grown will be extinct within ten years. In the United States, the replacement of new varieties of vegetable crops in the twentieth century has caused the extinction of 97 percent of 75 crops, according to a study conducted by the

Rural Advancement Fund International. With respect to new commercial crop varieties, once they are dropped from the catalogs of seed companies they disappear after one or two years (Fowler and Mooney 1990). There-fore the Green Revolution began the process of turning the centers of plant genetic diversity into centers of uniformity. While having a benefi-cial impact on yields, the homogeneity of crops associated with improved varieties of seeds has indeed occurred.

By the 1990s, the percentage of cropland dedicated to scientifically bred varieties was already high in developing countries: 90 percent for wheat; 74 percent for rice; and 62 percent for maize (WDR 2003, 11). Losses in genetic diversity also result from the direct loss of plant life through the clearing of forests and other ecosystems to make room for more cultivated crops. Conventional wisdom holds that poverty drives subsistence farmers to encroach on virgin lands, such as tropical forests. This is now being qualified by a more regionalized perspective, and while poverty does cause ecosystem degradation, it is not the sole cause it was once thought to be. On one hand, parts of Brazil, Peru, and Bolivia have faced sharp demographic increases coupled with very low levels of per capita food availability. Among the consequences of this mix are grow-ing "land degradation and fragmentation, growing landlessness, and out-migration . . . in search of cultivable land" (Dyson 1996, 191). On the other hand, deforestation in the Amazon Basin is attributed more to "cattle ranchers, logging companies, and the Brazilian military" than it is to the movement of subsistence farmers (Dyson 1996, 191). Even more significantly, the 2003 World Development Report concluded that "large-scale agriculture, including ranches and plantations, accounts for most deforestation in Latin America" (WDR 2003, 164).

National Food Security and Domestic Consumption Patterns

The transfer of the U.S. agricultural paradigm has involved profound changes not only in the patterns of crop production, but also in food con-sumption patterns in the receiving countries. Profound dietary changes for developing countries were first initiated by U.S. food aid, a strategy for disposing of agricultural surpluses of wheat (Burbach and Flynn 1980). Wheat was "both a change from most traditional dietary staples and an efficiently produced, often subsidized alternative to the marketed crops of domestic farmers" (Friedmann 1994, 182). Most Third World coun-tries have been unable to compete with cheap exports from the United

States, and their domestic agricultures have stagnated and declined: "Import policies created food dependency within two decades in countries which had been mostly self-sufficient in food at the end of World War II" (Friedmann 1994, 182).

The case of Mexico's agricultural collapse is illustrative of increasing food dependency in Latin America. In 1965, its population was about equally distributed between urban and rural areas, with agriculture contributing about 65 percent of foreign exchange. By 2000, the urban population was 75 percent of the total and 25 percent was rural. Yet only 20 percent of the economically active population had rural employment. The rural contribution to gross domestic product (GDP) was 7.3 percent in 1992 and a mere 3.5 percent by 2007. Under these conditions, 90 percent of all farmers produce primarily for self-subsistence, while 40 percent sell cash crops. Mexico's grain dependence on imports, primarily from the United States, had grown considerably by the end of the twentieth century. In corn, 23 percent of domestic consumption is imported, in wheat 50 percent, in sorghum 43 percent, in rice 60 percent, and in soybeans almost 100 percent (Bartra 2004, 22–25).

The import dependency and export demand dynamics of the global food regime have further entrenched these trends. Improvements in food consumption are now driven by food imports, and this dependency is increasing. Commercial imports in eleven Latin American and Caribbean countries accounted for 30 percent of domestic food supplies in the 1980s (USDA, ERS, 1997, 20). This figure went up to 44 percent in 1999, and is projected to increase to 50 percent by 2011 (Meade et al. 2002, 16). The flip side of food-import dependency is agricultural production for export, with exporters subject to external forces that dictate what is to be produced (Cabello 2003, 132). This has implications for local populations and national food security alike. Consequently, as indicated earlier, national dietary improvements are often not reflective of regional differentiations, particularly in rural areas. Rather, the economic development associated with a move from subsistence to commercial production can be associated with economic exclusion, decreased dietary diversity, and even increased malnutrition (see Teubal's chapter, this volume). As the capacity of peasants to produce their own livelihood on the basis of traditional crops decreases, these crops become replaced by more profitable cash crops (Cabello 2003, 131).

Traditional staples are substituted for high-yielding varieties, traditional mixed cropping is substituted for monocropping, and "industrial foods, which must be purchased, have taken the place of traditional staples and self-provisioning, often exacerbating the problems of hunger and malnutrition" (Gouveia and Stanley 1990). A good illustration of the impacts of such modernization is Whiteford's (1991) study of a Costa Rican community affected by the growth of the beef industry. Whiteford found that the change in land-use patterns was the major cause of undernourishment. Converting lands to pasture for cattle production reduces the land available for subsistence, while providing little employment for involvement in the cash economy. At the same time, by the mid-1970s, commercially packaged foods were introduced to the community for the first time, offering products such as infant formulas, packaged white bread, and "Jack's Snacks"; i.e., food for the few who could afford it (Whiteford 1991, 136).

The repercussions of these changing diets are not just limited to domestic social impacts; they are also global. From 1995 to 2050, the world's population is expected to increase by 72 percent (World Food Summit [WFS] 1996, 1), seemingly requiring another "revolution-like" production increase to meet food demands. However, while population growth remains the greatest factor for increased food demand, food pressures are exacerbated by two additional nondemographic factors: income growth and urbanization (Dyson 1996, 108, 101). Although these social factors are more difficult to predict than demographic ones, trends are discernable.

Latin America's food consumption pattern continues to move toward a U.S.-type diet based on wheat, meat, and milk, and away from local grains and cereals. Somewhat ironically, this is occurring at the same time that there is some movement by the North American public toward leaner meats and foods with more fiber and less fats and cholesterol. Consequently cereal demand in developed countries may actually decrease, as for health reasons people shift to less meat-based diets (Dyson 1996, 110). Nonetheless, given the population distribution between developed and developing countries, the "Americanization" of diets in developing countries will have a profound effect on food needs.

Currently, "low-value staple food products" (e.g., breads and cereals) account for 27 percent of the consumer food budget in low-income countries, but only 12 percent in high-income countries (Seale, Regmi, and Berstein 2003, 2). Further, "consumers in low-income countries . . . make greater adjustments to their household spending on food when incomes and/or prices change": for example, a 10 percent income increase would only produce a 1 percent increase in food spending in the United States, but an 8 percent increase in Tanzania (ibid.). Income increases produce higher demand for fruits, vegetables, and livestock (Weibe 2003, 8), and overall demand for high-value foods (e.g., meat and dairy) is growing (Seale et al. 2003, 2). Total income in the developing world is expected to increase at an average of 4.3 percent annually between 1995 and 2020 (Pinstrup-Andersen and Pandya-Lorch 2000, 7). Given that about 70 percent of world population concentrates in developing countries, this income increase will account for a proportionally much greater demand for food.

Dietary diversity also increases as people shift from rural to urban areas (ibid., 6), with urbanization tending to shift diets from coarse grains to rice or wheat, fruits, vegetables, animal products, and processed foods (Weibe 2003, 8). Rural populations are projected to remain constant from 1990 to 2020; however, because of "rural to urban migration, and the natural increase of urban populations, virtually all of the projected global population increase will occur in urban areas" (Dyson 1996, 105). Globally, then, the urban population should increase from 45.2 percent in 1990 to 62.0 percent in 2020; in Latin America it is projected to increase from 71.5 percent to 82.9 percent for the same years (Dyson 1996, 102).

In sum, in addition to food pressures from demographic trends, urbanization and rising incomes in developing countries are likely to have profound impacts on food demand, particularly high-value food demand. Given that the production of one calorie of beef requires eleven plant-derived calories (World Food Summit [WFS] 1996, 3), the dietary shift toward these high-value foods is globally significant. It is estimated that the food needs for those countries where the diet is dominated by maize, like most in Latin America, are expected to double by 2050 (WFS 1996, 4). Finally, there is a rather unsavory irony in further Americanization of Latin American diets. The shift to diets based on milk, meat, and wheat, in conjunction with the growing social polarization and rising incomes for some, is leading to situations where problems of over-nutrition and obesity can be found concurrently with malnutrition (Mancino, Lin, and Ballanger 2004).

Biotechnology and TNCs

Given the social and environmental impacts that modern agriculture has had on regions such as Latin America, it is reasonable to ask how these impacts will be affected by the addition of biotechnology to the modern agriculture package. Furthermore, how are these impacts affected by the central role that transnational corporations (TNCs) are playing in the context of neoliberal globalism since the mid-1980s?

Broadly speaking, biotechnology is the process of modifying living organisms for specific uses. We are particularly concerned with the products of new techniques developed after the 1970s: plant tissue culture, somatic hybridization or protoplast fusion, recombinant DNA, and the creation of transgenics. These technologies are frequently characterized as the "new biotechnologies." The main genetically modified crops are soybeans, maize, cotton, and canola. Globally, 56 percent of soybeans, 20 percent of cotton, 28 percent of maize, and 14 percent of canola were genetically modified as of 2004. The use of such technology is on an upward trend. Global adoption of transgenic crops increased from 1.7 million hectares in 1996 to 81 million hectares in 2004 (James 2003; Manjunath 2005, 5–6). The question that remains is whether this technology will be any less polarizing or more environmentally friendly than the Green Revolution technologies.

We argue that, given the trends identified above, the products coming from biotechnology will have an even more socially polarizing effect, and that they will further entrench the petrochemical era of food production brought about by modern agriculture. Socially, biotechnology as currently promoted can only exacerbate such trends (i.e., social polarization within and between regions, semi-proletarianization of traditional farmers, etc.). Its role in the intensification of high-input agriculture is not only consistent with Green Revolution technologies, but even furthers the centralization and concentration of production. These trends might be altered only by profoundly modifying the structures of political and economic power of the current agents of technological production and diffusion. The difficulties of achieving this, however, are inherent in the corporate structure leading technology development, which will be discussed with reference to its particularities from modern agriculture. Before doing this, however, we need to pay some attention to the specifics of biotechnology's environmental impacts.

Biotechnology's Environmental Impacts

Traditional plant breeding was based on the combination of traits among plants that were sexually compatible and, by definition, from the same species. That is to say, genetic improvement of plants came only from other plants. Genetic engineering changed all this by making it possible to mix genes from one species into another, eliminating previous barriers of sexual compatibility. Therefore, no one can honestly be certain of the long-term effects. Nonetheless, there are a host of concerns over biotechnology's potential environmental effects, some purely speculative, and others with significant empirical support. Food safety concerns, loss of genetic diversity, and unexpected repercussions from selected plant characteristics are a few of the main concerns.

Food safety concerns around biotechnology arise for a number of reasons. First, the transfer of genes from one species to another in consumable items raises the prospect for unexpected allergic reactions in humans. There is the associated risk of foods not approved for human consumption, but allowed for other uses, slipping into the human food production chain. This has already been seen in the StarLink® corn recall in 2000, where corn approved only for animal feed (containing a pesticide suspected to cause human allergies) found its way into a variety of human foods. Second, as genetic modifications get more adventurous, moving from the enhancement of existing plant characteristics to increasingly new applications, the health impacts become significantly less certain. Monsanto's NewLeaf® potato, for example, has been genetically engineered to fight potato beetles by incorporating a pesticide into each cell of the potato. U.S. firms and government have resisted labeling genetically modified foods on the basis that they are "substantially equivalent" to those naturally occurring. Yet the public no doubt has some reason to pause over the consumption of a food item that is required to be registered as a pesticide (Montague 2000, 181). In contrast to this lax approach used in North America, the European Union adopted the "precautionary principle" on whether to approve the marketing of products involving new technologies (Chap. 6, this volume).

Another environmental concern over biotechnology is the unexpected repercussions that can arise from selected plant characteristics. The exposure of "non-target" insects to insecticidal properties of modified plants—such as the impact of *Bt* corn on monarch butterflies—is only one such example. Because some of these insects play a critical role in transferring, for instance, pollen from one plant to another, entire ecosystems could be disrupted if non-target insect populations are affected. Concerns over the specific genetically modified traits are particularly salient because biotechnology creates a form of "living pollution," where its traits will continue to reproduce after it has been released, making "corrections" to any errors fairly difficult or perhaps impossible. Impacts are not limited to a genetically modified crop's potential to contaminate non-transgenic crops of the same type. Rather, some of the more serious concerns regard the potential of transgenic crops to cross with similar species and mutate. "Superweeds" are the potential result of such a cross, where herbicide-resistant plants transfer their properties to undesirable plants or weeds, thus requiring greater amounts of chemicals to eradicate the problem. The Organic Trade Association (OTA) cites that such a superweed was already discovered in Canada that was resistant to three different herbicides (OTA 2001).

The greatest environmental concern over biotechnology, however, is with respect to its impact on biodiversity. The effect of transgenic crops on biodiversity far extends the concerns already raised by monocropping under the Green Revolution. Not only is diversity decreased through the physical loss of species, but because of its "live" aspect, it has the potential to contaminate, and potentially to dominate, other strains of the same species. While this may be a limited concern with respect to the contamination of another commercial crop, it is significantly more worrisome when it could contaminate and eradicate generations of evolution of diverse and subtly differentiated strains of a single crop, such as the recently discovered transgenic contamination of landraces of indigenous corn in Mexico (see Chap. 3 by Kathy McAfee and Chap. 6 by Liz Fitting in this volume).

The above effects do not necessarily lead to a conclusion that the environmental effects of biotechnology are insurmountable, or that any such risk is always too great. Projected global population increases elevate biotechnology as a plausible response to food supply concerns. These dawning impacts, however, raise the prospect that the greatest environmental problem of all is that the speed and chief motives of biotechnology's development—greed and profits—prevent a careful social and public screening of benefits and risks. In this context, it would seem that a democratically determined direction of technology development would be necessary. The current corporate and institutional structure of biotechnology, however, makes this an increasingly challenging goal.

TNCs: The Corporate Structure of Biotechnology Development

The "genetics supply industry" has gone from its original public and private roots to be controlled by large transnational corporations (TNCs) from the chemical and pharmaceutical industries. This process of industrial concentration took place mostly in the 1980s and continued well into the 1990s. The economics behind agricultural biotechnologies are significant. By 2001, the value of the global transgenic seed market was $3.7 billion, and genetically modified crops made up approximately 13 percent of the $30 billion global commercial seed market. One year later, the value of the global transgenic seed market was already $4.0 billion (James 2003).

Given that Green Revolution technologies were developed and distributed by public agencies with apparently social intentions, the promotion of the new biotechnologies by a highly concentrated private sector seems more likely to run counter to social priorities. Indeed, the marketing strategies of these TNCs tend to enhance negative environmental impacts, as the preservation of biodiversity and other natural features offers no immediate economic benefit. Most likely, exactly the opposite is true: each successive "problem" raised by the technology will be responded to with a further technological solution, thus further enhancing the technological treadmill. This cycle is already in evidence in the issue of superweeds, and the increasingly toxic herbicides necessary to fight them.

In this context, the merger of chemical and pharmaceutical giants with seed companies takes on amplified social significance. Currently every major seed company has some form of direct link to a chemical company (Middendorf et al. 2000). A practical reason for this amalgamation is that they are all science-based industries. In the case of the chemical industry, however, amalgamation is also an important strategy for horizontal integration: seeds are the main vehicle through which companies can expand the technological package offered to farmers. These TNCs are reorienting the research in seed companies toward crops that can maximize their profit potential—such as through marketing a technological package—rather than their social utility. An early and consistent emphasis, for example, was to develop crop resistance to increased use of pesticides.

By 2002, "herbicide tolerance continued to be the most dominant trait, occupying 75 percent of the GM global area . . . followed by insect resistance (17 percent) and the stacked genes of herbicide tolerance and insect resistance, occupying 8 percent" (James 2003). On the face of it, it would seem like insect resistant plants will be environmentally beneficial, as they will help reduce the use of pesticides. Nevertheless, genetically inserting *Bacillus thuringiencis* or *Bt* within plants, a natural killer of certain pests, is creating insects resistant to it. This happens because *Bt* is in the plant at all times, instead of for short periods, as used by organic farmers. As a result of this constant presence and insect resistance, it is quite possible that *Bt* will be lost to organic farmers as a tool.

In contrast to the current mainstream approaches, a biotechnology program that emphasized nitrogen fixation could drastically reduce the environmental impact of intensive agriculture. Nitrogen exists in ample amounts in the environment, but it is not easily absorbed by plants. Thus, as a result of the advent of input-intensive modern agriculture, nitrogen fertilizer use increased from 3.5 million tons to 80 million tons in the period between 1950 and 1989 (Hardy 1993, cited in Board on Science and Technology 1994). Therefore, increased nitrogen fixation capabilities in plants could drastically reduce the amounts of energy required for fertilizer production, and eliminate the environmental repercussions of such fertilizers (e.g., groundwater pollution). But the short-term profit motivation of biotechnology corporations gets in the way: they find it easier to develop herbicide-tolerant transgenics and their associated high-input package.

Profitability also dictates that the majority of transgenic traits are developed with the needs of advanced industrial countries in mind. A comparative study of crops and crop character traits in transgenic field trials in the United States and developing countries found them to be highly similar (Arends-Kuenning and Makundi 2000). If the needs of farmers in developing countries were to be a consideration, however, these trials would emphasize different traits and crops; for example, plant hardiness, drought tolerance, and disease and insect resistance for regions where farmers cannot afford high inputs. Nonetheless, the top trait for field trials in developing countries remains herbicide tolerance, at 37 percent of trials (James and Krattiger 1996, cited in Arends-Kuenning and Makundi 2000, 323).

In the context of global corporate dominance, then, TNCs make the critical choice as to how to maximize profits. As the development of crops for subsistence farmers provides only a social, not a financial benefit, it is unlikely to attract the attention of the private sector. Similarly, as the short-term profit motivation of such corporations can be more easily met by environmental degradation than by environmental preservation, the development of biotechnology will proceed accordingly. Consequently, it is clearly a matter for civil-society organizations to rise up and demand appropriate state regulation to steer companies in a more socially oriented direction.

Conclusion

With the globalization of the world economy, national agricultures in Latin America have increasingly conformed to food production and consumption patterns of the United States. Because the industrialization processes of those countries have not been nearly as robust as those in their

Northern neighbor, social polarization resulting from the application of modern agricultural technologies has been more dramatic. Environmentally speaking, these modern technologies have also shown exacerbated effects in developing countries, through the erosion, degradation, and contamination of their agricultural resources. Developing countries do not have the resources or infrastructure to put in place preventative measures that could reduce the impact of these, as in more developed countries.

The irony, however, is that most of the plant varieties which today account for the world's food consumption had their biological origin and diversity in the countries of the Southern hemisphere. The expansion of high-yield varieties and transgenics, along with the rest of the technological package of modern agriculture, is posing a major threat to this plant genetic diversity—arguably to the increased benefit of those TNCs that promote it. Therefore, the new additions to modern agriculture as a technological paradigm—biotechnology and genetic engineering—threaten to exacerbate many of the existing dangers to long-term sustainable agriculture. Moreover, the growing vertical integration and linkages of TNCs threaten to further entrench corporate dominance to such an extent that democratization of the technological paradigm becomes increasingly difficult.

We have argued that the main economic actors reinforcing the current technological paradigm are the transnational corporations involved in the agricultural inputs–producing sector. These companies have largely determined the technological package that constitutes "modern agriculture." They have not done it on their own: their alliance with the U.S. government and the Land Grant Colleges was the crucial first stage of its development (Kloppenburg and Kenney 1983; Kloppenburg 1988). In the race not to lose out in the transition to an information era and a knowledge-based economy, the governments of developed nations have scrambled to facilitate the adoption of these technologies. This scramble includes not only the facilitation of research and development within their nations, but also the lobbying for international regulations that both broaden and strengthen private intellectual property rights, primarily to the detriment of developing nations.

In sum, the state-facilitated, TNC-dominated structure of technological development severely compromises biotechnology's potential to respond to environmental and social needs. While there can be no real argument with those who cite population growth estimates in conjunction with cropland limitations to justify support for biotechnology, it is less certain that without stronger social control these technologies will provide a net benefit to either environmental or food security concerns, particularly when long-term sustainability is considered. The Green Revolution has provided us with ample evidence to reconsider whether technologically induced production increases in themselves are sufficient for providing the intended benefits. While bottom-up directed technological development could lead the current technological paradigm into a truly revolutionary mode, one that puts social and public interests at the forefront, the increasing power of TNCs makes significant gains along these lines appear difficult but not impossible.

Given the dominant place of the U.S. economy in the world system and its ideology of neoliberal globalism, any significant and lasting solution to the direction of agricultural research and food production patterns would seem to rest within that country. The American public could push for a democratization of the agricultural research institutions and for a societalized strategy of product development for U.S.-based TNCs, ensuring that the technologies they develop are friendly to small producers, consumers, and the environment. Some sectors of the American public are speaking, although the voices are still faint. At the very least, the voices of dissent can be seen in the rising demand for organic and socially sustainable produce. For the most part, the peoples of Latin America still lack this power: they lack the scientific capability and sufficiently powerful economic actors to produce alternative agricultural technologies; and they lack the state structures and the public financial capability to pursue alternative paths of development. Perhaps equally significantly, the new individual or corporate economic "winners" in the agricultural modernization of developing countries are unlikely to relinquish their gains, despite the overall national losses.

There are of course a host of other actors between these poles—notably the European Union (EU). Even before the substantial geopolitical rift around the 2003 U.S. invasion of Iraq, the EU had a wealth of environmental and other nongovernmental organizations, many of them global in structure and reach. In new alliances with developing nations, a new EU hegemonic bloc could rise in the world sphere to challenge the U.S. technological paradigm in agriculture and food. Resistance exists in the EU, at various levels. The EU itself until recently maintained a de facto boycott of genetically modified or GM products, much to the anger of the United States, and insisted on the establishment of labeling regulations for genetically modified organisms or GMOs. More local forms of resistance mirror this national struggle. Many have hailed French farmer José Bové as a national hero, for example, after he bulldozed a

McDonald's restaurant under construction in 1999. Bové declared his actions to be an attack on "foul foods" (namely, American fast food and GMOs), and on the WTO, multinationals, and governments that promote them to the detriment of small producers (*AgBiotech Buzz* 2003). More conventional forms of protest can be found in the works of Greenpeace and other ENGOs (see Chap. 11 by Manuel Poitras, this volume). Equally significant, there is fair evidence that growing alliances between developing nations could counter U.S. pressure, as evidenced by the 2003 round of WTO negotiations in Cancún. This meeting failed to reach any agreements, thanks in large part to opposition by twenty-two developing countries, including Brazil and Mexico. They were opposed to the U.S. government's double standards: wanting to liberalize world trade in agriculture while keeping one of the largest per capita farm subsidies in the world, along with the European Union's and Japan's. On the other hand, if developing countries do get their way with the United States and the WTO finally imposes its trade-liberalization standards, the chief winners may not be Southern farmers. Rather, they will likely increase their exports, yes, but the main benefits would be concentrated by agricultural purchasing firms like Cargill and ADM, at the expense of U.S., EU, and Japanese farmers. This would be the logical result from a market structure with a large multiplicity of farmers, on one hand, and a mere handful of corporate buyers of their products, on the other.[2]

Nevertheless, resistance to U.S. trade domination holds some potential for more equitable distribution of the economic benefits of new technologies, if not a more socially sustainable agriculture. The Convention text of the UN Convention on Biological Diversity (CBD) (1992) and the Biosafety Protocol (2000) are examples of this slowly changing dynamic, even if the U.S. government has refused to become a signing member of the UN-CBD. Any structural solution, however, rests largely with the result of political and grassroots resistance to U.S. dominant power, both inside and outside of the United States.

Notes

1. "It is seriable in that it does not have to be used on a farm system as a whole; a piecemeal experiment is feasible. It is separable in that the package can be separated/unpacked. And it is single unit in that the adoption does not involve each unit in relationships with neighbours and/or authority structures; how much one gains from the technology does not depend on one's neighbours or authority structures" (Lipton 1989:316; 318–319, summarized by Das 2002:12).

2. This hypothesis was proposed by Friederich Buttel at a symposium held in his honor by the Rural Sociological Society in August of 2004 in Sacramento, California.

References

AgBiotech Buzz: News in Brief. 8 November 2003. "French Farmer to Receive Prison Time for Destruction of McDonald's." *AgBiotech Buzz* Website. http://pegwagbiotech.org/buzz/display.php3?StoryID=45. Accessed November 2003.

Altieri, M. 2001. *Genetic Engineering in Agriculture: The Myths, Environmental Risks, and Alternatives.* Oakland, Calif.: Food First Books.

Arends-Kuenning, M., and F. Makundi. 2000. "Agricultural Biotechnology for Developing Countries." *American Behavioral Scientist* 44(3): 318–350.

Bartra, Armando. 2004. "Rebellious Cornfields: Toward Food and Labour Self-sufficiency." pp. 18–36 in *Mexico in Transition: Neoliberal Globalism, the State and Civil Society,* ed. Gerardo Otero. London: Zed Books.

Board on Science and Technology for International Development. 1994. *Biological Nitrogen Fixation: Research Challenges—A Review of Research Grants Funded by the U.S. Agency for International Development.* Chapter 1: "Global Investment and Research Needs." National Research Council. National Academy Press. Washington, D.C. Available online (consulted on 27 November 2004): http://www.nap.edu/readingroom/books/bnf/.

Bryceson, Deborah, Cristóbal Kay, and Jos Mooij, eds. 2000. *Disappearing Peasantries? Rural Labour in Asia, Africa and Latin America.* London: Intermediate Technology.

Burbach, Roger, and Patricia Flynn. 1980. *Agribusiness in the Americas.* New York: Monthly Review Press and North American Congress on Latin America.

Cabello, G. 2003. "The Mexican State and the Agribusiness Model of Development in the Globalisation Era." *Australian Journal of Social Issues* 38(1).

Cartagena Protocol on Biosafety. 2000. Cartagena Protocol on Biosafety to the Convention on Biological Diversity. Montreal: Secretariat of the Convention on Biological Diversity. http://www.biodiv.org/biosafety/protocol.asp. Accessed November 2003.

Castles, Stephen, and Mark J. Miller. 2003. *The Age of Migration: International Population Movements in the Modern World.* New York and London: Guilford Press.

Convention on Biological Diversity. 1992. *Convention on Biological Diversity. Convention Text.* 5 June 1992. Convention on Biological Diversity Website. http://www.biodiv.org/convention/articles.asp. Accessed October 2003.

Conway, G., and E. Barbier. 1990. *After the Green Revolution: Sustainable Agriculture for Development.* London: Earthscan Publications Ltd.

Das, R. 2002. "The Green Revolution and Poverty: A Theoretical and Empirical Examination of the Relation between Technology and Society." *Geoforum* 33(1): 55–72.

Davies, W. 2003. "An Historical Perspective from the Green Revolution to the Gene Revolution." *Nutrition Reviews* 61(6): 124–131.

Delgado Wise, Raúl. 2004. "Labour and Migration Policies under Vicente Fox: Subordination to U.S. Economic and Geopolitical Interests." pp. 138–151 in Otero 2004.

Dosi, Giovanni. 1984. *Technical Change and Industrial Transformation: The Theory and an Application to the Semiconductor Industry.* London: Macmillan.

Dyson, T. 1996. *Population and Food: Global Trends and Future Prospects.* Environmental Change Programme. London: Routledge.

Eckstein, Susan. 1990. "Urbanization Revisited: Inner-City Slum of Hope and Squatter Settlement of Despair." *World Development* 18(2): 165–181.

Evenson, R., and D. Gollin. 2003. "Assessing the Impact of the Green Revolution, 1960–2000." *Science.* 2 May 2003. Vol. 300(5620): 758–763.

Food and Agriculture Organization of the United Nations, Statistical Database (FAOSTAT). Available online: http://apps.fao.org/. Accessed October 2004.

Fowler, Cary, and Pat Mooney. 1990. *Shattering: Food, Politics and the Loss of Genetic Diversity.* Tucson: University of Arizona Press.

Friedmann, H. 1994. "The International Relations of Food." In *Food: Multidisciplinary Perspectives.* Ed. B. Harriss-White and R. Hoffenberg. Oxford: Blackwell Publishers Ltd.

Gilbert, Alan. 1998. *The Latin American City.* 2nd Edition. London: Latin America Bureau.

Gouveia, Lourdes, and Kathleen Stanley. 1990. "Rural Crisis, North and South: The Role of Wheat in the United States and Venezuela," paper presented at the World Congress of Sociology, Madrid, Spain, July.

Greene, Catherine R. United States Department of Agriculture (USDA), Economic Research Service (ERS). 2001. *U.S. Organic Farming Emerges in the 1990s: Adoption of Certified Systems.* USDA, ERS AIB No. 770. Available online: http://www.ers.usda.gov/publications/aib770/.

Halweil, B. 2000. "Portrait of an Industry in Trouble." *Environmental Case Book: Genetically Modified Foods.* Ed. N. Miller. Boca Raton, Fla.: Lewis Publishers. 201–204.

Hardy, R. W. F. 1993. "Biological Nitrogen Fertilization: Present and Future Applications." pp. 109–117 in *Agriculture and Environmental Challenges.* Ed. J. P. Srivastava and H. Aldermans. *Proc. 13th Agric. Sector Symp.* Washington, D.C.: World Bank.

Heffernan, D., and M. Hendrickson. 2002. "Multi-National Concentrated Food Processing and Marketing Systems and the Farm Crisis." Paper presented at the Annual Meeting of the American Association for the Advancement of Science, Boston.

Hendrickson, Mary, and William Heffernan. 2005. *Concentration of Agricultural Markets.* Department of Rural Sociology, University of Missouri, February. www.agribusinesscenter.org/docs/Kraft_1.pdf.

Hewitt de Alcántara, Cynthia. 1978. *La modernización de la agricultura mexicana: 1940–1970.* Mexico City: Siglo XXI Editores.

ILO (International Labour Organization). 1988. Rural Employment Promotion, International Labour Conference, 75th Session, Report VII. Geneva: ILO.

James, C. 2003. "Global Review of Commercialized Transgenic Crops: 2002." Executive Summary of ISAAA Briefs 29. 4 November 2003. Full Report at http://www.isaaa.org/kc/CBTNews/ISAAA_PR/briefs29print.htm. Accessed November 2003.

Kloppenburg, J. 1988. *First the Seed: Political Economy of Plant Biotechnology, 1492–2000.* New York: Cambridge University Press.

Kloppenburg, Jack, Jr., and Martin Kenney. 1983. "The American Agricultural Research System: An Obsolete Structure?" *Agricultural Administration* 14: 1–10.

Lappé, F., J. Collins, and P. Rosset. 1998. *World Hunger: 12 Myths.* Second Edition. New York: Grove Press.

Lipton, M. 1989. *New Seeds and Poor People.* London: Unwin Hyman.

Mancino, Lisa, Biing-Hwan Lin, and Nicole Ballanger. 2004. "The Role of Economics in Eating Choices and Weight Outcomes." 24 pp. USDA, ERS Agricultural Information Bulletin Number 791, October. Available at: http://www.ers.usda.gov/whatsnew/ (consulted on 27 November 2004).

Manjunath, T. M. 2005. "A Decade of Commercialized Transgenic Crops—Analyses of Their Global Adoption, Safety and Benefits." The Sixth Dr. S. Pradhan Memorial Lecture, delivered at Indian Agricultural Research Institute (IARI), New Delhi, on 23 March. Available at: http://www.agbioworld.org/word/decade-commercialized.doc (consulted 8 May 2007).

Mares, David. 1987. *Penetrating the International Market: Theoretical Considerations and a Mexican Case Study.* New York: Columbia University Press.

Meade, B., S. Rosen, S. Shapouri, B. Lohmar, and M. Trueblood. United States Department of Agriculture (USDA), Economic Research Service (ERS). 2002. *Food Security Assessment.* ERS Outlook Report No. GFA 13.

Middendorf, G., M. Skladney, E. Ransom, and L. Busch. 2000. "New Agricultural Biotechnologies: The Struggle for Democratic Choice." *Hungry for Profit.* Chapter 5. 93–116.

Montague, P. 2000. "#695—Biotech In Trouble—Part 1, May 04, 2000." From Environmental Research Foundation Home. Rachel's Environment & Health News. *Environmental Case Book: Genetically Modified Foods.* Ed. N. Miller. Boca Raton, Fla.: Lewis Publishers. 201–204.

Murray, Douglas L., and Polly Hoppin. 1990. "Pesticides and Nontraditional Agriculture: A Coming Crisis for U.S. Development Policy in Latin America?" Paper presented at the 85th Annual Meeting of the American Sociological Association, 11–15 August, Washington, D.C.

National Research Council. 1989. *Alternative Agriculture: Committee on the Role of Alternative Farming Methods in Modern Production Agriculture.* Washington, D.C.: National Academy Press.

Organic Trade Association (OTA). 2001. "OTA's Congressional Education Day 2001. Issue: Genetically Engineered Crops and Food. Background for Congress." Available online: http://www.ota.com/pp/otaposition/papers/gmo.html (accessed 27 November 2004).

Otero, Gerardo. 1999. *Farewell to the Peasantry? Political Class Formation in Rural Mexico.* Boulder, Colo., and Oxford: Westview Press.

Otero, Gerardo, ed. 2004. *Mexico in Transition: Neoliberal Globalism, the State and Civil Society.* London: Zed Books.

Paarlberg, Robert L. 1988. "U.S. Agriculture and the Developing World: Opportunities for Joint Gains." Pp. 119–138 in John W. Sewell, Stuart K. Tuker, and contributors, *Growth, Exports and Jobs in a Changing World Economy: Agenda 1988.* New Brunswick (U.S.) and Oxford (UK): Transaction Books.

Petras, James, and Henry Veltmeyer. 2005. "Latin America's Social Structure and the Dynamics of Change." pp. 162–180 in Jan Kuuper Black, ed., *Latin America: Its Problems and Its Promise.* 4th edition. Westview Press.

Pichón, J., and J. Uquillas. 1997. "Agricultural Intensification and Poverty Reduction in Latin America's Risk Prone Areas: Opportunities and Challenges." *Journal of Developing Areas* 31: 479–514.

Pinstrup-Andersen, P., and R. Pandya-Lorch. 2000. "Meeting Food Needs in the 21st Century: How Many and Who Will Be at Risk?" Presented at AAAS Annual Meeting, February 2000, Washington, D.C. Accessible at http://www.ifpri .org/pubs/jhu/fed21century_chapter01.pdf.

Portes, Alejandro, 1989. "Latin American Urbanization in the Years of Crisis." *Latin American Research Review* XXIV(3): 7–44.

Portes, Alejandro, Manuel Castells, and Lauren A. Benton. 1989. *The Informal Economy: Studies in Advanced and Less Developed Countries.* Baltimore, Md.: Johns Hopkins University Press.

Restrepo, Iván (with Susana Franco). 1988. *Naturaleza Muerta: Los Plaguicidas en México.* Mexico City: Ediciones Océano.

Rubio, Blanca. 2001. *Explotados y excluidos. Los campesinos latinoamericanos en la fase agroexportadora neoliberal.* Mexico City: Plaza y Valdes/UACH.

———. 2004. "La fase agroalimentaria global y su repercusión en el campo mexicano." *Comercio Exterior* 54(11): 948–956.

Sanderson, Steven E. 1986. *The Transformation of Mexican Agriculture: International Structure and the Politics of Rural Change.* Princeton: Princeton University Press.

Schoonhoven, A., and D. Pachico. Centro Internacional de Agricultura Tropical (CIAT). 1998. "Rice and Beans in Latin America: A Summary Report on the Economic Impact of Improved Varieties." CIAT Website. http://www.ciat.cgiar .org/newsroom/release_10.htm. Accessed November 2003.

Seale, J., A. Regmi, and J. Berstein. United States Department of Agriculture (USDA), Economic Research Services (ERS). 2003. "International Evidence on Food Consumption Patterns." *ERS Research Briefs.*

Shapouri, S., and S. Rosen. USDA, ERS. 2000. "Food Security Assessments." *International Agriculture and Trade Reports.* Situation and Outlook Series. GFA12.

Thiesenhusen, William C. 1989. *Searching for Agrarian Reform in Latin America.* Boston: Unwin Hyman.

———. 1995. *Broken Promises: Agrarian Reform and the Latin American Campesino.* Boulder: Westview Press.

Thrupp, Lori Ann. 1991. "Social and Environmental Degradation from Pesticides in Central American Banana Plantations: Political Ecology of Agribusiness Structures and Organization." Paper presented at the Rural Sociological Society, Columbus, Ohio, 19–21 August.

United States Department of Agriculture (USDA). Economic Research Service (ERS). Website: http://www.ers.usda.gov.

———. 1997. "Food Security Assessments." *International Agriculture and Trade Reports.* Situation and Outlook Series. GFA9.

———. 2001. "Farm and Rural Communities: Hired Farm Labor: Comparing the U.S. and Mexico." *Agricultural Outlook.* D. Shields, Economics Ed. ERS, Market and Trade Economics Division. January–February 2001.

———. 2002. "Data: Organic Production." USDA, ERS Website. http://www.ers.usda.gov/Data/organic/. Accessed August 2004.

United States Department of Agriculture (USDA). Office of Communications. *Agriculture Fact Book 2001–2002.* Website: http://www.usda.gov/factbook/. Accessed November 2003.

United States Embassy Website, Tokyo, Japan. 2002. "U.N. Food Agency Adopts Updated Pesticide Standards: Stronger Code Will Reduce Associated Risks, FAO Says." http://usembassy.state.gov/tokyo/wwwhenv20021107a2.html. Accessed December 2003.

von Bertrab, Alejandro. 2004. "El efecto de la liberalización económica en los pequeños productores de maíz en México." *Comercio Exterior* 54(11): 758–769.

Weibe, K. Economic Research Service (ERS), U.S. Department of Agriculture (USDA). 2003. United States Department of Agriculture. "Trends in Food and Resources." *Linking Land Quality, Agricultural Productivity, and Food Security.* Agricultural Economic Report No. AER-823.

Whiteford, M. B. 1991. "From *Gallo Pinto* to 'Jack's Snacks': Observations on Dietary Change in a Rural Costa Rican Village." pp. 127–140 in *Harvest of Want: Hunger and Food Security in Central America and Mexico,* ed. S. Whiteford and A. Ferguson. Boulder: Westview Press.

World Development Report (WDR). 1992. *Development and the Environment. World Development Indicators.* Oxford: Oxford University Press.

———. 2003. *Sustainable Development in a Developing World: Transforming Institutions, Growth, and Quality of Life.* New York: Oxford University Press and the World Bank.

World Food Summit (WFS). 1996. "Food Needs and Population." *World Food Summit: Food for All.* Rome. 13–17 November 1996.

CHAPTER 3

Exporting Crop Biotechnology: The Myth of Molecular Miracles

KATHY MCAFEE

It is often asserted that genetically engineered crops can prevent a looming crisis of global agricultural productivity. Enthusiasts assert that these new, transgenic crops—varieties containing genes introduced in the laboratory—are essential to produce sufficient food for a burgeoning world population, and that they can avert ecological damage from the expansion of agriculture (Pardey 2001; Borlaug and Carter 2005; BIO 2005). The U.S. government, in cooperation with agribusiness interests, actively promotes this idea. Such arguments for a biotechnology-based solution to food insecurity can be dangerously misleading. The actual performance of transgenic crops has been mediocre, at best (see Chaps. 7 and 8, this volume). In the United States, their productivity has not generally been higher than that of conventional varieties, nor have they allowed reduced use of pesticides, as explained below.

Nevertheless, advocates of crop genetic engineering commonly assume that European and U.S. farm technologies, regulatory practices, and food-producing systems are not only superior but also universally applicable. As I have argued elsewhere, many proponents of a genetic-engineering solution to hunger make use of idealized conceptions of molecular biology and exceptional examples of genetic engineering successes (McAfee 2003a). Most contributions to international biotechnology policy literature do recognize that transgenic crops cannot be adopted easily and without risk in all parts of the world. Many authors, however, focus on what they see as deficits in the institutions and personnel of "less developed" countries. If these lacks can be remedied by means of scientific and legal training and other so-called capacity building, they reason, then Latin America and other regions will be able to share in the expected benefits of transgenic crops.

Proponents of a molecular-technology answer to hunger often fail to appreciate crucial differences between the ecological, cultural, institutional, and economic contexts of food systems in most developing countries and those of the United States, where most transgenic crops have been developed. Many forget to compare the hoped-for benefits of transgenics to the tremendous costs to developing countries of managing their risks and obtaining and enforcing the intellectual property rights that are required for their use. Most discussions of crop biotechnology for the global South fail to weigh the possible benefits of transgenics against the potential gains that could be obtained by more proven and promising uses of Southern-country expertise, institutions, and food-producing resources. And advocates of genetic-engineering responses to hunger rarely address the economic policies that discourage domestic food production in food-deficit countries.

The first section of this chapter places the controversy over transgenic crops in the context of larger questions of U.S. relations with Mexico and Latin America. The following section outlines the rationales for genetic engineering put forward by the U.S. government. It illustrates how these arguments are being used to promote the globalization of intellectual property regimes and changes in multilateral trade and environmental rules that favor transnational corporations (TNCs) that have invested heavily in biotechnology. The third section explains why, even if transgenic crops were performing well in the United States, one could not extrapolate from this experience to predict net benefits from crop genetic engineering for Africa, Asia, and Latin America. The chapter concludes by pointing toward more promising approaches to improving Latin American agriculture.

Transgenic Crops: A New Technological Paradigm?

Recent conflicts over transgenic crops in Mexico illustrate why it is misleading to rely on the U.S. experience to assess the likely effects of transgenic crops in countries where self-provisioning agriculture and crop genetic diversity remain important today. These conflicts have embroiled peasant farmers, scientists, government agencies, lawmakers, nongovernmental organizations, the media, and foreign experts in conflicts over how genetic engineering should be regulated and whether transgenic crops should be allowed in Mexico at all (McAfee 2003b; Jansen and Roquas, this volume; Fitting, this volume).

The wider significance of this controversy becomes clearer if one re-members that the export of U.S. agricultural models did not begin with transgenic crops. While working in Mexico during the 1940s, the re-nowned geographer Carl O. Sauer questioned the fundamentals of the U.S. project that was soon to become the Green Revolution. In his ca-pacity as consultant to the Rockefeller Foundation, Sauer warned against using "agricultural science to recreate the history of U.S. commercial agriculture in Mexico" (Sauer 1941, quoted in Bebbington and Carney 1990, p. 35). He worried that

> a good aggressive bunch of American agronomists and plant breeders could ruin the native resources for good and all by pushing their Ameri-can commercial stocks. The little agricultural work that has been done by experiment stations here [in Mexico] has been making that very mistake, by introducing U.S. forms instead of working on the selection of ecologi-cally adjusted native items. . . . Mexican agriculture cannot be pointed toward standardization on a few commercial types without upsetting native economy and culture hopelessly. (Letter from Sauer to Joseph Willits, Director of the Rockefeller Foundation Division of Social Sci-ence, quoted in F. Apffel-Marglin and S. Marglin 1996, p. 212)

Sauer's faith in the acumen of small-scale farmers did not convince the Green Revolutionaries. The latter focused instead on achieving net food production increases, intended mainly for urban and rural consumers, by means of irrigation, mechanization, and chemical fertilizer applied in large amounts to new, standardized crop varieties designed to produce well under just those conditions (see Chap. 2 in this volume). But Sauer's warning of the damage that could be done in Mexico by "a good bunch of aggressive American[s]" would ring true to many peasant producers in the Southern Mexican highlands today. Farmers there learned in 2001 that their local maize varieties may be "contaminated" with genetic ma-terial from U.S. transgenic corn bioengineered to produce insecticides (McAfee 2003b). Whether or not this transgene flow endangers maize biodiversity is not yet known, but the farm communities whose maize plots are affected are already vulnerable.

Trade-policy reforms have weakened local markets for maize and bean producers (Nadal and Wise 2004). "Free trade" reforms have led to a threefold increase in Mexican imports of U.S.-grown corn, both con-ventional and transgenic, and account for about 20 percent of Mexico's domestic corn consumption (USDA, cited in Vaughan 2002; Ackerman et al. 2003; Nadal and Wise 2004). The cheaper U.S. corn has not reduced hunger in Mexico. The inflation-adjusted price of the tortillas, Mexico's main staple food, rose by about 40 percent during the ten years after the advent of the North American Free Trade Agreement and soared again in the context of high oil prices in 2006 (Ackerman et al. 2003; Bartra 2004; Said 2007). By 2007, the drive to produce biofuels, such as ethanol, from corn and sugar sent tortilla prices soaring again by another 30 to 40 percent. Economic liberalization policies have cut state support for farm prices, inputs, and markets, increasing pressure on peasants to give up farming and migrate Northward, as I learned from interviews with farmers in Southern Mexico in 2003 and 2004 (also see Weiner 2002; Rodarte 2003; Fitting in this volume).

U.S. trade negotiators have been pressing Mexico to permit unrestricted planting of genetically engineered corn, which was banned there in 1998 (Jansen and Roquas, this volume). Advocates of stricter regulation, however, received support from a 2004 report by the NAFTA Commission on Environmental Cooperation, based on a two-year study and public hearings (CEC 2004).[1] The three-country commission pointed out that, while significant damage to human health and maize biodiversity from transgene flow appears unlikely based on present knowledge, the effects of transgene introgression have not been studied in Mexican ecosystems. Over strong U.S. objections, the report recommended that maize imported by Mexico be milled to prevent its being planted. This recommendation has not been acted upon.

In December 2004, Mexican legislators approved a new law allowing commercial planting of transgenic maize. It requires that such products be labeled and that transgenic seeds must be declared risk-free before they are released, and further restricts releases of transgenic maize in states where maize biodiversity is concentrated. The measure was opposed by activist Mexican farmers and environmental organizations that fear the loss of valued maize traits and food self-reliance as a consequence of increased imports, the demise of Mexico's own seed companies, and increased dependence on patented seeds, which farmers would be forbidden to share or replant (Bartra et al. 2005). Disputes over the new law split Mexico's scientific community, and many of its details remain unresolved. As of late 2006, these conflicts were still blocking the law's implementation.

As the Mexican controversy illustrates, assessing the promise or peril of crop genetic engineering requires consideration of ecological, economic, political, and cultural factors. A strictly technological analysis, or a dualist approach that talks of "technology's impact on society" as if the two were separable, will not suffice. Like all technologies, crop varieties are co-creations by people and nature, a dynamic nexus of agrarian cultures and the ecosystems of which they are part. This relationship is little appreciated by most proponents of transgenics. Instead, the assumption of cultural superiority and the hubris of Green Revolution science that worried Carl Sauer also characterize much of contemporary biotechnology discourse.

The Mexican biotechnology disputes have dramatized the contrast between different types of agrifood systems and divergent ways of understanding the role of food and farming in rural life and development. The choice facing Mexico and other Latin American countries is often reported as *not* a choice between the past and the future, nor between "modern" and "backward" agriculture, but it is not this. The food security challenge, rather, is to find a third path, or multiple paths, through the particular ecological, social, and political-institutional conditions of different Latin American countries.

Although the cause of hunger in Latin America is not lack of sufficient food, increased food production for local consumption and regional markets could help alleviate malnutrition. But the main obstacles to this are not technological. They cannot be overcome by improved seeds, whether genetically engineered or not. Greater obstacles for most farmers are their lack of control over food-producing resources (land, water, planting materials, fertilizer and other inputs, and credit); inadequate storage, transport, and marketing infrastructure; the lack of political power among small- and medium-scale producers and poor consumers; and economic policies that promote agro-exports and imports at the expense of domestic food and employment needs.

Some genomic technologies may yet contribute to sustainable food production in Latin America. Even small-scale farmers might benefit from crop varieties produced with the aid of genetic markers and, perhaps, recombinant-DNA methods, if these are developed with farmer input, adapted to local conditions, and rigorously tested. But transgenic crops are not the key to increasing productivity or to ending hunger in Latin America. Rather, emphasis on transgenics is distracting attention and resources from more promising investments in farmer-centered agricultural research and extension and from the agricultural and economic policy changes that are urgently needed to support food security by raising productivity for the majority of farmers, not just a few.

The Myth of Molecular Miracles

The U.S. state has actively promoted the worldwide adoption of transgenic crops, a globally standardized regime of intellectual property rights, and changes in multilateral trade and environmental rules that favor the transnational corporations that are involved in agrochemical and food processing and trade as well as genetic engineering. Much of the rationale put forward in favor of these policies does not stand up to informed scrutiny.

The Role of the United States Government in Biotechnology Promotion

The United States is the world's largest producer and exporter of transgenic seeds and crops: mainly corn, soy, cotton, and rapeseed (canola), and pesticides meant to be used on those crops. Argentina, which grows transgenic soy and corn, is second (see Teubal, this volume), followed by Canada, which exports genetically engineered rapeseed and grains. China has experimented with transgenic tobacco, produces *Bt* cotton, and is working on other crops, but placed a moratorium on transgenic plantings in 2002. By 2004, thirteen other countries had permitted limited planting and field testing of transgenics: Brazil, Paraguay, India, South Africa, Uruguay, Australia, Romania, Mexico (cotton and soy), Spain, the Philippines, Colombia, Honduras, and Germany (ISAAA 2004).

U.S. officials have worked zealously to convince publics and governments that genetic engineering is the only rational response to hunger (Becker 2003). In 2002, then–U.S. Agriculture Secretary Ann Veneman proclaimed that new biotechnologies will "make agriculture more environmentally sustainable" and offer "the opportunity of economic self-sufficiency for subsistence farmers in developing countries" (FAO 2002). In 2003, U.S. Trade Representative Robert Zoellick condemned as "Luddite" and "immoral" the policies of European governments that supported African decisions to decline donations of U.S. transgenic maize. U.S. President George W. Bush told African presidents that Africans are starving unnecessarily because they are not using the science of genetic modification (Fleisher 2003).

Such hyperbolic statements are part of a broader strategy to promote worldwide exports of U.S. grains, processed foods, and agricultural inputs. This pleases politically influential agribusiness interests and helps slow the growth of U.S. balance-of-payments deficits (McAfee 2003a). Although U.S. officials assert that biotechnology will save Africa's poor, the industry's actual focus is on heavily traded crops for large farms in food-exporting regions. U.S. trade and agricultural officials have long maintained that in a system of global free trade, "less efficient" producers will import grains and other staples from countries and firms with more advanced agricultural systems. The U.S. Agency for International Development has argued since the 1970s that developing countries should grow tropical fruits, cocoa, coffee, tea, and winter vegetables so they can import staples such as rice, maize, and wheat, and milk powder.

In contrast, other proponents of crop genetic engineering contend that Southern countries need to develop their own biotechnology for domestic production of staple foods: a publicly funded "second Green Revolution" (Juma et al. 1994; Conway 1997; De Young 2006). Some favor greater precaution in the application of biotechnology and more research on its unknown effects, especially in tropical regions (NAS 2000; Serageldin and Persley 2000; Royal Society 2002). Exemplary of this approach is a 2003 study by leading UK scientists, ethicists, and industry analysts commissioned by the Nuffield Council on Bioethics (NCB 2004, 93). Recognizing that the bulk of agro-biotechnology research and development has focused on varieties and traits for commercial agriculture in developed countries, the group called for "a major expansion of GM-related research into tropical and sub-tropical staple foods," with substantial public funding (NCB 2004, 90). The Nuffield report acknowledges that some highly touted transgenic products, such as "golden rice," engineered to produce the nutrient ß carotene, have yet to be fully developed, much less proved beneficial. The Nuffield report also takes seriously the possible health and environmental risks of transgenic products. It recommends that, rather than being given blanket approval, GE applications be assessed on a case-by-case basis, taking into account "a variety of factors, such as the gene, or combination of genes, being inserted, and the nature of the target crop" as well as "local agricultural practices, agro-ecological conditions and trade policies of the developing country in which GM crops might be grown" (NCB 2004, xiv).

Some of the euphoria that accompanied introduction of transgenic crops in the 1990s has faded as their predicted benefits have proved modest or nonexistent, especially for farmers and consumers. Ecological hazards which had been discounted by industry publicists, particularly the migration of genetic material from transgenic to conventional crops, and the rapid evolution of herbicide-resistant weeds, are now more widely recognized. The UN Food and Agricultural Organization has stressed the need for "responsible deployment" of transgenics "to protect agro-systems, rural livelihoods and broader ecological integrity" (FAO 2005). These caveats notwithstanding, the FAO, World Bank, United Nations Development Program, and Consultative Group for International Agricultural Research (CGIAR) place crop genetic engineering high on their agendas (IISD 1999, UNDP 2001).

The U.S. government has worked to reshape international trade regimes and global-governance institutions, especially the World Trade Organization, to support biotechnology exports. WTO sub-agreements help to establish legal and discursive frameworks, under the rubric of "free trade," for international markets in genetic resources and biotechnology products. The most important of these are the Agreement on Agriculture (AoA) and the Trade-Related Intellectual Property Rights Agreement (TRIPs). Other WTO sub-agreements with ramifications for food and biotechnology trade are the Codex Alimentarius, the accord on Sanitary and Phytosanitary Measures, and the General Agreement on Trade in Services (GATS).

The WTO AoA calls for the phased elimination of most subsidies, quotas, and tariffs that many countries have used to maintain their domestic farming sectors, and, in some cases, to reward agricultural elites. U.S. agribusiness firms want a version of the AoA that will open foreign markets to their products without eliminating those categories of state subsidies and export credits that the U.S. defines as "non-trade-distorting." Such subsidies barely keep U.S. farmers afloat, but they enable transnational firms to export farm commodities at prices below the economic costs of production (Murphy, Lilliston, and Lake 2005). Of the subsidized crops exported from the United States, about half of the corn, 76 percent of the soy, and 73 percent of the cotton are genetically engineered (Gersema 2003; USDA 2005).

The TRIPs Agreement was initiated by a coalition of biotechnology firms and introduced by the United States. It requires WTO member countries to enforce private property rights "in all fields of technology." It would prevent public agencies and private enterprises from distributing their own versions of drugs, therapies, research tools, and crop varieties, as well as other inventions "protected" by patents, trademarks, trade secrets, or plant-breeders' rights. The U.S. government has sponsored even stronger intellectual property rights (IPR) requirements in its regional and bilateral trade treaties (GRAIN 2001; Drahos 2003). African governments, especially Kenya and Ethiopia, have led a broad alliance of developing countries in opposing globalized IPR (e.g., Africa Group 2003).

The United States has influenced the international Convention on Biological Diversity (CBD) in support of biotechnology interests. Early in the CBD negotiations, the U.S. delegation insisted on a provision recognizing intellectual property rights (UNEP/CBD 1994, Article 16; McConnell 1996; Drahos 1999). Developing countries opposed it, but then conceded. They accepted the IPR provision in exchange for CBD wording that recognizes national sovereignty over medicinal plants and other genetic-resource raw materials for biotechnology and additional CBD language calling on users of genetic resources to "share the benefits" of "commercial utilization of genetic resources" with the providers of those resources (UNEP/CBD 1994, Articles 1, 3, and 15).

The United States also tried for a decade to block the Cartagena Protocol on Biosafety, a sub-agreement of the CBD established at the urging of developing countries and NGOs (UNEP/CBD 1994, Article 19). Its effects in particular Latin American countries are discussed by Jansen and Roquas and by Fitting in this volume. The United States then tried unsuccessfully to insert language subordinating the Protocol to the WTO (McAfee 2003c). The United States failed to convince Southern and European governments to drop Protocol language that permits a "precautionary approach" to biotechnology regulation and was unable to defeat a provision that allows countries to take account of the socioeconomic consequences of transgenics (Article 26), a potentially important precedent in biotechnology regulation. The Protocol became international law on 11 September 2003.

The United States also promotes biotechnology through "capacity building" programs. Biotechnology capacity building is meant to help Southern scientists and administrators to understand genetic engineering, learn about the latest technology packages, manage the risks of transgenics, or (less often) to engineer plant and livestock varieties appropriate to their soils, climates, and markets. Capacity building often includes training in intellectual property instruments: licenses and material transfer agreements that are negotiated when patent holders permit restricted use of their property.

Most capacity-building programs portray partnerships with private enterprises as the best way to gain access to advanced biotechnology tools, data, and germplasm. This is because private corporations have patented or hold exclusive licenses to many of the enabling technologies and genetic data needed to engineer new crop varieties (Wright 2000; Barton and Berger 2001; Boyd 2003). Many Southern NGOs and officials have challenged the legitimacy of these proprietary claims. In many cases, the "inventive" activity on which patent applications are based consists of no more than characterizing and purifying a plant extract and proposing a novel use for it, or consistently reproducing a crop variety obtained from peasant farmer-breeders. Such "innovation" can be sufficient grounds for obtaining a utility patent in the United States. Biotechnology capacity builders must therefore convince skeptics of the morality and practical value of these private, exclusionary property rights.

Biotechnology capacity building has become a minor growth industry for international consultants and lawyers. By June 2003, participants from 117 countries were enrolled in a joint effort of the U.N. Environment Program and the Global Environment Facility (UNEP/GEF 2003). The U.S. Foreign Agriculture Service sponsors seminars and short courses worldwide "to build regulatory and institutional capacities and educate a variety of foreign audiences on issues surrounding agricultural biotechnology production, consumption, and trade" (USFAS 2003). USAID monitors developing-country policies on GM imports and sponsors a $14.8 million program "to enhance biosafety policy, research, and capacity" in Bangladesh, India, Indonesia, the Philippines, and East and West Africa, and other countries (USAID 2002).

Capacity building for managing the environmental hazards of transgenic organisms typically teaches methods modeled on those used in the United States or Europe. WTO and Biosafety Protocol provisions require that countries must justify any decisions not to accept imports of particular products on the basis of "sound science." The WTO TRIPs agreement allows exceptions in cases of dangers to public order and morality, and the Cartagena Protocol on Biosafety (CPB) permits countries to decline to import genetically modified living organisms for socioeconomic reasons, although thus far, no countries have attempted to use the latter provision. Countries and companies that export transgenic crops therefore have a stake in convincing Southern scientists and civil servants that their task is to bring their domestic laws and practices "up" to U.S. and E.U. standards and that the latter are unassailably rigorous and universally applicable.

Thus, substantial resources are being spent to influence Southern perceptions and restructure Southern-country institutions to favor the importing of proprietary biotechnology products. A closer look at the actual performance of transgenic crops to date, however, suggests that even if agriculture in developing countries differed little from that of the United States, there would be reason to doubt the wisdom of such a policy.

Promises and Perils of GM Crops

While industry public relations materials stress the value of transgenic crops to the hungry, existing commercially planted GMOs have not been designed to produce more food. Yields from these crops have generally been about the same and sometimes lower than those of their conventional counterparts, except in some places and some growing seasons when certain pest infestations have been unusually heavy (Benbrook 1999, 2000, 2003; Carpenter 2001; Elmore et al. 2001; Hyde et al. 1999). GM crops have not resulted in reduced use of pesticides, except in the case of *Bt* cotton in some regions, nor have they generally saved farmers money (Benbrook 2001, 2004).[2]

Industry publicity therefore highlights anticipated new crops with nutritional or disease-resistance traits that—it is hoped—will benefit farmers and consumers, not just agrochemical and seed firms. Most of these applications, such as the highly touted "golden rice" containing ß carotene (pro-vitamin A), are still in the development stage or are merely hypothetical. It is unlikely that private firms will have much incentive to develop many specialized varieties for small-scale or subsistence producers because these farmers rarely provide profitable markets. Nevertheless, industry spokespeople and agencies such as the World Bank routinely refer to GM organisms as "genetically enhanced," creating the impression that genetic engineering is already providing superior crops.

Meanwhile, a decade's experience with GMOs has cast doubt on early, reassuring endorsements of their safety and predictability. Regulators on both sides of the Atlantic have revised their recommendations for planting and field testing of transgenics in light of evidence that pollen from crops such as rapeseed (canola) and maize can travel farther than they had assumed, transferring synthetic DNA to related crops and sometimes to other species. Other unresolved issues concern the effects of transgenics on "non-target" organisms, including beneficial insects and soil microorganisms that keep agro-ecosystems healthy.

Because research on transgenic crops focuses mainly on short-term yields and on one crop at a time, scant attention has been paid to the agro-ecosystemic interactions of transgenic organisms or their longer-term effects. Scientific critics have been concerned about possible hazards of the antibiotic marker genes and bacterial and viral constructs that are used to engineer transgenic crops and that are reproduced along with them (McAfee 2003a). Another poorly understood phenomenon is the multiple, unpredicted outcomes of genetic engineering (Meyer et al.

1992; Halsberger 2003). These include "silencing" of the expression of apparently unrelated genes and the appearance of unexpected traits in response to changes in growing conditions (McAfee 2003a). Recent scientific evidence about ecological risks of transgenic crops is evaluated in a 2005 report by the Ecological Society of America (Snow et al. 2005).

One likely negative consequence of transgenics, pest resistance, is widely recognized. Weeds and insects can evolve in just a few years to tolerate pesticides, including those used in tandem transgenic plants or produced by the plants themselves. Approximately three-fourths of currently marketed GM crops are herbicide-tolerant (HT), engineered to withstand spraying with glyphosate. Glyphosate kills non-HT crops and most other plants it reaches, including weeds. This can reduce the need for farmers to till erosion-prone soils. Several species of weeds have already become resistant to glyphosate, however, causing serious problems for farmers and threatening to make glyphosate useless in the future (Pollack 2003).

Another 23 percent of planted GM crops carry genetic instructions from a bacterium, *Bacillus thuringiensis* (*Bt*), that causes them to produce insecticides in their tissues. About 8 percent have both *Bt* and HT traits. To slow the development of insect pests that are resistant to *Bt* toxins, the U.S. Environmental Protection Agency tells farmers to plant 20 percent of their corn acreage in non-*Bt* crops, but this requirement has been laxly enforced (Jaffe 2003). Genetic engineers have developed varieties that produce more than one form of toxin, a tactic that may postpone but is unlikely to solve resistance problems (Mellon and Rissler 1998).

Thus far crop genetic engineering represents a continuation of, rather than an alternative to, the technological paradigm of conventional U.S. agriculture (see Chap. 1 by Otero, this volume). Modern monocultures, whether transgenic or conventional, are designed to maximize short-term crop yields, but they often create and worsen pest and disease problems. Their high concentrations of nutrients, biophysical uniformity, and large scale make them ideal grounds for pest organisms to move, breed, and evolve. Lack of genetic diversity in monocrop agro-ecosystems makes them particularly vulnerable when pests gain an edge, as happened during the 1970 U.S. corn blight epidemic and the Irish potato famine 120 years before. These greatly simplified ecosystems often lack the beneficial microbes, insects that prey on pests, and other organisms involved in the eco-regulatory mechanisms that help keep more complex farming systems productive. Use of insecticides in the United States rose tenfold over forty-four years, but the proportion of crops lost to insects nearly doubled in the same period (Wargo 1996, p. 7). This pesticide treadmill is equally problematic in developing countries, especially in regions of Latin America where the U.S. agricultural paradigm has been adopted wholeheartedly (see Chap. 2 in this volume).

Biotechnology and the Problem of Genetic Erosion

Technology-centered agriculture and genetic engineering in particular add momentum to the trends of agro-biodiversity loss, genetic homogeneity, and diminished farmer choices of crops and varieties. This happens in several ways, three of which are as follows. First, biotechnology has spurred the further consolidation of seed, agrochemical, and food-processing enterprises. Mergers and buy-outs among chemical, pharmaceutical, and seed firms were driven in the 1980s and 1990s by the race to acquire genetic data, transformation tools, and intellectual property rights to use them. A small number of mega-firms now dominate agro-industry and trade worldwide (OECD 2000; Murphy 2002; Heffernan and Hendrickson 2002; Hendrickson and Heffernan 2005). Many genetic-engineering technologies are not particularly costly to apply but licenses to use these proprietary methods for plant breeding are often expensive. IPR portfolios are the most valuable asset of some major biotechnology firms (Barton and Berger 2001; Boyd 2003).

A second way that genetic engineering can speed genetic erosion is by contributing to social differentiation among farmers. This is likely to occur for the same reason that Green Revolution technology has frequently had this effect. Only some producers are able to afford higher-priced seeds and accompanying technology packages, while many poorer, part-time, or older farmers and farmers in more isolated places or varied ecological and cultural settings will be unable to continue. As this happens, many more crop types and local land races will no longer be cultivated.

A third likely threat to agro-biodiversity from genetic engineering is more direct. Most of the major seed/agrochemical firms are developing methods to prevent farmers from saving seeds. Various genetic-use restriction technologies (GURT)—"terminator" technologies to their critics—are meant to ensure that farmers must either purchase new seed for each planting or buy chemical keys to activate bio-engineered crop traits (Jefferson et al. 1999). The rationale advanced by GURT defenders is that companies need to recoup their high expenditures on research, development, regulatory requirements, acquisitions, and IP management. If widely employed, GURTs could block the continued creation of crop biodiversity, which now occurs as farmers select seeds from their best plants, exchange seeds, and allow modern varieties to interbreed with local ones. Should genetic constructs that cause seed sterility find their way into populations of wild crop ancestors and crop varieties grown for subsistence, the results could be devastating. As the GURT example illustrates, differences between conventional modern agriculture and the smaller-scale, more diversity-based farming that prevails in much of the South make the risks of transgenic crops much greater in many parts of Africa, Latin America, and Asia.

Geographies of Difference in Farming and Food Systems

For the sake of argument, let us presume that with new discoveries, future crop genetic engineering will indeed increase crop productivity in the large-scale agrifood systems for which it has been developed. Let us assume also that with greater attention to ecological processes and better-enforced regulation, future transgenic crops may ameliorate the environmental damage caused by high chemical input modern agriculture in advanced capitalist countries. On the basis of these suppositions, can we logically conclude that genetic engineering is the foundation of a strategy to end hunger and strengthen food security in developing countries? The answer, unfortunately, is no. There are important geographies of difference that make it impossible to extrapolate, even from a best-case scenario for advanced capitalist countries, to a positive prognosis for the effects of GM crops in most developing nations.

Differences in the Risk to Agro-Biodiversity in the Global North and South

The risks of negative consequences from the release of transgenic organisms are greater in many parts of the global South than in the United States, Argentina, Canada, and China, where most genetically engineered crops have been grown. None of these countries are centers of genetic diversity for the crops now grown there in transgenic forms. About 40 percent of the world's traded staple crops were first domesticated in Latin America; others came from Asia and Africa (Kloppenburg and Kleinman 1987; FAO 1997). A vast wealth of traits is conserved in these regions in the genomes of local crops and the wild plant types from which early farmers derived these food crops. Also important are secondary centers of agro-biodiversity, where farmers have developed varieties suited to their particular growing conditions and farming practices: for example, maize varieties in parts of Africa and rice landraces in South Asia.

Farmers who save and exchange seeds still draw upon this genetic diversity to maintain the vigor of their crops. Crop breeders in formal institutions, whether using conventional or molecular means, employ genetic material from farmer varieties and crop wild relatives to develop improved or transgenic varieties. The plant samples and seeds conserved in CGIAR and other gene banks are precious potential sources of traits that may help with pest and plant-physiology problems and climatic stress. Local varieties cultivated by small- and medium-scale farmers are at least as valuable, and not only to the farmers themselves. While little is known about many of the samples stored in seed banks, information about how different living varieties perform is available from farmers who grow them.

Is this vital agricultural genetic diversity threatened by genetically altered crops? Many researchers believe that the possibility is great enough to warrant postponing the release of transgenics, especially in regions of genetic diversity, until much more is known about their effects. There is no doubt that engineered genetic material can be transferred via pollen, even from plants that are usually self-pollinating, to fields of conventional crops and wild relatives of these crops (Ohio State 2002). Some studies suggest that genetic engineering itself may, in some cases, cause plants to produce more seed or spread their pollen more widely (Ellstrand 2001; Snow et al. 2002). There is evidence that in ecosystems and in animal guts, engineered genetic constructs may also be transferred "horizontally," from one species to another, with the aid of naturally occurring microbes or viral vectors. Horizontal gene transfer is even less understood than gene flow via pollen.

In November 2001, University of California researchers reported in *Nature* that they had detected transgenic material in kernels taken from mountainside maize plots in Oaxaca, Mexico, where planting transgenic corn was illegal (Quist and Chapela 2001; McAfee 2003b). Other scientists, and a concerted campaign by the industry leader, Monsanto, cast doubt upon some of the diagnostic methods used in the initial *Nature* study (Kaplinksi et al. 2002; Matthews 2002). A later study failed to find evidence of transgene introgression in specific locations where it was reported in 2001 (Ortiz-García et al. 2005). Nevertheless, few researchers doubt that transgenic material probably has been or soon will be incorporated into the genomes of local maize varieties wherever transgenic corn is planted. Many believe that engineered DNA might also be taken up by teosinte, the ancestor of maize (*Zea mays*).

Is transgene introgression likely to cause the loss of genetic or varietal diversity in corn or teosinte or in different crops and wild plants? This could occur if engineered genetic material confers a survival advantage to some varieties, which might out-compete related plants that carry potentially valuable characteristics. Some researchers think that such effects will not be significant. This is the official position of the CGIAR's International Center for Maize and Wheat Improvement. Nevertheless, CIMMYT has taken pains to ensure that its seed collections are not "contaminated" by transgenic varieties (CIMMYT 2001). What is indisputable is that *if* gene flow from transgenic crops causes any of these problems, the risk of harm is far greater in centers of genetic diversity and regions where crop ancestors still thrive, and where farmers still depend upon inter- and intra-species diversity, than it is in the major food-exporting countries, where most commercial crops now grown were unknown four hundred years ago and there is little diversity to be lost.

Differences in the Role of Agro-Biodiversity in Different Kinds of Agro-Food Systems

Inter-species and intra-species genetic diversity is economically inefficient in some agrifood systems but it is critically important in others. In capitalist agriculture, the values of crop varieties, like other factors of production, depend upon their contribution to the profitability of the enterprise, usually calculated over one harvest cycle. For large farming operations and agribusiness firms, genetic uniformity has advantages related to the exigencies of scale, mechanization, and marketing. Identical plants, whether transgenic or conventional, that ripen simultaneously can be harvested, quality-checked, transported, and processed in bulk.

In contrast, in more "traditional" farming systems in developing countries and elsewhere, crop variety and genetic diversity may be advantageous for multiple reasons, not all of them directly economic. The values of different crops and traits are related to the manifold functions of agricultural landscapes. They are place-specific and particular to different eco-social systems. Particular varieties may be preferred for different purposes, especially where some foods are grown for subsistence, where food, labor, seeds, and inputs are exchanged among family and community members and migrants, and where local markets and festivities are important. Traits that affect local marketability, tastes and colors, cooking and storage qualities, and symbolic significance can be important to cultural cohesion as well as to nutrition.

Genetic homogeneity may be dangerous, especially where peasants must use marginally arable land. It is common to plant different varieties at different altitudes or in soils and microclimates with different characteristics, or to plant landraces with different traits, in order to increase the chances of an adequate harvest if the growing season turns out to be particularly dry, wet, hot, or cold. Moreover, as noted above, small-scale farmers not only select their best seeds for planting and exchange, but also may allow their own varieties to hybridize with modern or wild varieties to improve their crops. If they have access to fewer traditional and new varieties, and especially if the new varieties were to carry genetic instructions for seed sterility (GURT), this economically and culturally vital process will be stifled.

Commodity Relations and Greater Economic Risks for Self-Provisioning Farmers and Tropical Smallholders

The effects of crop genetic engineering are substantially different for fully capitalist agriculture than for farming systems in which the processes, inputs, and products of production are not primarily based on commodity relations. One reason is that farmers in "modern," fully commercialized farming systems purchase new seed frequently, often annually, while the majority of Southern-country farmers save seeds for planting. As noted, seed-sterility technologies could disrupt this cycle of agrarian life and survival.

Moreover, in places where seeds are saved and multiple varieties are planted, transgenic traits are more likely to be dispersed, with unpredictable consequences. Just as pesticides more generally can speed the evolution of resistant weeds and pests, so can transgenic crops that require pesticides or produce pesticides, as noted above. Potentially resistant predators are often more numerous and varied in tropical zones and in regions of crop genetic diversity, where pests have co-evolved with domesticated crops and their wild relatives. If resistant pests cause partial or complete crop failure, the consequences are more likely to be severe in regions where farmers lack other sources of incomes and food.

Contrasts in Regulatory Needs and Effectiveness in Different Kinds of Agrifood Systems

Most agencies that facilitate biotechnology capacity-building take U.S. or European regulatory institutions as models. EPA rules and USDA guidelines underestimate or ignore the dispersal of transgenic artifacts by natural means such as wind-borne pollen and horizontal (interspecies) gene transfer. Some worry that regulatory bodies in developing countries are subject to political pressures or are excessively compliant (Cohen 1999; Jansen and Roquas, this volume). While this may often be true, there is ample evidence of pro-industry bias and laxity in biosafety testing and resistance-management regulation in the United States itself (NAS 2002; Jaffe 2003). Faith in industry self-monitoring and government transparency is as ill founded in advanced economies as it is in developing countries.

Even if U.S. safety measures were stringent, thorough, and honestly enforced, models developed for industrialized farming are not appropriate for numerous regions in developing nations. The meanings and practices that surround the seed are different in agrifood systems that are fully integrated into national and international markets, compared to those in which a significant proportion of farming is for subsistence and local exchange. The model that U.S. agencies are trying to globalize presumes a system of commercialized agriculture in which varieties are uniform, seed is a commodity sold for planting, and harvested grain is a different commodity sold for consumption. But in partially self-provisioning peasant economies in many regions of Latin America, agricultural production and consumption are phases of a cycle that is both more local and more closed. The same seed may be the source of life in at least two senses: it is the next day's meal *and* the next season's planting material.

In modern capitalist agrifood systems, society-nature relations are mediated by markets and conceptual dualism prevails. Seed for planting, feed for animals, and food for people are perceived and regulated differently, even when the same grain is the source of all three. The contamination in 2001 of U.S. corn food products with transgenic StarLink® corn, approved for animal feed but not for humans because of its potential allergenicity, showed that such distinctions are hard to maintain. Nevertheless, U.S. negotiators have insisted upon those distinctions. They were adamant about excluding from the Biosafety Protocol's purview any GM organisms not meant "for intentional introduction into the environment," including those meant "for direct use as food or feed, or for processing" (Article 7). This makes it more difficult for countries to decline to import U.S. agricultural products on environmental grounds.

The controversy over gene flow in Oaxaca illustrates the folly of modeling developing-country biotechnology management on regulations designed for fully capitalist agrifood systems. As we have seen, seeds are social as well as natural. Those who would base biosafety rules on Northern models fail to appreciate the meanings and mobility of seeds as cultural objects and as coactants in eco-social systems. Recall the 1998 Mexican moratorium on transgenic corn. Such a proscription might work in a fully commodified agrifood system, but Mexican agriculture does not fit that description. Despite the efforts of the last two governments to eliminate the country's "inefficient" peasant food producers, many Mexican *campesinos* continue to raise corn for household use even when this appears to be irrational from a narrowly economic point of view (Fitting, this volume; McAfee, forthcoming). Few Mexican observers were surprised to hear that "escaped" transgenic constructs were detected in the Oaxacan sierra, meaning that the official GM moratorium had been futile. The origin of the traveling transgenes is almost certainly whole-grain (unmilled) U.S. *Bt* corn, which is widely marketed as grain to be fed animals but which can also be planted or used to make tortillas and other foods. *Bt* corn seeds may also be brought home by seasonal migrant workers following the time-honored peasant practice of seed exchange and experimentation. Laws against use of untested transgenics will not stop this.

Differences in the Ability of Northern and Southern Food Producers and Enterprises to Profit from Biotechnology and Property Rights

In an idealized world of globalized property rights, anyone anywhere has equal opportunity to innovate and exclude others from the use of his or her invention. In the real world, there are historical and structural differences in the abilities of transnational firms, domestic companies, public-sector agencies, indigenous peoples' organizations, and small-scale food producers to make profitable use of property rights. Only the financially well endowed can afford to establish and defend proprietary claims to genetic and agriculture resources.

Accumulation of capital in biotechnology is based on the enclosure of the intellectual commons: private ownership of scientific knowledge for commercial purposes and privatization of genes and organisms that were once open-access resources (Press and Washburn 2000; Barton 1998). The great majority of patents on products and processes used around the world are held by Northern corporations, with the United States well in the lead. The agrochemical giant Monsanto has had proprietary rights to most of the transgenic crop applications (mainly herbicide tolerance) now in commercial use. A few private firms control a large proportion of the rights to genetic-engineering enabling and platform technologies as well as the rights to make use of natural or induced genetic variations. Recent efforts by private foundations and international research agencies to "invent around" privatized technologies or to persuade TNCs to donate them have had limited success. Academic and public-sector plant breeders often find it impossible to use proprietary germplasm and technologies to produce public goods, such as crop varieties crafted for poor farmers (Herdt 1999; Barboza 2001).

NGOs, farmer activists, and governments of Southern states point to the injustice of "biopiracy." This charge is usually aimed at Northern firms and research institutions when they patent pharmaceuticals, crop varieties, or other products derived from materials and information obtained from Southern ecosystems, farmers, or healers, sometimes with little or no added innovation by the patent holder. In response, defenders of universalized IPR point out that proprietary claims on medicines and crops cannot prevent the original cultivators or healers from continuing to use crops or natural products in their original form, so long as they do not sell them. To those whose resources have been "pirated," that message reads: "It's fine to keep benefiting from these valuable genetic resources—as long as you agree to remain poor."

A consortium of Mexican small- and medium-scale farmers got such a message when they took NAFTA's free-trade promise at face value. In 1994, the Rio Fuerte producers' union began increasing their exports to the United States of a type of yellow bean they had been growing and selling since the 1970s. Their U.S. market suddenly disappeared when their distributor was sued for patent infringement by a U.S. bean broker, Pod'ner LLC. Pod'ner's CEO claimed to have invented the "Enola" bean in 1996, using beans he bought in Mexico in 1994. In 1999, he was granted a U.S. patent of dubious validity on beans with the same color and other traits also found in the Mexican growers' mayacoba bean (Pallottini et al. 2004). U.S. and EU patents have been granted for other varieties developed and marketed in developing countries, including basmati rice, neem plant extracts, quinoa and beans from the Andes, and kava from the South Pacific.

Some defenders of the property-rights paradigm urge governments and communities in gene-rich regions to assert their own proprietary claims on genetic resources and then trade them internationally. When the CBD was drafted, hopes were high that the sale of such genetic-resource rights under the terms of biodiversity prospecting contracts with pharmaceutical firms would generate incentives and incomes for biodiversity conservation and for indigenous peoples and other local resource providers. These hopes were ill founded; market prices for natural-product samples are minimal and no community has earned significant royalties under the terms of a bioprospecting contract.

The benefits which have actually been provided to genetic-resource providers have typically been modest, up-front payments for community-development projects. Such benefits depend upon the goodwill or the public-relations agendas of pharmaceutical firms and other bio-buyers, not on the legal force of IP claims by Southern communities or governments. Bioprospecting deals can aid some communities, but it is wrong to imagine that they can be the basis of a strategy for the "equitable sharing" of the "benefits of biodiversity" that is called for in the CBD. The power relations between buyer and sellers are too hugely asymmetrical. Should a firm develop a highly profitable use for a substance obtained through such arrangements, it could use its immensely greater legal and technological resources to interpret the contract or to alter the product in such a way as to limit profit transfers to the raw-material providers. Its shareholders would expect nothing less.

Vast differences in the capacity of economic actors to employ biotechnology and to accumulate capital from property rights do not fall strictly along a North-South divide. State agencies and private firms in India, Brazil, Cuba, and some other Southern nations may be able to find export niches for specialized, proprietary biotechnology products. This is unlikely to alter the trend toward consolidation of food-producing resources in fewer hands worldwide. In the absence of national- and global-level policy changes, the combination of food-trade and investment liberalization and IPR globalization will speed displacement of domestic food producers, agro-industries, and seed companies. One example is Brazil, where, pending the legalization of transgenic crops, Monsanto positioned itself to take over some of the country's major seed firms (Reuters 2003; Rocha 2003; Monsanto 2004).

All the dimensions of difference discussed above are consequences of the place-specificity of agro-eco-social systems and of the biodiversity they contribute to and depend upon. All three dimensions of difference are simultaneously ecological, economic, and cultural. All three are also related to the degree to which agrifood systems have become capitalist, modernized, and integrated into wider markets. However, this does not imply an either-or choice between agrifood systems that are "modern," market-oriented, technology-centered, and highly productive (but probably not ecological sustainable) on one hand and farming systems that are "traditional," subsistence-oriented, and less productive (even if more sustainable) on the other.

Alternative Agricultural Futures

A huge amount of attention and resources has been directed toward developing, denouncing, celebrating, privatizing, regulating, and scrutinizing crop genetic engineering. This is remarkable in light of the fact that, even if transgenic varieties can be made to perform as their enthusiasts predict, and even if their risks prove minimal, their promised productivity, environmental, and nutritional benefits are quite modest. For these hoped-for gains, developing countries would need to take on "a major financial and logistical challenge" (NCB 2004, p. 94).

Therefore, the response to the issues outlined above should not be to eliminate the differences that make transgenic crops irrelevant or hazardous for most farmers in developing countries, at least in their present form. Instead, we can build upon the positive aspects of these differences. Truly "modern" agricultural science can then combine knowledge from genomics, ecology, and local experience to pursue multiple technological options, experimental methods, and learning models. These approaches need to be developed from the bottom up, with active farmer participation, and adapted to greatly diverse landscapes and ecologies.

Let us assume that the goal is to produce adequate food and fiber in ways that are socially and ecologically sustainable and that contribute to equitable economic development. If that is the goal, then there are strong reasons to avoid replicating the model of high-chemical-input, heavily mechanized, monocrop plantations for which transgenic crops, so far, have been designed. These reasons include energy inefficiency, waste and degradation of water and soil, and ever-escalating pesticide requirements, even with transgenic varieties. Added to that are profound social costs: the ruin of small- and medium-scale farmers throughout North America and Europe, and for those farmers who remain, virtual indenture to giant agrochemical and food-processing firms (see Chap. 1 in this volume).

These few firms have extensive control over crop prices, farming methods, and varieties (Murphy 2006). They monopolize most technologies used in genetic transformation. Currently, crop genetic engineering is doing more to perpetuate the modern agri-technological paradigm discussed in Chap. 1 above than it is doing to help create more sustainable and adaptable options. This has partly to do with the techno-science itself; most crop biotechnology is focused on germplasm as a laboratory object: denatured, decontextualized, and disembedded from its eco-social habitats (McAfee 2003a). It is also the consequence of the concentration of agricultural and scientific resources in the hands of transnational agrochemical oligopolies.

Such a path for genomic sciences and rDNA technology is not inevitable. As experience with crop genetics and transformation has increased, useful guides for ecologically informed and prudent use of these powerful tools have begun to emerge (Benbrook 2003). Potentially, molecular biotechnology can contribute to decentralized ex situ and in situ genetic-resource conservation, wider exchange of plant genetic resources and knowledge about them, crop breeding and evaluation aided by farmer-led experimentation, molecular-genetic markers, farmer-focused and participatory livestock breeding, and integrated resource management. But this approach will not prevail so long as profitability criteria set research agendas, budget cuts and property claims block production of public goods, and policy makers are mesmerized by the mirage of molecular wonders.

Insights into more sustainable and more productive agriculture are emerging from farmer-centered agronomic research, agroecology (both as science and as a social movement), and networks of farmer-scientists, peasant activists, and their allies in nongovernmental, multilateral, and academic institutions (Altieri 1995; Brookfield 2001; Uphoff 2002; Hisano and Altoé, this volume). Mexican peasant organizations are part of a growing international social movement of farmers and other food producers organizing under the banner of food sovereignty (Food Sovereignty Forum 2002; Wise, Salazar, and Carlson 2003; NFFC 2004; McMichael 2004; McAfee forthcoming). Latin American small-scale farmers do not want to be poor or to survive on self-provisioning alone, nor do they reject new knowledge. But many are struggling for the right to remain farmers, or at least to maintain some agricultural activities and productive land in their extended families and communities. Their choices demand respect, as a principle of human rights and democracy, and because productive rural communities can enhance food security, biodiversity and environmental sustainability, and social well-being.

It is clear that cultural, economic, and political factors are at least as critical as technological factors to overcoming hunger. While crop biotechnology debates rage on, far less public and scientific attention is being paid to policy changes and activities that could do more, on a larger scale, to increase staple food production and food security: thoroughgoing land reform, backed by farmer-centered research and extension services, low-interest credit, and assistance with storage, transport, and marketing for low-income farmers. In most policy circles, it seems, obstacles to such changes appear far more daunting than the prospect of curing hunger by manipulating molecules. Bottom-up mobilization by peasant and other grassroots movements may yet encourage policy makers to see prospects differently and become more responsive.

Notes

1. I was one of twenty-seven external peer reviewers of the CEC report.

2. Dr. Charles Benbrook, former Executive Director of the Board on Agriculture for the U.S. National Academy of Sciences, is one of the most thoughtful and objective evaluators of the results of biotechnology applications in the United States. While not opposed to genetic engineering, Benbrook is critical of those who treat it as a panacea. He points out that in the limited number of cases where it has been plausible to claim that transgenic varieties have proved beneficial, the studies involved have compared transgenic performance to the option of doing nothing, rather than to the options of employing methods to address the causes, not just the symptoms, of the problems that plague high-chemical-input, monocrop farming (personal communication, November 2005).

References

Ackerman, F., T. A. Wise, K. P. Gallagher, L. Ney, and R. Flores. 2003. "Free Trade, Corn, and the Environment: Environmental Impacts of US-Mexico Corn Trade under NAFTA." Global Development and Environment Institute Working Paper No. 03–06. Tufts University, Medford, Mass. http://ase.tufts.edu/gdae/Pubs/wp/03-06-NAFTACorn.pdf.

Africa Group. 2003. "Taking Forward the Negotiation of Article 27.3B of the TRIPs Agreement." Communication from the Africa Group to the WTO TRIPs Council.

Altieri, M. 1995. *Agroecology: The Science of Sustainable Agriculture*. Boulder, Colo.: Westview.

Apffel-Marglin, F., and S. Marglin. 1996. *Decolonizing Knowledge: From Development to Dialogue*. Oxford: Clarendon Press.

Barboza, D. 2001. "As Biotech Crops Multiply, Consumers Get Little Choice." *New York Times* (Late Edition, Final), 9 June, §1, 1.

Bartra, Armando. 2004. "Rebellious Cornfields: Toward Food and Labour Self-sufficiency." pp. 18–36 in *Mexico in Transition: Neoliberal Globalism, the State and Civil Society*, ed. Gerardo Otero. London: Zed Books; Nova Scotia: Fernwood Publishing.

Bartra, A., et al. 2005. *Transgénicos, ¿quién los necesita?* DF: Centro de Producción Editorial, Grupo Parlamentario del PRD, Cámara de Diputados, Congreso de la Unión, LIX Legislatura.

Barton, J. 1998. "The Impact of Contemporary Patent Law on Plant Biotechnology Research." pp. 85–97 in *Global Genetic Resources: Access and Property Rights*, ed. S. A. Eberhart et al. Madison: Crop Science Society of America.

Barton, J., and P. Berger. 2001. "Patenting Agriculture." *Issues in Science and Technology* 7(4): 43–50.

Bebbington, A., and J. Carney. 1990. "Geography in the International Agricultural Research Centers: Theoretical and Practical Concerns." *Annals of the Association of American Geographers* 80(1): 34–48.

Becker, E. 2003. "U.S. Threatens to Act against Europeans over Modified Foods." *New York Times*, Late Edition—Final, Section A, Page 4, Column 3.

Benbrook, C. M. 1999. "Evidence of the Magnitude and Consequences of the Roundup Ready Soybean Yield Drag from University-Based Varietal Trials in 1998." Ag BioTech InfoNet, Technical Paper Number 1. http://www.biotech-info.net/RR_yield_drag_98.pdf.

———. 2000. "Who Controls and Who Will Benefit from Plant Genomics?" Paper presented at the Annual Meeting of the American Association for the Advancement of Science, Washington, D.C.

———. 2001. "Do GM Crops Mean Less Pesticide Use?" *Pesticide Outlook* (now renamed *Outlooks on Pest Management*), October: 204–207.

———. 2003. "Principles Governing the Long-Run Risks, Benefits, and Costs of Agricultural Biotechnology." Paper presented at the Conference on Biodiversity, Biotechnology, and the Protection of Traditional Knowledge, Saint Louis, Mo., 5 April. http://www.biotech-info.net/biod_biotech.pdf.

———. 2004. "Genetically Engineered Crops and Pesticide Use in the United States: The First Nine Years." Technical Paper #7, Northwest Science and Environmental Policy Center, Sandpoint, Idaho, 25 October. http://www.biotech-info.net/technicalpaper7.html.

BIO (Biotechnology Industry Organization). 2005. http://www.bio.org/foodag/.

Borlaug, Norman, and Jimmy Carter. 2005. "Food for Thought," in *Wall Street Journal* (Opinion Editorial), http://www.cartercenter.org/search/viewindexdoc.asp.

Boyd, W. 2003. "Wonderful Potencies? Deep Structure and the Problem of Monopoly in Agricultural Biotechnology." pp. 24–62 in R. A. Schurman and D. D. T. Kelso, eds., *Engineering Trouble: Biotechnology and Its Discontents*. Berkeley: University of California Press.

Brookfield, H. C. 2001. *Exploring Agrodiversity*. New York: Columbia University Press.

Carpenter, J. E. 2001. *Comparing Roundup Ready and Conventional Soybean Yields*. Washington, D.C.: National Center for Food and Agricultural Policy.

CEC (North American Commission on Environmental Cooperation). 2004. *Maize and Biodiversity: The Effects of Transgenic Maize in Mexico: Key Findings and Recommendations*. http://www.cec.org/maize.

CIMMYT (Centro Internacional Mejoramiento de Maíz y Trigo). 2001. "Response to Discovery of Transgenic Maize Growing in Mexico." Mexico City, Mexico: CIMMYT. http://www.cimmyt.org/english/wps/transg/transgenic.htm.

Cohen, J. I., ed. 1999. *Managing Agricultural Biotechnology: Addressing Research Program Needs and Policy Implications*. Biotechnology in Agriculture Series, 23. Wallingford, England: CABI.

Conway, G. 1997. *The Doubly Green Revolution: Food for All in the Twenty-First Century*. Ithaca, N.Y.: Comstock/Cornell University Press.

De Young, K. 2006. "Gates, Rockefeller Charities Join to Fight African Hunger." *Washington Post*. 13 September, page A01. 125 (4): 1543.

DeVries, J., and G. Toenniessen. 2001. *Securing the Harvest: Biotechnology, Breeding and Seed Systems for African Crops*. Wallingford, England: CABI.

Drahos, P. 1999. The TRIPs Review. Paper presented at the Conference on Strengthening Africa's Participation *GR941*: in the Review and Revision of the TRIPs Agreement of the World Trade Organization, February. Nairobi: African Centre for Technology Studies.

———. 2003. "Expanding Intellectual Property's Empire: The Role of FTAs." Australian National University, Research School of Social Sciences, Regulatory Institutions Network. http://www.grain.org/rights_files/drahos-fta-2003-en.pdf.

Ellstrand, N. C. 2001. "When Transgenes Wander, Should We Worry?" *Plant Physiology* 125(4): 1543.

Elmore, R. W., F. W. Roeth, L. A. Nelson, C. A. Shapiro, R. N. Klein, S. Z. Knezevic, and A. Martin. 2001. "Glyphosate-Resistant Soybean Cultivar Yields Compared with Sister Lines." *Agronomy Journal* 93(2): 408–412.

FAO (Food and Agriculture Organization of the United Nations). 1997. *State of the World's Plant Genetic Resources for Food and Agriculture.* Rome: FAO.

———. 2002. U.S. Secretary Of Agriculture: "The World Must Unite to Fight Hunger and Poverty; Success Will Require an International Coalition." Press Release 01/85 C9. http://www.fao.org/WAICENT/OIS/PRESS_NE/PRESSENG/2001/pren0185.htm.

———. 2005. Monitoring the Environmental Effects of GM Crops: FAO Expert Consultation Recommends Guidelines and Methodologies. Rome: FAO Newsroom. http://www.fao.org/newsroom/en/news/2005/89259/index.html.

Fleisher, A. 2003. "West Africa: Background Briefing on the President's Meetings with President Wade and with the Leaders of the West African Democracies." Transcript. Washington, D.C.: White House.

Food Sovereignty Forum. 2002. *Food Sovereignty: An Action Agenda.* Rome: International NGO/CSO Planning Committee. http://www.cenesta.org/projects/FoodSovereignty/plan_actioning.pdf.

Gersema, E. 2003. "Soybean Planting at Lowest since 1998." Associated Press. 30 June. http://www.agbios.com/main.php?action=ShowNewsItem&id=4551.

GRAIN (Genetic Resources Action International). 2001. "TRIPS-plus through the Back Door: How Bilateral Treaties Impose Much Stronger Rules for IPRs on Life Than the WTO." http://www.grain.org/briefings/?id=6.

Halsberger, A. G. 2003. "Codex Guidelines for GM Food Include the Analysis of Unintended Effects." *Nature Biotechnology* 21(7): 739–742.

Heffernan, D., and M. Hendrickson. 2002. "Multi-National Concentrated Food Processing and Marketing Systems and the Farm Crisis." Paper presented at the Annual Meeting of the American Association for the Advancement of Science, Boston.

Hendrickson, Mary, and William Heffernan. 2005. *Concentration of Agricultural Markets.* Department of Rural Sociology, University of Missouri, Columbia, Mo. February. www.agribusinesscenter.org/docs/Kraft_1.pdf.

Herdt, R. W. 1999. "Enclosing the Global Plant Genetic Commons." Paper delivered at the China Center for Economic Research, 24 May, based on a paper presented at the Institute for International Studies, Stanford University, 14 January. http://www.rockfound.org/display.asp?Collection=3&context=1&DocID=220&Preview=0&ARCurrent=1.

Hyde, J., M. A. Martin, P. V. Preckel, and R. Edwards. 1999. "The Economics of Bt Corn: Adoption Implications." Purdue University Agricultural Communications Online Publications, ID-219, 1–18. www.biotech-info.net/bt_economics.pdf.

IISD (International Institute for Sustainable Development). 1999. "A Summary Report of the Workshop on Agricultural Biotechnology and Rural Development." *Sustainable Developments* 26(1): 1–8.

ISAAA (International Service for the Acquisition of Agri-biotech Applications). 2004. Preview: Global Status of Commercialized Biotech/GM Crops: 2004. www.isaaa.org.

Jaffe, G. 2003. *Planting Trouble: Are Farmers Squandering Bt Corn Technology? An Analysis of USDA Data Showing Significant Noncompliance with EPA's Refuge Requirements.* Washington, D.C.: Center for Science in the Public Interest. http://www.cspinet.org/new/pdf/bt_corn_report.pdf.

Jefferson, Richard A., Carlos Correa, Gerardo Otero, Don Byth, and Calvin Qualset. 1999. "Genetic Use Restriction Technologies: Technical Assessment of the Set of New Technologies Which Sterilize or Reduce the Agronomic Value of Second Generation Seed, as Exemplified by U.S. Patent 5,723,765, and WO 94/03619. Montreal: United Nations Convention on Biological Diversity." Technical Report. Available at http://www.biodiv.org/sbstta4/HTML/SBSTTA4-9-rev1e.html.

Juma, C., J. M. Mugabe, and P. Kameri-Mbote, eds. 1994. *Coming to Life: Biotechnology in African Economic Recovery.* Nairobi: ACTS Press and Zed Books.

Kaplinski, N., D. Braun, D. Lisch, A. Hay, S. Hake, and M. Freeling. 2002. "Biodiversity (Communications Arising): Maize Transgene Results in Mexico are 1746 Artifacts." *Nature* 416(6881): 601–602.

Kloppenburg, J. R., Jr., and D. L. Kleinman. 1987. "The 1751 Plant Germplasm Controversy." *BioScience* 37(3): 190–198.

Matthews, J. 2002. "Amaizing Disgrace." *Ecologist.* 30 May.

McAfee, K. 2003a. "Biotech Battles: Plants, Power, and Intellectual Property in the New Global Governance Regimes." pp. 174–194 in R. A. Schurman and D. D. T. Kelso, eds., *Engineering Trouble: Biotechnology and Its Discontents.* Berkeley: University of California Press.

———. 2003b. "Corn Culture and Dangerous DNA: Real and Imagined Consequences of Maize Transgene Flow in Oaxaca." *Journal of Latin American Geography* 2:18–42.

———. 2003c. "Neoliberalism on the Molecular Scale: Economic and Genetic Reductionism in Biotechnology Battles." *GeoForum* 34(2): 203–222.

———. 2006. "Sostenibilidad y Justicia Social en el Sistema Alimentario Global." In A. Cohn, J. Cook, M. Fernandez, and R. Reidor, eds., *Nuestras Tierras, Nuestros Alimentos, Nuestras Vidas: Movimientos de Campesinos, Comercio, y Medio Ambiente en las Américas.*

———. Forthcoming. Yale School of Forestry and Environmental Studies. http://www.yale.edu/environment/publications.

McConnell, F. 1996. *The Biodiversity Convention: A Negotiating History.* London: Kluwer.

McMichael, P. 2004. "Global Development and the Corporate Food Regime." Agribusiness Accountability Initiative Clearinghouse: http://www.agribusinessaccountability.org/page/231?order_by=Abstract&opts=.

Mellon, M., and J. Rissler, eds. 1998. *Now or Never: Serious New Plans to Save a Natural Pest Control*. Cambridge, Mass.: Union of Concerned Scientists.

Meyer, P., F. Linn, I. Heidmann, H. Meyer, I. Niedenhof, and H. Saedler. 1992. "Endogenous and Environmental Factors Influence 35S Promoter Methylation of a Maize A1 Gene Construct in Transgenic Petunia and Its Colour Phenotype." *Molecular Genes and Genetics* 231(3): 345–352.

Monsanto Corporation. 2004. *2003 Annual Report*. http://www.monsanto.com/monsanto/layout/investor/financial/archive/2003AnnualReport.asp.

Murphy, S. 2002. *Managing the Invisible Hand: Markets, Farmers and International Trade*. Minneapolis, Minn.: Institute for Agriculture and Trade Policy.

———. 2006. "Concentrated Market Power and Agricultural Trade." EcoFair-Tade Discussion Paper 1, Heinrich Böll Foundation. http://www.iatp.org/iatp/publications.cfm?accountID=451&refID=89014.

Murphy, Sophia, Ben Lilliston, and Mary Beth Lake. 2005. *WTO Agreement on Agriculture: A Decade of Dumping*. Minneapolis, MN: Institute for Agriculture and Trade Policy.

Nadal, Alejandro, and Timothy A. Wise. 2004. "The Environmental Costs of Agricultural Trade Liberalization: Mexico–U.S. Maize Trade under NAFTA Working Group on Development and Environment in the Americas." *Discussion Paper Number 4*. Medford, Mass.: Tufts University.

NAS (National Academy of Sciences). 2000. *Transgenic Plants and World Agriculture*. Washington, D.C.: National Academy Press. http://www.nap.edu/books/NI000227/html.

———. 2002. *Environmental Effects of Transgenic Plants: The Scope and Adequacy of Regulation*. Washington, D.C.: National Academy Press. http://www.nap.edu/books/0309082633/html.

NCB (Nuffield Council on Bioethics). 2004. "The Use of Genetically Modified Crops in Developing Countries." Follow-up Discussion Paper. http://www.nuffieldbioethics.org/go/ourwork/gmcrops/page_218.html.

NFFC (National Family Farm Coalition). 2004. "Fair Trade Policy: Promoting Food Sovereignty." http://www.nffc.net/issues/fair/fair_1.html.

OECD (Organization for Economic Cooperation and Development). 2000. "Economic Impacts of Genetically Modified Crops on the Agri-Food Sector: A Synthesis." Working Document 1838, OECD Secretariat.

Ohio State University. 2002. "Scientific Methods Workshop: Ecological and Agronomic Consequences of Gene Flow from Transgenic Crops to Wild Relatives." 5 and 6 March, Columbus, Ohio. http://www.biosci.ohio-state.edu/~asnowlab/gene_flow.htm.

Ortiz-García, S., E. Ezcurra, B. Schoel, F. Acevedo, J. Soberón, and A. A. Snow. 2005. "Absence of Detectable Transgenes in Local Landraces of Maize in Oaxaca, Mexico (2003–2004)." *Proceedings of the National Academy of Sciences*, *Early Edition*, August 2005.

Paarlberg, Robert. 2001. *The Politics of Precaution: Genetically Modified Crops in Developing Countries*. Baltimore: Johns Hopkins University Press.

Pallottini, L., E. Garcia, J. Kami, G. Barcaccia, and P. Gepts. 2004. "The Genetic Identity of a Patented Yellow Bean." *Crop Science* 44(3): 968–977.

Pardey, P. G., ed. 2001. *The Future of Food: Biotechnology Markets and Policies in an International Setting.* Washington, D.C.: International Food Policy Research Institute.

Pollack, A. 2003. "Widely Used Crop Herbicide Is Losing Weed Resistance." *New York Times.* Late Edition, Final, 14 January, §C, 1.

Press, E., and J. Washburn. 2000. "The Kept University." *Atlantic Monthly.* March, 33(285): 39–53.

Quist, D., and I. Chapela. 2001. "Transgenic DNA Introgressed into Traditional Maize Landraces in Oaxaca, Mexico." *Nature* 414(6863): 541–543.

Reuters News Service. 2003. "Monsanto to Double Brazil sales by 2003." By Phil Stewart. Sao Paulo, 29 October.

Rocha, V. De O. 2003. "Biotechnology in Brazil: Legal and Economic Aspects." Harvard University, David Rockefeller Center for Latin American Studies. 4 April. http://www.veirano.com.br/html/english/conteudo_articles.cgi?ARTIGO=13.

Rodarte, O. A. 2003. "Toward an Equitable, Inclusive, and Sustainable Agriculture: Mexico's Basic Grain Producers Unite." 129–148 in T. A. Wise, H. Salazar, and L. Carlsen, eds., *Confronting Globalization: Economic Integration and Popular Resistance in Mexico.* Bloomfield, Conn.: Kumarian Press.

Royal Society of London. 2002. *Transgenic Plants and World Agriculture.* Report under the auspices of the Royal Society of London, U.S. National Academy of Sciences, Brazilian Academy of Sciences, the Chinese Academy of Sciences, Indian National Science Academy, the Mexican Academy of Sciences, and Third World Academy of Sciences. Washington, D.C.: National Academy Press.

Said, C. 2007. "Nothing Flat about Tortilla Prices." *San Francisco Chronicle.* Section C, page 1.

Serageldin, I., and G. J. Persley. 2000. *Promethean Science: Agricultural Biotechnology, the Environment, and the Poor.* Washington, D.C.: Consultative Group on International Agricultural Research.

Snow, A. A., D. Pilson, L. H. Rieseberg, M. J. Paulsen, N. Pleskac, M. R. Reagon, D. E. Wolfe, and S. M. Selbo. 2002. "A Bt Transgene Reduces Herbivory and Enhances Fecundity in Wild Sunflowers." *Ecological Applications* 13(2): 279–286.

Snow, A. A., D. A. Andow, P. Gepts, E. M. Hallerman, A. Power, J. M. Tiedje, and L. L. Wolfenbarger. 2005. "ESA Report: Genetically Engineered Organisms and the Environment: Current Status and Recommendations." *Ecological Applications* 15(2): 377–404.

UNDP (U.N. Development Programme). 2001. *Making New Technologies Work for Human Development. Human Development Report 2001.* New York: Oxford University Press.

UNEP/CBD (U.N. Environment Programme/Convention on 1905 Biological Diversity). 1994. *Convention on Biological Diversity: Text and Annexes.* Châtelaine, Switzerland: Interim Secretariat for the Convention on Biological Diversity.

UNEP/CBD/ICCP. (U.N. Environment Programme/Convention on Biological Diversity/Intergovernmental Committee for the Cartagena Protocol on

Biosafety). 2002. Intergovernmental Committee for the Cartagena Protocol on Biosafety, Third Meeting, The Hague, 22–26 April 2002. Capacity-Building (Article 22, Article 28). Note by the Executive Secretary. http://www.biodiv .org/doc/meeting.aspx?mtg=ICCP-03.

UNEP/GEF (U.N. Environment Programme/Global Environment Facility). 2003. *Biosafety News*, 4 June.

Uphoff, N., ed. 2002. *Agroecological Innovations: Increasing Food Production with Participatory Development.* London: Earthscan.

USAID (U.S. Agency for International Development). 2002. Collaborative Agriculture Biotechnology Initiative: Mobilizing New Science and Technology to Reduce Poverty and Hunger. Press release. http://www.usaid.gov/press/ releases/2002/fs020612.html.

USDA (U.S. Department of Agriculture). 2003. U.S. and Cooperating Countries File WTO Case against EU Moratorium on Biotech Foods and Crops. Press release. www.ustr.gov/Document_Library/Press_Releases/2003/May/ U.S._Cooperating_Countries_File_WTO_Case_Against_EU_Moratorium _on_Biotech_Foods_Crops.html.

USDA (U.S. Department of Agriculture) Economic Research Service. 2005. Adoption of Genetically Engineered Crops in the U.S. http://www.ers.usda .gov/Data/BiotechCrops/ExtentofAdoptionTable1.htm.

USFAS (U.S. Foreign Agricultural Service). 2003. Biotechnology and U.S. Agricultural Trade: Training and Capacity Building. www.fas.usda.gov/itp/ biotech/development.html.

Vaughan, S. 2002. Economic Valuation and Trade-Related Issues. Montreal: Secretariat of the Commission for Environmental Cooperation of North America. www.cec.org/files/pdf/Vaughan-e.pdf.

Wargo, J. 1996. *Our Children's Toxic Legacy: How Science and Law Fail to Protect Us from Pesticides.* New Haven, Conn.: Yale University Press.

Weiner, T. 2002. "In Corn's Cradle, U.S. Imports Bury Family Farms." *New York Times.* Late Edition, Final, 26 February, §A, 4.

Wise, T. A., H. Salazar, and L. Carlson, eds. 2003. *Confronting Globalization: Economic Integration and Popular Resistance in Mexico.* Bloomfield, Conn.: Kumarian Press.

Wright, B. 2000. "Intellectual Property Rights: Challenges and International Research Collaboration in Agricultural Biotechnology." pp. 289–314 in M. Qiam, A. F. Krattinger, and J. von Braun, eds. *Agricultural Biotechnology in Developing Countries: Toward Optimizing Benefits for the Poor.* Dordrecht, Netherlands: Kluwer Academic Publishers.

CHAPTER 4

Biosafety Regulation and Global Governance: The Problem of Absentee Expertise in Latin America

KEES JANSEN AND ESTHER ROQUAS

Global developments are putting pressure on Latin American countries to construct regulatory frameworks that open the path for a smooth introduction of transgenic crops. Prompting demands for regulations are a series of factors, including technological uncertainties; potential high risks such as biosafety, human health, and changes in agrarian structures; the ethics of modifying nature; and the huge economic interests involved. The latter interests stem from high investments and projected future profits in the so-called "life sciences" industries. Three major issues are prominent: First, the regulation of intellectual property rights. Concerns in this regard include the interests of transnational "life science" companies to patent their technological innovations and monopolize markets, and prepare developing countries (or indigenous communities) to commodify their local knowledge and control of genetic resources through bioprospecting and benefit sharing with foreign companies. A classical and well-publicized example of the latter is the INBio-Merck bioprospecting agreement in Costa Rica (Aguilar-Støen and Dhillion 2003; Artuso 2002). In this agreement the commercial utilization of Costa Rica's biodiversity was made possible in exchange for up-front monetary compensation, training, technology transfer, and future royalties.

A second regulatory issue concerns risk assessment procedures addressing biosafety (and to a lesser extent food safety) when transgenic seeds or other biotechnology products are being introduced (e.g., Burachik and Traynor 2002; Newell and Mackenzie 2000). Third, there are societal concerns about the socioeconomic consequences and the direction of modern agriculture and agro-industrialization once biotechnology use is spreading. Brazil is one of the few countries where this issue has become a real topic in public debates about the admission of transgenic crops and is being discussed along with biosafety issues (Pelaez and Schmidt 2004; Jepson et al., this volume; Hisano and Altoé, this volume). In most other Latin American countries the possible impact of biotechnology on agrarian change has not yet been an issue of wider public concern. Ethical concerns about the rights of humankind to manipulate nature have not received such prominent attention in Latin America as in some European countries, at least in the field of plant biotechnology—to which we limit our account.

The topic of biosafety regulation, in particular, has gained importance in recent years. Social movements against genetically modified organisms (GMOs) have pushed for more regulation to halt uncontrolled introductions of transgenic crops. They point to the continuous pressure of the U.S. government to open up their countries for doing field trials with GMOs, a pressure that is difficult to contest. Some pro-GMO coalitions consider overabundance of regulation to be a hindrance for public sector involvement in biotechnology (FAO 2003). Other pro-GMO coalitions, however, regard a lack of biosafety regulation as a potential source for future controversies and a hindrance to the introduction of transgenic crops. Harmonization of biosafety laws should enhance trade opportunities, they believe. An international consensus emerged to push for proper biosafety regulation in developing countries, mainly through the vehicle of the Cartagena Protocol on Biosafety, which is an outcome of the 1992 UN Convention on Biological Diversity. The Cartagena Protocol was entered into force on 11 September 2003 and stipulates that a participating country develops a biosafety regulatory framework to deal with GMO release in the environment and its potential adverse consequences (Art. 1).

This chapter explores the interactions evolving between Latin American countries and international organizations to establish biosafety regulatory frameworks in the shortest possible time. A series of critical questions will be raised about the current trend to harmonize biosafety regulations. We will argue that the problem of developing biosafety regulatory frameworks cannot be reduced to the relatively simple problem of how to increase local scientific and regulatory capacity. The issue is much broader and requires proper attention to controversies and contrasting views around biotechnology, the heterogeneity of national political cultures and socio-economic conditions, and the complexities of regulation making and implementation in weak and developmental states. The making of biosafety regulation is thus basically a sociopolitical issue and not a technocratic issue. The latter receives ample attention of international organizations; the former receives hardly any.

Biosafety Regulation and Global Governance

Biosafety risks cross national boundaries and are thus by their very nature an international issue. This section discusses who participates in expertise building and political negotiations about internationally proposed regulatory models and how developing countries build up an apparent lack of capacity.

Setting the Standards

A look at participation lists of some international events where standards and guidelines are being developed suggests that only experts of "larger" Latin American countries participate. For example, in March 2002 the Ad Hoc Intergovernmental Task Force on Foods Derived from Biotechnology—which advises the Codex Alimentarius Commission (FAO/WHO)—met in Japan to develop standards and guidelines for foods derived from biotechnology or traits introduced into foods by biotechnology. There were 245 participants from 34 countries, but from Latin America only Argentina, Brazil, and Mexico had sent delegates (2, 2, and 4, respectively). The U.S. delegation was composed of 17 members (FAO/WHO 2002). The June 2000 meeting of the Interim Commission on Phytosanitary Measures, Exploratory Working Group on Phytosanitary Aspects of Genetically Modified Organisms, Biosafety and Invasive Species had 62 participants from 29 countries, but only one representative from Mexico and one from Chile (IPPC 2000). Looking at lists of participation of a series of meetings, we have the impression, but not the full evidence, that smaller Latin American countries are generally not represented in expert meetings where standards are being set. This may be partly a result of available capacity and partly a result of available resources. Even a country like Argentina may find it difficult to free up resources to send a representative (e.g., Burachik and Traynor 2002, 41). The evidence we could collect suggests that in meetings where agreements on standards and guidelines have to be formalized by politicians, smaller countries seem to be more often represented compared to the expert meetings. It is quite unlikely that politicians and diplomats from developing countries can develop any influence on the content of standards at this stage. Moreover, developing-country experts who are able to participate in these international arenas are largely trained in the North, under the scientific and technological paradigms that currently dominate biotechnology development (see Chap. 1 in this volume).

Capacity Building

Although Latin America may be underrepresented in the international undertaking of standard setting, it is heavily involved in international activities for capacity building. The establishment of REDBIO (Red de Cooperación Técnica en Biotecnología Vegetal; that is, Technical Cooperation Network on Plant Biotechnology in Latin America and the Caribbean) is an example of this. REDBIO's main activities in the 1990s were the compilation of a database of plant biotechnology laboratories (with 619 member organizations as of 18 June 2001) and the organization of a series of conferences to bring together scientists from participating countries. The regional office of FAO in Santiago, Chile, acts as the REDBIO technical secretariat. What is termed as biotechnology research in Latin America implies in most cases working with relatively simple technologies such as plant tissue culture (micropropagation) and biopesticide production (Goldstein 1995; Solleiro and Castañón 1999). Only a few domestic firms or government laboratories are involved in transgenic plants, most of them located in Brazil. REDBIO coordinates its activities with the United Nations University Programme for Biotechnology in Latin America and the Caribbean (UNU/BIOLAC).

The emergence of REDBIO in Latin America illustrates that capacity building is high on the political agenda of biotechnology development in Latin America. Several studies identify knowledge gaps and propose capacity building (training for expert knowledge). In its first phase, biotechnology training in Latin America was primarily focused on training scientists in doing biotechnology. Issues of concern were the coordination of training activities by international agencies and the pooling of resources, the level at which training should take place (short courses, B.Sc., M.Sc., or Ph.D.), the adjustment of training courses and curricula to industrial needs, the issue of whether training should focus on core biotechnologists or on scientists from other disciplines such as agronomy and medicine, in order to expose their specialties to biotechnology, and, last but not least, the mobilization of funds for all this work (Daza 1998; Jaffé and Rojas 1994). More recently, it seems that the training of regulators has received growing attention, and activities are organized on the social aspects of biotechnology, with emphasis on bio-ethics and bio-safety (e.g., UNU/BIOLAC, n.d.). These activities intend to make the development of local regulation possible and to facilitate participation in international epistemic communities. Trainees may consist of people already educated in biotechnology, but also social scientists, policy makers, legislators, and regulators. The history of REDBIO and BIOLAC shows that international efforts to build capacity around biotechnology in Latin America have shifted from purely scientific, disciplinary training and scientific/technical cooperation to the enhancement of regulatory capacity. International interests in the implementation of proper risk assessment procedures and patent laws that protect biotechnology innovations stood at the cradle of this shift.

The major push for capacity building around the biosafety issue, however, is an outcome of international concerns about biodiversity. Most Latin American countries signed the 1992 Convention on Biological Diversity (CBD) that left room for further negotiating an international framework for biosafety. There was broad consensus that the international community needed some kind of international instrument for dealing with biosafety issues: the promise of use of biotechnology for developing new agricultural crops had to be reconciled with potentially damaging effects on the environment and human health.

The result of several years of negotiations was the emergence of a protocol of the CBD: the 2000 Cartagena Protocol on Biosafety. A protocol is a binding international instrument, separate from, but related to, another treaty (Mackenzie et al. 2003). It is a separate instrument that must be negotiated, signed, and eventually ratified by individual countries. It thus has its own parties, and creates separate rights and obligations for them. The Cartagena Protocol specifically deals with transboundary movements of GMOs.[1] It establishes an Advanced Informed Agreement procedure that obliges the exporter of a GMO to notify the importing country and provide certain information about the GMO and its intended use. This gives the importing country the opportunity to review this information and to decide on approving or prohibiting the import of the GMO, or to determine the conditions for importation.

Latin America has not acted as a unified bloc in the Protocol negotiations. Originally, all Latin American countries negotiated together as part of the so-called Like-Minded Group of developing countries. At a later stage, however, Argentina, Chile, and Uruguay left the Like-Minded Group to join the Miami Group—which further consisted of the United States, Australia, and Canada (Vogler and McGraw 2000). While the negotiating was going on, these three Latin American countries had decided to approve the release of some specific GMOs, which subsequently changed their attitudes toward the issue at stake. It evidently makes a difference whether the biosafety issue is being discussed from the perspective of biodiversity or consumer wishes, or from the perspective of

Table 4.1. UNEP-GEF Drafting Biosafety Frameworks: Latin American Participants

Country	Phase of project in 2003
Bolivia	— (participated in project's pilot phase)
Argentina	3 (drafting national biosafety system)
El Salvador	2 (identification of stakeholders, training, discuss framework with UNEP-GEF staff)
Chile	2
Guatemala	2
Honduras	2
Nicaragua	2
Peru	1 (surveys and inventories)
Ecuador	0 (project design)
Costa Rica	0
Panama	0
Paraguay	0
Uruguay	0
Venezuela	0
Belize	0

Source: Extracted from UNEP-GEF 2003.

having, for example, a considerable acreage of transgenic soybeans. Below we will explore more extensively the different factors that compel Latin American countries to work on biosafety regulation.

The most important effort at the international level to shape the making of biosafety regulation is the worldwide UNEP-GEF (United Nations Environmental Programme–Global Environment Fund) program to develop national biosafety systems (Briggs 2003). This project makes an inventory of existing regulatory instruments for biosafety and leads to the drafting of a national biosafety framework and the outline of an institutional structure for biosafety regulation. Characteristic of this program, according to UNEP-GEF, is a country-driven perspective. Below we will discuss what this means. Many Latin American countries participate in this project (see Table 4.1), although not Mexico, Brazil, or Colombia. So far none of the participating Latin American countries has become enrolled in a UNEP-GEF follow-up project for the actual implementation of the drafted biosafety framework.

One of the most eye-catching characteristics of biosafety regulation in

Latin American countries is its intertwining with global governance. The question then emerges as to what extent such international shaping of regulation triggers domestic discussions about it, instead of endogenously developed needs and arguments. Before addressing this question, we will first explore a UNEP-GEF follow-up project for a more complex set of factors leading to different responses of countries to the actual implementation of the drafted push for biosafety framework regulation.

Factors Driving the Discussion on Biosafety Regulation in Latin America

This section briefly presents some aspects of the development of biosafety regulation in Argentina, Brazil, Mexico, Honduras, and the Andes region. On the basis of these examples we identify several patterns in the process of making biosafety regulation in several countries or regions and we assess specificities of their regulatory frameworks.

The Alliance of Foreign and National Economic Interests: Argentina

Argentina took a leading role in accepting GM soybean and other crops after 1991, in response to domestic interests and research and the desire by the United States and TNCs to use Argentina as a location for off-season GM seed production and field trials (Cohen et al. 2003). In 2002 Argentina grew 23 percent of the global GM crop area (James 2002; Teubal's Chap. 8 in this volume). The Argentine government as well as producers were open to experimenting with the new GM varieties. The alliance of national and international interests thus quickly led to the spread of GM crops and to putting biosafety on the regulatory agenda.

Argentina developed a biosafety system in the early 1990s when it did not have clear policy goals about biotechnology use and safe handling of GMOs. Consequently, the system has many shortcomings (Burachik and Traynor 2002). Cohen et al. (2003) remark that a major problem for developing a reliable biosafety system in Argentina is that government officials disappear every time general elections are held and are replaced by others—a problem that is evident in many other Latin American countries as well. Moreover, policy makers in Argentina, according to Burachik and Traynor (2002), need to be made aware of the consequences of signing the Cartagena Protocol and the importance of a biosafety system to counteract the activities of the biotechnology critics. The current system is institutionally unclear with regard to mandate and decision-making power of reviewing institutions and the Agricultural Directorate. The procedure for reviewing the acceptability of GMO release furthermore shows a lack of timeliness and transparency. Reviewers, although sufficiently qualified, are limited in number and there is a danger for potential conflict of interest as many of them have ties with biotechnology companies (Cohen et al. 2003). A new biosafety law developed through the UNEP-GEF project is supposed to clear away these shortcomings.

Hence the willingness of Argentina to indulge the U.S. pressure to admit GM soybean resulted in the instantaneous creation of a biosafety system. The government developed the system without taking the time for a thorough public debate about the aims and principles of such a system. The biosafety system is existent but neither efficiently working nor extensively debated among the Argentine people.

Domestic Visions of Agricultural Modernization: Brazil

The controversy within Brazilian society about the regulation of transgenic crops is a highly publicized case (Hisano and Altoé, this volume; Jepson et al., this volume; Paalberg 2001; Pelaez and Schmidt 2004). Brazil passed its first biosafety law in 1995, long before other Latin American countries even started thinking about biosafety issues. Among other things, this law created the National Technical Commission on Biosafety (CTNBio), which oversees the risk assessments of GMO introductions on a case-by-case basis. The commission's intention to approve the marketing of Monsanto's Roundup Ready® soybeans in 1998 pulled the trigger of a still ongoing controversy on the appropriateness of transgenic soy in Brazil. Court cases initiated by the consumer organization IDEC and by Greenpeace led to verdicts by federal judges, who ruled that CTNBio could not permit the introduction of GMOs without the required environmental impact assessments and the required collaboration with and authorization of other authorities in the Brazilian state system. The biosafety regulatory framework is not the single responsibility of CTNBio but also involves a series of other ministries and state organizations (Fontes 2003). To make the situation more complex, the southern state of Rio Grande do Sul declared itself a GMO-free zone, an example followed later by several other Brazilian states. The possibility of selling GMO-free soy to European countries and Japan at a premium price aroused the interest of certain economic sectors in GMO-free production areas. The importance of the controversy went beyond the confrontation between pro- and anti-GMO interest groups, and played a role in exploring the boundaries between the judicial and executive powers in Brazil, and between the states and the federal government (Hisano and Altoé, this volume; Jepson et al., this volume).

The Brazilian case makes clear that a concern about biosafety issues can be driven from within. Two different pillars with contrasting interests sustained the controversy. First, public agricultural research in Brazil had focused on biotechnology from the early 1980s onward. It had invested substantial amounts of money on training for its scientists in Brazil and abroad (Valadares and Monte-Neshich 1996). EMBRAPA, the largest state agricultural research organization—which alone swallows up approximately two-thirds of the total budget of all public research institutes of Latin America combined (Echeverría 1998)—was heavily involved in biotechnological research in association with the National Research Center on Biotechnology and Genetic Resources (CENARGEN). Substantial benefits were expected from biotechnology. A second pillar was the evolution of social movements in Brazil. The possible introduction of transgenic soy by Monsanto confronted a network of organizations which had opposed the appropriation of seed by transnational companies for more than a decade (Pelaez and Schmidt 2004). Not only the transgenic content of Roundup Ready® soy, but even more important, the next step made by foreign companies to control seed supply of Brazilian farmers, had stirred the opposition. The ideas held by this network resonated with growing middle-class consumer awareness about the "right to choose." Although the pillars may have confronted each other at certain moments, they both originated from and were rooted in democratic and economic change in Brazil.

Characteristic of the evolution of the debate on biosafety in Brazil is the importance of these internal actors. Their importance manifests itself in the way Brazilian diplomats had to operate in the negotiations that led to the adoption of the Cartagena Protocol on Biosafety. They had to amalgamate the various and frequently opposing opinions: Brazil stayed within the Like-Minded Group of developing countries, which strove for a precautionary regulation of transboundary movements of GMOs, while at times Brazil supported positions much closer to the Miami Group and pursuing a promotional regulation, afraid that the Protocol would erect new barriers to trade (Nogueira 2002). The precautionary principle establishes that a lack of scientific certainty about negative side effects of a new technology is a legitimate reason to take preservative action. At that time, Brazil thus rejected the contrasting principle of substantial equivalence advocated by the Miami Group, which establishes that negative side effects should be scientifically proven before any preventive measures are justified.

Images of a Beckoning Future versus Images of Good Peasant Agriculture: Mexico

As in Brazil, scientists in Mexico considered biotechnology an opportunity to leap-frog into modern agriculture, human welfare, and national food security, thus becoming "an equal partner in the commercial pact" with the United States, Canada, and Europe (Possani 2003; see also Bolívar Zapata 2001, Otero 2001). Mexico had a considerable number of scientists working in the field of biotechnological research and development, and several transgenic crops were cultivated. But unlike the situation in Brazil, there was little governmental activity in creating a supporting infrastructure and regulatory environment. During the 1990s, Mexico's major policy mantra was deregulation and neoliberal free trade, which left little room for government support for specific developments or the strengthening of regulatory frameworks. Regulation of GMO introductions was superficially mentioned in a wide range of laws (including health law, phytosanitary law, plant variety and seed certification laws, consumer law, environmental law, sustainable rural development law, and forestry law) and, paradoxically, involved formally a whole series of government agencies in the GMO issue (including the ministries of Health, Environment, Agriculture, Education, Finance, and Economy). The result was that companies' requests for field trials were evaluated on a case-by-case basis, but no full biosafety system was put in place with proper risk assessment and full monitoring and evaluation of introduced GMOs. The situation changed when the Interministerial Commission for Biosafety and Genetically Modified Organisms (CIBIOGEM) was formed in 1999 to coordinate biosafety issues as part of the implementation of the Cartagena Protocol in Mexico (Glover et al. 2003). The project for a new Biosafety Law on Genetically Modified Organisms was presented to and approved by the Senate on 24 April 2003. Known to its critics as the "Monsanto law," it was passed on 14 December 2004 (Hernández Navarro 2005).

One recurrent theme in the discussions about biosafety regulation in

Mexico is the potential gene flow from transgenic maize to native varieties and wild relatives. Mexico is the biological center of origin of maize (and center of genetic diversity); wild relatives of maize (*Zea mays*) are endemic. Scientists as well as civil-society organizations discussed the possible risks of introducing transgenic maize (e.g., Serratos et al. 1997). In 1998, Mexico announced a moratorium on the cultivation of transgenic maize. The debate became passionate when Quist and Chapela (2001) published their controversial article in *Nature*, which argued that landraces in Oaxaca already contained genes from transgenic varieties. In various publications scientists disputed its conclusions, but more or less implicitly acknowledged that transgenic maize had hybridized with native varieties (cf. Butler 2002, Ho 2002, Kaplinsky et al. 2002, Metz and Fütterer 2002, and the reply by Quist and Chapela 2002).[2] In January 2002, NGOs, farmers, and indigenous organizations organized the seminar "In Defense of Maize" and demanded to halt imports of GM corn. The Ministry of Environment announced the results of new studies which confirmed that maize in the states of Oaxaca and Puebla was contaminated (ETC 2002). The most likely source of contamination is imported corn for animal feed, which is used as seed by farmers (McAfee, this volume; Fitting, this volume).

Another possible source is maize seed shipped by relatives in the United States or brought back from the United States by Mexican immigrants. The concerned civil-society organizations started to carry out their own research and presented the results of a large-scale study of 2,000 sampled plants in 138 farming communities where farmers generally do not buy maize seed but use locally available seed stocks (Comunidades 2003). The results confirmed the initial Quist-Chapela thesis: 24 percent of the samples, from nine Mexican states, contained transgenes, sometimes from two, three, or four GM types. The involved organizations repeated their demand to halt GM corn imports. They also called for a rejection of the Biosafety Law in preparation, which they considered as a legalization of the unwanted gene flow. This movement emphasized that it was indigenous farmers in Mexico who have developed the existent diversity of maize races and that the "culture of maize" has to be safeguarded. Although rather late in the process, the official making of biosafety law, mainly a result of obligations to international agreements, confronts a national social movement that does not believe in the official vision of Mexico's agricultural future (Bartra 2004).

Jumping on the Bandwagon of Global Capacity Building Projects and Harmonization Projects: Honduras

Honduras is listed as one of the Latin American countries with an operating biosafety regulation (Jaffé 2003). It had a sort of biosafety "law" issued as early as 1998. But the biosafety "Acuerdo" of 1998 is basically not more than a short indication that "proper measures have to be taken to prevent negative effects of GMOs on human health and the environment" (Chap. III, art. 6) without any further specification. The Ministry of Agriculture is the designated authority to regulate GMOs and to evaluate requests for field trials and releases. No criteria for evaluation are specified. Hence, it cannot be called a fully operational regulatory system for dealing with biosafety issues.

As with other complex regulatory matters of high-risk technologies, the main drive for developing regulation in Honduras is international agreements and activities of international (regional) organizations. Motivated by the CBD and the Central American Alliance for Sustainable Development (ALIDES, to which Honduras is a party since 1994), it has installed a National Biosafety Commission (CONABIOH) since 1998, but little activity has been developed since then. The responsibility for approving GMO introductions remained with the Ministry of Agriculture. Another factor initiating thinking about biosafety regulation is the efforts of CORECA, the Regional Council for Co-operation on Agriculture (consisting of the ministers of agriculture of the seven Central American countries, Mexico, and the Dominican Republic as well as the Inter-American Institute for Co-operation on Agriculture (IICA). CORECA is involved in developing a harmonized regulatory framework on GMOs for Central America (CORECA 2002a). In CORECA, Mexico is taken as an example, as a country where transgenic crops are produced and with a regulatory system in place which is more advanced than those of other countries in the region (CORECA 2000, 2002b). Furthermore, Mexico is working with crops and varieties of importance to other countries in the region, and different from the varieties for temperate climates. Unlike the situation in the United States—the major producer of GMOs—Mexico's biological diversity and environmental conditions bear resemblance to those in the other countries of CORECA.

The first official commercial planting of transgenic crops in Honduras took place in 2002 when the government approved Monsanto's request to introduce MON-810 *Bt* maize cultivars (scientists had warned some years before that illegal plantings of GMOs could be expected and that policy makers should prepare for proper regulatory frameworks; cf. Hruska and Lara Pavón 1997). The transnational fruit corporations with plantations in Honduras (Chiquita, Dole) work on transgenic bananas on an experimental level.

There are only two incidental events in which opposition to transgenic crops was voiced. The first was a press meeting of several networks of farmer organizations in October 2001, in which they rejected the import of GMO seed, arguing that it would replace farmers' seed with seed from transnationals, which would affect human health.

After making international commitments, the Honduran government was forced to consider biosafety issues, not because there was an internally felt need for doing so. This case is an example of how global governance can steer lawmaking at the national level.

Byproduct of Biodiversity and Bioprospecting Interests: Andean Countries

The entry of Andean countries in the biotechnology discussion focused initially not so much on the introduction of GMOs but on the access to, and property of, genetic resources. At the XI Andean Presidential Council held in May 1999, the presidents declared that the importance of the Andean region for the world's biological diversity "is one of the greatest strengths of the Andean sub-region and source of opportunities for the development of its Member Countries." In Bolivia, for example, some people and organizations imagined enormous benefits from bioprospecting. The scanty biosafety regulation that has been issued in the Andean countries so far tends to restrict the trade and use of GMOs (GTZ/FUNDECO/IE 2001). In June 2002, the XIV Andean Presidential Council approved the "Regional Biodiversity Strategy for the Andean Tropical Countries" (Decisión 523), which sets out the basic principles for dealing with biosafety issues (development of research capacity in biosafety management, application of the precautionary principle, development of risk assessment, and monitoring of risk assessment).

In contrast to Central America, where more attention is being asked at the ministerial level for consultation with civil-society organizations critical to GMOs, a more open and participatory approach was built from the start into the Andean initiative in order to generate regional consensus about biosafety issues. Experts from the agricultural sector (ministries of agriculture) set the agenda for the development of biosafety regulation in Central America, whereas environmental experts play a key role in the Andean countries. A central organ is the Andean Committee on Environmental Authorities (CAAAM), which was assigned the task to update and strengthen the Regional Biodiversity Strategy. The German governmental corporation for international cooperation (GTZ) played an important role in the realization of a regional strategy. Further support came from the Inter-American Development Bank.

Diversity of Factors That Push for Biosafety Regulation

This overview of work on biosafety regulation in various Latin American countries points to a range of stimuli: foreign and national interests in selling and cultivating transgenic crops (Argentina), reconciliation of domestic controversies over possible roads to agricultural modernization (Brazil), concern about possible threats to, or "pollution" of, the tradition of peasant agriculture (Mexico), the idea that a smaller country cannot afford to negate international agreements and the presence of global capacity-building projects and harmonization projects (Honduras), and the imagined benefits from biodiversity and bioprospecting interests (Andean countries). To some extent these factors may play out in all their complexities in each Latin American country. However, some factors appear as very dominant in a particular country, and it is in this sense that they are used here as an illustration.

There are crucial differences between the countries with regard to the level at which GMOs are considered important in the short term. Brazil and Argentina had to take decisions when GMOs had entered or were about to enter on a large scale. In most countries, however, it was the CBD, and later on the Cartagena Protocol, that triggered the need for policy makers to start considering how to deal with GMOs. Smaller roles were played by IICA, who at an early stage developed some study activities and workshops about biotechnology and biosafety and (Jaffé 2003), the mentioned GTZ-Andean Community activities, the REDBIO network of FAO, and the World Bank, together with CIAT, which are discussing biosafety issues in Colombia (Pehu 2003). The recent work on regional harmonization of biosafety regulations basically builds on, and is inspired by, other ongoing activities for regional harmonization.

Absentee Expertise

Experts play a key role in the development of biosafety regulation and in the transfer of global biosafety models to the national situation. Emphasizing the lack of participation of (smaller) Latin American countries in the global making of biosafety regulation and demanding more capacity building tend to overlook another crucial issue, which we have conceptualized elsewhere with the concept "absentee expertise" (Jansen and Roquas 2005). A key characteristic of absentee expertise is that the expert is operating rather detached from local concerns and contexts. The few national experts that do participate in international expert communities may well be as much tied to the international expert community as to their home country. Participation in the international network provides them with the highly valued status attached to such a position, access to international project funds, and the benefits of international travel. It could be argued that such participation contributes to transnational social learning (Haas and McCabe 2001), but the question has to be raised as to whether this also leads to strong national social learning. The absentee expert shares some major characteristics with the absentee landlord, who has figured so often in past debates on agrarian structures in Latin America. An absentee landlord is a landowner not resident on the estate from which he/she derives income.

In the case of absentee experts, the major source of power is not ownership of land but monopolized access to expert knowledge. The power of absentee expertise tends to be invisible or hidden. It is the power laid down in objective standards, model laws, training manuals, and so on. The power is being sustained through the public trust in scientific institutions and the authority of sound science. For many people it remains obscure how absentee experts arrive at their models for biosafety regulation, even though internally the procedures and discussions may be quite transparent and subject to peer review. The role of absentee expertise is also hidden because, in the dominant model of science, there is little acknowledgement of the political character of the so-called scientific consensus on problematic knowledge issues. Absentee experts operate in a cosmopolitan world of international scientists in which certain disciplines, causal principles, definitions of problems, and solutions predominate. Absent information about local contexts, a universal truth of the potentials of global technologies and regulatory frameworks is developed in conferences, meetings, and workshops. As discussed above, although flexibility and local adaptation and development of regulation are much-promoted elements of the global making of biosafety regulation, the present social construction of biosafety regulatory frameworks is driven primarily by global standards.

Hence, although global governance, international capacity building, and harmonization may result in a gain in efficiency and the development of capacity at low costs, the flip side may be that it only leads to the building of absentee expertise. Global predominance may mean that it only builds experts and knowledge which hardly relate to local conditions and cannot deal properly with local complexities, diverging political-economic interests, and the vagaries of weak states to implement laws and regulations. The next section explores further the general contradictions arising from the strengthening of absentee expertise.

To Harmonize or Not to Harmonize?

International agreements aim to harmonize regulation related to all kinds of issues that cross boundaries and involve more than one country. To share the same regulatory system with the same aims, principles, and procedures enhances efficiency and makes international trade easier. Most international agreements, however, do not carry through a completely harmonized system, but only set a general framework, leaving enough room for a country's individual interpretation.

The Cartagena Protocol—being an international instrument striving at harmonized biosafety systems—orders the establishment of biosafety systems but creates room for individual countries to develop their own idiosyncratic biosafety framework. The Cartagena Protocol seeks harmonization of biosafety frameworks because it focuses on the unwanted consequences of uncontrolled—either spontaneous or deliberate—*transboundary* movements of GMOs. A major discussion point during the negotiations was the magnitude of precaution that should be built in the Protocol: a discussion that emerged as a consequence of a lack of scientific evidence and consensus about the long-term impact of GMO release in the environment.

Principle 15 of the Rio Declaration (1992) defines the precautionary approach as: "Where there are threats of serious or irreversible damage, lack of full scientific certainty shall not be used as a reason for postponing cost-effective measures to prevent environmental degradation." The Protocol negotiations led to disagreement about whether or not precaution is indeed an international legal principle that might be used to claim rights or create uncompromising obligations (Mackenzie et al. 2003, 13). Concerns about introducing the principle of precaution are often driven by the idea that the principle might run into conflict with international trade agreements.

It is precisely the lack of consensus about how to refer to the precautionary principle in the Protocol which created the space for individual member countries to make their own choices. The view that the Protocol as a whole should be an instrument of precaution without detailed prescriptive precautionary principles became dominant. Precaution became an issue treated in the preamble. The subsequent prescriptions about precaution leave countries considerable room for choosing between different possibilities to perform risk assessments and making decisions on the extent to which scientific uncertainty should be decisive in permitting or forbidding GMO release in the environment.

There are thus many choices to make for a country after signing the Protocol, but these require a great deal of expertise and considerable financial means. It is here where international experts start to play an important role as transmitters of knowledge and models, and as trainers of domestic experts. Between inventing the wheel yourself and literally copying an existing format for a biosafety system, a third way has become dominant in less developed countries. This third way emerges in cases in which developing countries receive support and assistance from international organizations and international experts to draft their home-made regulatory frameworks. This appears to be a two-way process: Countries might be convinced that they need international expertise to deal with the matter, and international organizations encourage the use of international models and standards for biosafety regulation.

The UNEP-GEF program to develop national biosafety systems—in which many Latin American countries participate—is a good example of such a supportive program. It uses a "country-driven" approach, but for the outsider it is difficult to see what this means. The program design has four fixed phases which—by making use of a toolkit—lead to a draft National Biosafety Framework in the briefest possible period of about eighteen months. The UNEP-GEF program sets in motion a technocratic drafting process which offers little space for endogenous reflection and extended in-country discussions about aims and principles. A salient aspect is the importance of "sound science"—only recognizing risks when they are scientifically proven—in national biosafety systems, in spite of the Protocol's precautionary nature. It is the principle of sound science—and not the precautionary approach—that seems to dominate international biosafety frameworks. The sound science principle is basically disseminated through training and support provided by international experts.

The consequences of the international support to build national biosafety systems are less visible. In countries like Honduras, such support efforts tend to suppress emerging national debate, deliberations, and arguments. A country's choice between "doing things right" along international expert–defined phases and standards, and "doing the right thing" while considering the particular history, needs, and wants of a country, is thus never categorically made.

Some supporting international organizations characterize opponent views as nonscientific and deceiving the public. The ISNAR Country Study of Argentina (Burachik and Traynor 2002), for example, reports that Greenpeace has been active in campaigning against GMOs. Without paying any attention to the particular *content* of the arguments that Greenpeace uses, the report concludes:

> Some NGOs, having sharply raised their visibility by astute use of the media, are particularly effective in getting their message heard. Echoing the opposition arguments arising in Europe, some of the "information" disseminated about biotechnology and GM foods suffers from gross inaccuracies, false assumptions, and unsupportable extrapolations. Not surprisingly, though, given the technical nature of the subject, most reporters and editors are incapable of distinguishing fact from fiction, and give broad exposure to the NGOs' claims. Providing accurate information to the public thus becomes the responsibility of research institutes and government agencies. (2002, 45–46)

The message in this ISNAR report is that the arguments of the opponents do not deserve any attention, as they are not serious or scientific and they are copied from international biotechnology opponents in Europe. The opponents are shrewd in using the media in contrast to the naiveté of the objective scientific experts. Both public and press are not capable enough to understand and value complicated technological issues, and the monopoly on accurate information about biotechnology should be in the hands of the government and scientific institutes. The ISNAR report provides a quite faithful example of the "deficit model" (Collins and Evans 2002). Public rejection of biotechnology is caused by the public's lack of knowledge; a deficit that should be removed by providing objective and truly scientific information. The report thus reproduces the image of science as superior, non-ideological knowledge. Deviant opinions are ideological, paranoid, and non-scientific.

The widely distributed message that boosting the knowledge of the public will automatically increase acceptance seems to efficiently silence all debate in society. There is no room left for debating the issue of biosafety and "doing the right thing." Turning to international models and standards to comply with the international obligations concerning biosafety seems thus the only way and has become the common practice in many Latin American countries.

Biosafety Regulation: Absentee Expertise and Public Debate in Latin America

The problem of how to regulate biosafety is inherent to the use of plant biotechnology and—with or without international pressure—it is therefore a prominent issue in all countries. The interests of Latin American countries differ and the ways they are driven to address the issue of biosafety regulation are also quite diverse. Alliances of foreign and national economic interests, democratic participation, domestic views on agricultural development and biodiversity, and international capacity building projects may play a role in the choices that countries have made in relation to biosafety. The issue of difference is not so much whether or not a country has issued biosafety laws, but how thoughtful, critical, and authentic the discussion on biosafety regulation has been.

Conclusion

This chapter has argued that major problems regarding biosafety regulation are a lack of public debate and a lack of country-level, endogenous processes of lawmaking. It is precisely the lack of debate and endogenous judgment that makes it difficult to contest international pressure from international organizations, powerful countries, and transnational corporations. Regulation of biosafety is internationally presented as a "must," but many Latin American countries lack the financial and institutional resources to keep control over the development of such regulation. In spite of the aims of international projects to support domestic processes to involve stakeholders, the importance of regulating biosafety has been pushed forward merely in technocratic regulatory terms. The necessary incentives to promote the public debate that should precede norm-settling processes have not been provided. In this process legitimacy is being created by appealing to science advisers. This paper calls for a better rethinking of the role of absentee expertise.

"Absentee expertise" is used here as a heuristic concept to draw attention to the separate worlds of science advice for global governance and day-to-day realities of agrarian change, state formation, and shifting governance in Latin American countries. Looking at absentee expertise at a more concrete level, it becomes clear that the roles of both international and national experts vary in different countries. The most extreme cases are Brazil on one hand, where different views on biosafety regulation are expressed by different groups of experts (expertise and counter-expertise are involved in a continuous controversy), and countries like Honduras on the other hand, which sign international agreements with hardly any citizen or expert having seriously considered the consequences. Such countries are guided by international "absentee expertise" in copying internationally developed guidelines and regulatory models. We could say that there is a pattern by which "semiperipheral countries," i.e., resource-rich dependent and developing countries, have a greater ability for internally driven legislation than resource-poor countries. Even then, however, there are substantial differences between resource-rich countries such as Brazil and Mexico.

To establish a link with the international absentee expertise has become a goal in itself. It has become the goal of training and capacity-building activities aimed at expertise development in Latin American countries. This chapter suggests that alternative ways of looking at expertise development may be possible. In this view, expertise development should originate in a societal debate on priorities and goals and not be reduced to a technocratic transfer of knowledge about existing regulatory models. An essential element is that Latin American countries start thinking of biosafety in the framework of their particular societal development, economic plans, the role of agriculture, and the preservation of biodiversity. Controversies and oppositional views can then be taken as an essential source for creative thinking about the complexity of the issue and the necessary political choices. These do not threaten progress but are essential for democratically shaping common futures.

Our ideas on absentee expertise also imply that the quality of international modeling of biosafety regulation can only be assessed by looking at how the formal model holds out in the concrete world. Much modeling tends to forget the "weak state" character of many Latin American governments in its ambitious attempts to set up a comprehensive biosafety system. Many countries in Latin America face severe problems in generating their own lawmaking processes and institutional development. They have great difficulties in enforcing laws once these have been issued. The inadequacy of state performance does not cease to exist with a heavily subsidized development of a biosafety framework. The internationally promoted regulatory models have not found a way to improve state performance. The latter entirely depends on endogenous political processes that only Latin American peoples themselves may be able to change in time.

Notes

1. The Cartagena Protocol does not use the term *GMOs* but *LMOs*: Living Modified Organisms (see Gupta 2002:242, for the reasons for this term shifting).

2. The scientific debate intermingled with mud slinging and other elements of science politics. One of the major scientific controversies was about whether Quist and Chapela had detected hybridization (i.e., some cross-pollination) or introgression (i.e., the permanent incorporation of genes from another population). Only the latter, as some have argued, could really be a matter of concern (Stewart et al. 2003). Although Quist and Chapela used the term *introgression*, they had not proven the stabilization of the transgene in the new host genome. A remaining unresolved issue is whether gene flow as such is a problem, or whether only potentially harmful traits should be considered as a risk (see McAfee's Chap. 3 in this volume for further discussion of this issue).

References

Aguilar-Støen, M., and S. S. Dhillion. 2003. "Implementation of the Convention on Biological Diversity in Mesoamerica: Environmental and Developmental Perspectives." *Environmental Conservation* 30(2): 131–138.

Artuso, A. 2002. "Bioprospecting, Benefit Sharing, and Biotechnological Capacity Building," *World Development* 30(8): 1355–1368.

Bartra, Armando. 2004. "Rebellious Cornfields: Toward Food and Labour Self-sufficiency." pp. 18–36 in Gerardo Otero, ed., *Mexico in Transition: Neoliberal Globalism, the State and Civil Society.* London and New York: Zed Books.

Bolívar Zapata, F. 2001. *Biotecnología moderna para el desarrollo de México en el siglo XXI: Retos y oportunidades.* Mexico: CONACYT and Fondo de Cultura Economica.

Briggs, C. 2003. "Planning for the Development of Nearly 100 National Biosafety Frameworks," in M. A. Mclean, R. J. Frederick, P. L. Traynor, J. I. Cohen, and J. Komen, eds., *A Framework for Biosafety Implementation: Report of a Meeting,* pp. 31–34. Washington: ISNAR.

Burachik, M., and P. L. Traynor. 2002. "Analysis of a National Biosafety System: Regulatory Policies and Procedures in Argentina. ISNAR Country Report 63." The Hague: International Service for National Agricultural Research.

Butler, D. 2002. "Alleged Flaws in Gene-transfer Paper Spark Row Over Genetically Modified Maize." *Nature* 415 (28 February 2002): 948–949.

Cohen, J. I., P. L. Traynor, M. Burachik, M. Madkour, and J. Komen. 2003. "Biosafety Studies in Egypt and Argentina: Two Pathways to Implementation," in

M. A. Mclean, R. J. Frederick, P. L. Traynor, J. I. Cohen, and J. Komen (eds.), *A Framework for Biosafety Implementation: Report of a Meeting*, pp. 21–25. Washington: ISNAR.

Collins, H. M., and R. Evans. 2002. "The Third Wave of Science Studies: Studies of Expertise and Experience." *Social Studies of Science* 32(2): 235–296.

Comunidades (Comunidades indígenas y campesinas de Oaxaca, Puebla, Chihuahua, Veracruz, CECCAM, CENAMI, Grupo ETC, CASIFOP, UNOSJO y AJAGI). 2003. "La contaminación transgénica del maíz campesino en México." http://www.biodiversidadla.org/article/articleview/3621/1/23/. Accessed 15 October 2003.

CORECA (Consejo Regional de Cooperación Agrícola). 2000. "Producción y Comercialización de Productos Transgénico: Consideraciones para el Sector Agropecuario en los Países del CORECA." Managua, Nicaragua: CORECA, IICA.

———. 2002a. "Informe de la reunión extraordinaria del consejo de ministros realizada en San José, Costa Rica, el 26 de septiembre, 2002." San José, Costa Rica: CORECA.

———. 2002b. "Informe de la reunión extraordinaria del consejo de ministros realizada en ciudad de Panamá, Panamá, el 6 de diciembre, 2002." Panamá: CORECA.

Daza, C. 1998. "Scientific Research and Training in Biotechnology in Latin America and the Caribbean: The UNU/BIOLAC Experience." *Electronic Journal of Biotechnology* 1(2): 1–6.

Echeverría, R. G. 1998. "Agricultural Research Policy Issues in Latin America: An Overview." *World Development* 26(6): 1103–1111.

ETC (Action Group on Erosion, Technology and Concentration). 2002. "The Great Containment: The Year of Playing Dangerously." www.etcgroup.org (accessed 11 October 2003).

FAO (Food and Agriculture Organization). 2003. "Regulating GMOs in Developing and Transition Countries." Rome: FAO.

FAO/WHO (Food and Agriculture Organization of the United Nations/World Health Organization). 2002. "Report of the Third Session of the Codex *Ad Hoc* Intergovernmental Task Force on Foods Derived from Biotechnology, Yokohama, Japan, 4–8 March 2002." Rome: FAO.

Fontes, E. M. G. 2003. "Legal and Regulatory Concerns about Transgenic Plants in Brazil," *Journal of Invertebrate Pathology* 83: 100–103.

Glover, D., J. Keeley, P. Newell, and R. McGee. 2003. "Public Participation and the Biosafety Protocol: A Review for UNEP-GEF and DFID." Brighton: IDS.

Goldstein, D. J. 1995. "Third World Biotechnology, Latin American Development, and the Foreign Debt Problem," in N. P. Peritore and A. K. Galve-Peritore, eds., *Biotechnology in Latin America: Politics, Impacts, and Risks*, pp. 37–56. Wilmington: Scholarly Resources Inc.

GTZ/FUNDECO/IE. 2001. "Estrategia regional de biodiversidad para los países del trópico andino." La Paz, Bolivia: Fundación para el Desarrollo de la Ecología.

Gupta, A. 2002. "When Global Is Local: Negotiating Use of Biotechnology," in F. Biermann, R. Brohm, and K. Dingwerth, eds., *Proceedings of the 2001 Berlin*

Conference on the Human Dimensions of Global Environmental Change, "Global Environmental Change and the Nation State," pp. 238–247. Potsdam: Potsdam Institute for Climate Impact Research.

Haas, P. M., and D. McCabe. 2001. "Amplifiers or Dampeners: International Institutions and Social Learning in the Management of Global Environmental Risks," in the Social Learning Group, ed., *Learning to Manage Global Environmental Risks*, Volume 1, pp. 323–348. Cambridge: MIT Press.

Hernández Navarro, Luis. 2005. "El PRD y la Ley Monsanto." *La Jornada.* 8 March. Available at: http://www.jornada.unam.mx/2005/03/08/023a1pol. php (last accessed 24 June 2005).

Ho, M. W. 2002. "Astonishing Denial of Transgenic Pollution." www.i-sis.org .uk: ISIS Report, ISIS.

Hruska, A., and M. Lara Pavón. 1997. *Transgenic Plants (*Bacillus thuringiensis*) in Mesoamerican Agriculture.* Tegucigalpa: Zamorano Academic Press.

IPPC—Working Group on Phytosanitary Aspects of Genetically Modified Organisms, Biosafety and Invasive Species. 2000. "Meeting Statements Presented to the Interim Commission on Phytosanitary Measures." Rome: IPPC, FAO, Rome.

Jaffé, W. 2003. "Cooperation for Biosafety in Latin America and the Caribbean: Towards Shared Capabilities." Latin America and the Caribbean Regional Consultative Meeting, Brasília, Brazil, 22–25 July 2003.

Jaffé, W., and M. Rojas. 1994. "Human Resources for Biotechnology: Latin America." *Biotechnology and Development Monitor* 21: 21–22.

James, C. 2002. "Preview: Global Status of Commercialized Transgenic Crops: 2002." Ithaca, N.Y.: ISAAA Briefs No. 27, ISAAA.

Jansen, K., and E. Roquas. 2005. "Science Advice for Biotechnology Regulation in Developing Countries," in M. Leach, I. Scoones, and B. Wynne, eds., *Science and Citizens: Globalization and the Challenge of Engagement.* London: Zed.

Jepson, W. E. 2002. "Globalization and Brazilian Biosafety: The Politics of Scale over Biotechnology Governance." *Political Geography* 21(7): 905–925.

Kaplinsky, N., D. Braun, D. Lisch, A. Hay, S. Hake, and M. Freeling. 2002. "Maize Transgene Results in Mexico are Artefacts." *Nature* 416 (11 April 2002): 601.

Mackenzie, R., F. Burhenne-Guilmin, A. G. M. La Viña, J. D. Werksman, J. Kinderlerer, K. Kummer, and R. Tapper. 2003. *An Explanatory Guide to the Cartagena Protocol on Biosafety.* Gland, Switzerland, and Cambridge, UK: IUCN.

Metz, M., and J. Fütterer. 2002. "Suspect Evidence of Transgenic Contamination." *Nature* 416 (11 April 2002): 600–601.

Newell, P., and R. Mackenzie. 2000. "The 2000 Cartagena Protocol on Biosafety: Legal and Political Dimensions." *Global Environmental Change* 10: 313–317.

Nogueira, A. H. V. 2002. "Brazil," in C. Bail, R. Falkner, and H. Marquard, eds., *The Cartagena Protocol on Biosafety: Reconciling Trade in Biotechnology with Environment and Development*, pp. 128–137. London: Earthscan.

Otero, G. 2001. "Book Review of *Technological Capabilities in Developing Countries: Industrial Biotechnology in Mexico* by Ruby Gonsen." *Journal of Latin American Studies* 33(3): 629–632.

Paalberg, R. L. 2001. *Politics of Precaution: Genetically Modified Crops in Developing Countries.* Baltimore: Johns Hopkins University Press.

Pehu, E. 2003. "Biosafety Capacity Building: A World Bank Perspective," in M. A. Mclean, R. J. Frederick, P. L. Traynor, J. I. Cohen, and J. Komen, eds., *A Framework for Biosafety Implementation: Report of a Meeting,* pp. 27–30. Washington: ISNAR.

Pelaez, V., and W. Schmidt. 2004. "Social Struggles and the Regulation of Transgenic Crops in Brazil," in K. Jansen and S. Vellema, eds., *Agribusiness and Society: Corporate Responses to Environmentalism, Market Opportunities and Public Regulation,* pp. 232–260. London: Zed.

Possani, L. D. 2003. "The Past, Present, and Future of Biotechnology in Mexico." *Nature Biotechnology* 21 (May 2003): 582–583.

Quist, D., and I. Chapela. 2001. "Transgenic DNA Introgressed into Traditional Maize Landraces in Oaxaca, Mexico." *Nature* 414 (29 November 2001): 541–543.

———. 2002. "Reply." *Nature* 416 (11 April 2002): 602.

Serratos, J. A., M. C. Willcox, and F. Castillo-González. 1997. *Gene Flow among Maize Landraces, Improved Maize Varieties, and Teosinte: Implications for Transgenic Maize.* Mexico: International Maize and Wheat Improvement Center.

Solleiro, J. L., and R. Castañón. 1999. "Technological Strategies of Successful Latin American Biotechnical Firms." *Electronic Journal of Biotechnology* 2(1): 26–35.

Stewart, C. N., M. D. Halfhill, and S. I. Warwick. 2003. "Transgenic Introgression from Genetically Modified Crops to Their Wild Relatives." *Nature Reviews Genetics* 4 (October 2003): 806–817.

UNEP-GEF. 2003. "Biosafety Newsletter No. 4." Geneva: UNEP.

UNU/BIOLAC. n.d. "UNU Programme for Biotechnology in Latin America and the Caribbean (UNU/BIOLAC)." http://www.unu.edu/capacitybuilding/Pg_biolac/pg.html (accessed 12 June 2002).

Valadares, M. C. C., and D. C. Monte-Neshich. 1996. "Responding to Product Development Needs: Funding and Prioritizing Biotechnology Research at EMBRAPA." The Hague: Biotechnology Seminar Paper, ISNAR.

Vogler, J., and D. McGraw. 2000. "An International Environmental Regime for Biotechnology," in A. Russell and J. Vogler, eds., *The International Politics of Biotechnology: Investigating Global Futures,* pp. 123–141. Manchester: Manchester University Press.

CHAPTER 5

Unnatural Growth: The Political Economy of Biotechnology in Mexico

MANUEL POITRAS

Biotechnology has been said to hold unique promises both for peasants and for the hungry. A number of *ex ante* studies considering the technological potential of biotechnology have proposed that this set of technologies, including genetic engineering, is distinct from Green Revolution technologies and should lead to more positive development outcomes (Barajas 1991). Notably, in contrast to the Green Revolution package of hybrid seeds and agrochemicals, biotechnologies have been said to be scale-neutral, and thus potentially beneficial to producers of any size. However, the analysis of the actual political economy of the introduction and use of genetically engineered technologies in a developing country such as Mexico, as will be presented here, leads to different conclusions. I will first introduce recent modifications to the intellectual property rights regime and to the regulation of this most crucial agricultural input, the seed. I will then analyze state responses to the new agricultural biotechnologies by reviewing the regulation of the use of these technologies in Mexican agriculture, and the position of public research on this matter. Finally, I will present the private sector's outlook regarding these technologies by looking at the transgenic products that have been commercialized in Mexico so far, as well as the strategies of the only life science company of Mexican capital, Pulsar. What will come out of this analysis is that the mode of introduction of genetic engineering technologies into the Mexican countryside in the current context of neoliberal political economy does not favor their use by peasant farmers.

What characterizes the general political economic context for the introduction of agricultural biotechnology in the Mexican countryside is a situation in which technological change in agriculture is the preserve of capital-rich producers. This situation is reinforced by the current credit crunch suffered by all sectors of the economy, except for finance capital–held corporations, and by the withdrawal of the state from agricultural and technological support. Moreover, the deepening process of transnationalization of Mexican agriculture has driven and is most likely to continue to drive the process of technological change, responding not to the needs of an impoverished peasantry or of a large portion of the Mexican working class, but rather to those of giant agro-industrial corporations and of affluent consumers worldwide.

Capitalist Disciplining of Biotechnology in Mexico Reforming IPRs: Toward the Commodification of Life

A major line of reform affecting the way biotechnology is introduced into Mexico is the changes regarding intellectual property rights (IPRs). The emergence of biotechnology and the increasing concentration of capital have led to a rush for patents on genetic material, made possible by significant changes to patent laws in the United States and internationally. Mexico is no exception to this trend, as IPRs were a crucial item on the GATT negotiation agenda (in what later became the TRIPs agreement), which Mexico joined in 1986. The United States had also been exerting pressure on Mexico since the passing of the protectionist 1978 IPR law, with threats of trade sanction by the inclusion of Mexico on the infamous article "Special 301" of U.S. trade law (which lists countries "under observation" before imposing trade sanctions) (López and Orozco 1995, 594–596). The reform of Mexico's IPR law then became one of the conditions for the United States to conduct free trade negotiations at the beginning of the 1990s (Solleiro and Coutiño 1998, 23; Rodríguez 1994, 285).

Article 27.3b of the TRIPs agreement states that countries can exclude from their patent regime plants and animals (except microorganisms) as well as the essentially biological processes to produce them. But all signatories should adopt a system of legal protection of plant varieties, either patent-based or with an effective sui generis system, no later than by the year 2000. This deadline has now been extended indefinitely, with a number of developing countries wanting to modify the agreement. In the 1991 reforms to Mexico's IPR law, plant varieties were included as patentable while genetic material was excluded, which runs contrary to international practice. This was changed in 1994 in order to comply with the TRIPs agreements. Under the current law, plant varieties cannot be patented, and are protected instead by a new plant variety law, promulgated in 1996. The law is based extensively on the 1978 version of the International Union for the Protection of New Varieties of Plants (UPOV) Convention, which Mexico signed in 1979 but did not ratify until 1995. The 1996 law grants exclusive rights to the breeder of a variety for a period of fifteen to eighteen years. But the seeds protected under this agreement can be reproduced and multiplied by farmers without license for personal sowing and consumption (the "Farmers' Rights" clause). It can also be used, with permission from the breeder, for research purposes (Solleiro and Coutiño 1998, 24–27).

As for genetic resources, they are also exempted from private ownership. However, the law makes an important distinction when it states that the exemption applies to "biological and genetic material *as encountered in nature*" (Solleiro and Coutiño 1998). The last part of this sentence implies that any material that is *genetically modified* can be patented, a distinction that in turn can be interpreted more or less broadly. In this law, no recognition is given to farmers for the work of selective breeding that has happened over generations, nor for the preservation of natural and traditional biological and genetic resources (Aboites and Martínez 1995). The introduction of biotechnology in such a legal context can thus easily lead to scenarios of biopiracy, whereby traditionally known properties of plants are "discovered" in a corporate laboratory and then appropriated by the corporation.

The alleged aim of these reforms to the IPR law is to encourage investment in research and development of quality planting material. According to Solleiro, who conducted a series of questionnaires with a number of actors in the seed sector, better quality planting material is likely to be available since the implementation of the reforms, though it is more likely to come from transnational corporations than from domestic seed research and production (Solleiro 1996). In fact, domestic agricultural research has become increasingly costly: the proliferation of processes and gene sequences that have been patented has made research so expensive that only the huge research budgets of TNCs can provide for it. As a director of the Biotechnology Research Unit of the International Centre for Maize and Wheat Improvement (CIMMYT) confided in an interview, the cost of royalties to be paid in order to conduct research is becoming prohibitive, and public institutions like CIMMYT have to enter into agreements with private companies in order to access the material or funds necessary to conduct research (Savidan 1999). Increased protection of seed varieties and of genetically modified products is thus likely to benefit transnationally based capital rather than the domestic seed industry (López and Orozco 1995; Aboites and Martínez 1995).

Technology Encapsulated: The Seed

Another central line of reform that affects the meaning of the introduction of biotechnology in Mexico is the seed sector. While there are other applications of biotechnology and genetic engineering, notably in the animal sector, the seed is the main recipient of agricultural genetic engineering. It is the actual location of the application of the technology and thus the carrier of such technology from the laboratory to the field. The main interaction between biotech and Mexico's agricultural producers is thus through the seed. Like other sectors of the agricultural economy, the seed sector has been radically altered during the neoliberal restructuring process. In peasant economies, seeds for the next planting season are traditionally selected from the harvest or exchanged with other peasants through a variety of means. The hybrid revolution of the 1930s in the United States and 1950s in Mexico brought about a new dynamic in the provision of seeds. Hybrids do not breed true, and farmers using hybrids must go back to the market periodically. The hybrid revolution thus brought with it the development of a new agricultural input industry, which has made the fortune of transnational corporations such as Pioneer, which is now owned by DuPont, and Cargill, whose seed operations have since been split and acquired by Dow and Monsanto.

In Mexico, hybrid seed production and distribution was, as with other agricultural inputs, conducted through a state agency. The first hybrid seeds were developed by the Oficina de Estudios Especiales (OEE) and the Instituto de Investigación Agrícola (IIA). The dissemination of the resulting maize seeds was imparted to the Comisión del Maíz (Maize Commission) from 1947, and then to the Productora Nacional de Semillas (PRONASE) from 1961. The first Seed Law also took effect in 1961, creating other institutions to coordinate and oversee the breeding, production, and commercialization of seeds. Together with the rapid internationalization of Mexican agriculture from the 1960s on, however, came the beginnings of the transnational private sector involvement in the seed sector. PRONASE had the advantages of an exclusive mandate to produce and commercialize all the varieties developed by the national public agricultural research institutes, and of its seeds being included in the Green

Revolution package pushed onto the agricultural producers through the development banks (Solleiro and Coutiño 1998). But even with these advantages, the state agency's share of the certified seed market diminished from 41 percent in 1970 to 27 percent in 1977 (Barkin and Suarez 1983). This situation could only worsen with the 1982 crisis. PRONASE had accumulated a huge surplus inventory during the last attempt at regaining food self-sufficiency, under the Sistema Alimentario Mexicano (SAM), and thus became bankrupt when the subsidy program of SAM was terminated in the wake of the first debt crisis in 1982. In 1990, two-thirds of PRONASE's production units were liquidated and the rest were kept functioning on the condition of financial self-sufficiency. Government transfers to it were stopped in 1992 (Solleiro and Coutiño 1998; OECD 1997, 92).

In addition, a new seed law was proclaimed in 1991, liberalizing the lines of seed research (previously controlled by the Ministry of Agriculture), and removing PRONASE's monopoly over the distribution of the seeds developed by the public research system (Aboites and Martínez 1995; Solleiro and Coutiño 1998). The private sector share of the seed market thus grew tremendously from the 1970s on, from 13 percent of the maize seed market in 1970 to 22 percent in 1980, 54 percent in 1990, and 93 percent in 1993 (Solleiro and Coutiño 1998, 11). Noteworthy also is the fact that the deregulation of the seed market included the removal of the requirement for all commercialized seeds to be certified by the state, as was the case previously with the National Seed Inspection and Certification Service (ibid., 10, 13). There has thus been also a partial deregulation of seed quality monitoring, as well as of the biosafety of new seeds introduced (see Jansen and Roquas, this volume). The latter element is most crucial when considering transgenic seeds.

The State's Response to the Challenge of Biotechnology: Incomplete Regulation of Transgenic Material

As a result of the deregulation of the seed sector, seed quality and biosafety are now dealt with by the same corporations and producers' associations that use the seed. The agricultural engineers or technicians employed by these corporate entities are often the same ones who are supposed to implement state-mandated quality monitoring programs, with a potential conflict of interest (González and Chauvet 1999). But exceptions to this modus operandi have been set up for transgenic seeds. Three areas of concerns are dealt with by three separate institutions: the

Ministry of Agriculture (SAGAR) deals with the question of agricultural and animal biosafety, the Ministry of the Environment (SEMARNAP) with the preservation of biodiversity, and the Ministry of Health (Secretaría de Salubridad y Asistencia Pública) with questions of human health and nutrition. Moreover, the Ministry of Commerce (SECOFI), bolstered by the free trade agenda, has a very large influence on what is decided in regard to biotechnology generally. It was also the administrator of the animal health law until the mid-1990s (when the dossier was transferred to SAGAR) (Solleiro et al. 1999; Quintero 1999).

Most attention and action concern the question of agricultural biosafety. So far, human health and biodiversity have received comparatively little institutional attention. A National Agricultural Biosafety Committee (CNBA) was set up by SAGAR in 1988 to deal with the introduction of biotechnology in Mexican agriculture, mandated with the evaluation of requests for the commercial liberation of transgenic crops on a case-by-case basis, considering the potential impact of a new crop on the biodiversity of the country (Solleiro et al. 1999). The CNBA is made up of government representatives and academics. In contrast to similar institutions in the United States or Canada, it does not include industry representatives, not even as observers (Gálvez and González 1998, 20). Given the complexity of the task, the CNBA is largely underfunded and understaffed. Also, since there are no experimental procedures or models to test the interaction between a new transgenic crop and specific Mexican agricultural ecosystems, the decisions are based largely on speculative considerations (Serratos 1998, 81).

In 1995, SAGAR set up a *Norma Oficial Mexicana* (NOM056-FITO-1995) to help the work of the CNBA, making more explicit the conditions and procedures to follow in order to import or produce genetically engineered products in Mexico. It regulates the allocation of field test licenses and decides when a new agricultural product can be deregulated. Furthermore, in the 1991 seed laws, it was planned that research with "high risk transgenic material" can be undertaken only after having obtained permission from SAGAR (Aboites and Martínez 1995, Solleiro et al. 1999). Concerning animal biosafety, the use and import of genetically engineered products for animals (hormones, vaccines, monoclonal antibodies) is dealt with by a new animal health committee (the National Animal Health Technical Consultative Council, or CONASA), set up in 1991 and now under the aegis of SAGAR. Under planning is a NOM to deal expressly with the introduction of genetically engineered products destined for animal use or consumption.

These developments provide a general framework for the regulation of transgenic products. But more specific norms and procedures to oversee the introduction of transgenic crops are still lacking (Solleiro et al. 1999). As a result, the track record of the government in regulating the introduction of transgenic products in the country has been somewhat inconsistent and at times even contradictory. On one hand, the state appears to support the introduction of agricultural biotechnology. Monsanto's dairy hormone rBST was allowed in Mexico four years before it was allowed in the United States, after a review process by SECOFI that involved Monsanto inviting high-level government officials to travel to the United States at their expense, and a subsequent speedy approval of the hormone without any independent review process of the human and animal health aspects of the problem (Quintero 1999; see Otero, Poitras, and Pechlaner, this volume). The relatively high level of milk imports at the time was enough to justify its deregulation. Another example of government promotion of imported biotechnology is the direct financial support given to farmers for the introduction of transgenic seeds through the *Alianza para el Campo*, which handed out 300 pesos per hectare to some cotton producers at the end of the 1990s to introduce Monsanto's insect resistant (*Bt* cotton) seeds, accounting for almost half of the extra costs of the seeds (González and Chauvet 1999; Solleiro et al. 1999).

But on the other hand, the state has approved only a few licenses for field trial or deregulation. In fact, only one genetically modified crop has been deregulated to date, a tomato genetically engineered to have a longer shelf life. *Bt* cotton, which accounts for a high proportion of the transgenic crops planted, is only at a "pre-commercial level" of authorization (Gálvez and González 1998). In terms of animal products and besides rBST, there currently seems to be only one transgenic vaccine being registered (Chavez 1999). In addition, only a few transgenic products have been permitted to enter as imports: an herbicide-tolerant canola, imported as grain for oil production, and a pest-resistant (*Bt*) potato, used to make frozen fries (Gálvez and González 1998).

Part of the contradictory behavior of the Mexican state regarding biotechnology is the fact that biotechnology touches on multiple political economic constituencies: environment and biodiversity, industry and commerce, agriculture, human health—each with its own ministries, institutional agenda, and political forces to respond to. Together with other factors such as the lack of human, technical, and financial resources, this has led to a weak regulatory environment, full of loopholes and lacking in coordination and regulatory efficiency (González and Chauvet 1999,

58; Quintero 1999), fully protecting neither the interests of the Mexican population nor those of the corporations involved.

Biotechnology and Public Research

Biotechnology has caught the Mexican public agricultural research system in a time of deep restructuring and wide-ranging budget cuts. At the beginning of the 1990s, Mexico was poised to become quite active in biotechnology research. Many governmental and academic documents identified it as a promising avenue for Mexican agriculture, with adequate human and scientific resources, as well as successful and extensive productive experience with traditional biotechnology such as fermentation (UNIDO 1991; SRE 1991; Quintero 1996). The only potential barrier envisioned was the initial large-scale investment for genetic engineering research equipment. An intergovernmental program, supported by the UNDP, to promote a comprehensive biotechnology development plan in Mexico was also designed during that period (SRE 1991). While this plan would have allegedly been directed at the needs of Mexican agricultural producers, it was soon abandoned (Quintero 1999).

Instead, a number of cuts were imposed on the public agricultural research system, with no exception made for the fledging biotechnology research agenda. Probably as a result of the cuts endured, INIFAP has allegedly been very slow and ineffective in introducing the new biotechnologies in its research program (Pedraza et al. 1998). This is not to say, however, that there is no public biotechnology research in the country. Since the mid-1980s, there has been an important growth of biotechnology research groups in public institutions, making Mexico one of the leading Latin American countries in this regard (Quintero 1999). But most of the research being done at this point does not involve genetic engineering, but rather micropropagation and tissue culture. Only two research centers reportedly have an ongoing genetic engineering program: CINVESTAV-Irapuato (the Irapuato unit of the Advanced Research and Studies Center), which has a strong biotechnology research orientation and twenty-two researchers, and the Biotechnology Institute of the UNAM in Cuernavaca (UNIDO 1991; Pedraza et al. 1998; Quintero 1996 and 1999).[1] However, only CINVESTAV has had concrete results so far, with pre-commercial field trials of three virus-resistant potatoes and with many other projects under way, such as virus-resistant tomatoes, papaya, and asparagus, and aluminum-tolerant wheat and maize.

What has permitted these developments is the diversification of the sources of financing, mostly from international private foundations and from international organizations (primarily the UN, the FAO, and the OAS), but also from the corporate sector. The latter's participation, however, has been scarce and usually oriented toward the satisfaction of its own interests (Pedraza et al. 1998). Monsanto, for instance, granted CINVESTAV the royalty-free use of some of its technology for its transgenic potato project and offered to train some of its scientists in its own laboratories in St. Louis, Missouri. Monsanto drew some publicity from this and tried to counter the waves of discontent against its use of genetic engineering. According to some observers, Monsanto was also trying to force the question of regulation of transgenic crops into the government's hand. While the first field trials occurred in the United States, they then moved to Mexico in 1993. The government had to take action in regulating this and eventually drew the NOM mentioned earlier. While Monsanto does not charge royalties on this research and its products, it explicitly forbids CINVESTAV to commercialize the products of the research in the United States. In fact, Monsanto filed for patents on the varieties developed at CINVESTAV with the U.S. patent office in 1995. It is thus in a position to recoup any money this royalty-free license agreement might have cost the corporation. Also, CINVESTAV is forbidden to use the technology in any industrial potato varieties except one. Nevertheless, CINVESTAV did benefit from this project, with new technology, training for some of its scientists, international exposure, and the first experience of using genetic engineering on Mexican varieties (Commandeur 1996; Chauvet et al. 1998; Gálvez and González 1998).

Public research on genetic engineering is likely to face the same problem as with previous public research: that of not reaching the intended beneficiaries. This has been one of the main problems of the Green Revolution, which was allegedly intended to raise the productivity of all developing country producers and thus end rural poverty as well as hunger, but which instead benefited a few large commercial producers and exacerbated polarization in the countryside. Similarly, the lack of interaction between public agro-biotechnology research and producers has been the focus of a number of critics on the topic (Solleiro and Rocha 1996; Quintero 1999). Rodolfo Quintero, one of the prime biotechnology researchers in the country, estimates that part of the problem is that public research institutes do not encourage the development of products adapted to Mexican producers, but rather scientific achievement, measured by the number of publications rather than by farmers-based results (Quintero 1999).

The case of the CINVESTAV virus-resistant potato project is a good example of this de-linking. While it is said in the objectives of this project that the intended beneficiaries of the project are small producers (Gálvez and González 1998, 29; Chauvet et al. 1998), they have never been consulted on their needs. Only commercial producers have been approached for field trials. In fact, the virus the new potatoes are resistant to is not considered a problem by the small producers in question (Chauvet et al. 1998). In addition, the project has no specific provisions for the dissemination of the new seeds. With the dismantling of PRONASE in 1992, and the end of public technical assistance programs, the production and distribution of improved seeds would now be left to market forces (Appendini 1994; OECD 1997). This could be a major problem, as most small-scale farmers use informal seed markets for the purveyance of their needs, and use new seeds only sporadically, when financial resources are available (Chauvet et al. 1998).

This problem of de-linking could eventually be solved by using a participatory method of technology development and use, such as those developed and sponsored by the Tropical Agriculture Research Center in its approximately 250 local agricultural research committees (CIALs) active in eight Latin American countries (Humphries et al. 2000; Ashby et al. 2000). But such participatory methods have not yet been used in Mexico, at least not in conjunction with biotechnology. In sum, therefore, even if there are some public research projects in agro-biotechnology oriented toward small-scale producers, and even if there are academic achievements in biotechnology research in general, a number of elements combine to make it difficult for the products of such research to reach their intended beneficiaries, such as the general crisis of the countryside, the persistent de-linking of public research goals with producers' needs, the cuts to assistance programs, and the dismantlement of the socialized technology transfer infrastructure.

Is the Private Sector Up to the Task? Limited Commercialization of Agricultural Biotechnology in Mexico

The commercialization of agricultural biotechnologies around the world has been slower than expected, due most of all to the reaction of civil society organizations and a series of moratoria on the introduction of genetically engineered products by a number of governments, notably in the European Union but also in many developing countries. Nonetheless, their use has progressed tremendously in other parts of the world, especially in the United States, Argentina, Canada, and China, which together account for most of the 25-fold increase of the total area where transgenic crops are grown in only five years, from 1.7 million hectares in 1996 to 44.2 million in 2000 and 81 million in 2004. The United States and Argentina alone account for 90 percent of total transgenic crop acreage, and with Canada and China, 99 percent of total transgenic acreage is accounted for (James 2001; Manjunath 2005, 5).

Despite the limitations imposed on the commercialization of biotechnology by the regulatory problems already mentioned as well as by the crisis affecting the countryside, there are more than 100,000 hectares of transgenic crops cultivated in Mexico, which places it in the top ten list of countries with a significant acreage of transgenic crop cultivation. But only a few transgenic products have been used or cultivated on a significant scale in Mexico: the dairy hormone rBST (cf. Chap. 6), Calgene's (now a subsidiary of Monsanto) extended shelf life tomato, Monsanto's *Bt* cotton, transgenic flowers for export, and a number of transgenic vaccines.

The McGregor tomato, developed by Calgene, is the first and so far only genetically engineered crop to have been deregulated in Mexico, in 1994. The tomato is genetically modified to inhibit the production of the enzyme responsible for the decay of the vegetable, considerably extending its shelf life. This tomato was one of the first commercial products of genetic engineering in the United States, and caused a potent wave of consumer protests against biotechnology. Such protests, combined with higher retail prices than for conventional tomatoes, led to its commercial failure. In Mexico, this crop is of significant importance, as it is the most important export crop, with some 20–25 percent of all agricultural exports to the United States (Massieu Trigo 1999). However, the deregulation status conferred on the McGregor variety implies that no official record is kept on the extent of its cultivation. It is thus difficult to know the current extent of its cultivation, although a government official working on the regulation of genetic engineering at SAGAR claims that it is either very little or has stopped (Bélez 1999). It was cultivated by the relatively prosperous export-oriented producers of the Northwest of the country (notably in Sinaloa). Since the tomato is harvested when it is ripe, its harvest cannot be mechanized and more labor must be used. With its vast sources of cheap labor, Mexico had a comparative advantage over Florida producers in the cultivation of this crop, and most probably marginally increased seasonal employment in the regions where it was cultivated (Massieu Trigo 1999; Chauvet 1999 and 2000).

Monsanto's transgenic cotton, engineered to produce its own biotoxin (*Bacillus thuringiensis, Bt*), was cultivated from the mid-1990s in the northern irrigation districts of Mexico. It is now the single largest transgenic crop in Mexico, with between 100,000 and 120,000 hectares cultivated each year. According to some estimates, this would account for some 40 percent of all cotton harvested in the country (González and Chauvet 1999; Bélez 1999). Cotton is a traditional export crop, reaching back to the beginning of the century. It attained its zenith in 1956, when 1.1 million hectares were cultivated (González and Chauvet 1999). As a "vanguard" crop of the postwar agricultural political economy (Rubio 1998), cotton was grown on the best-irrigated land in the country, mostly in the irrigation districts of the North. After a prolonged crisis, due to the increased production prices and declining international commodity prices when synthetic fabrics were introduced, cotton production was reduced considerably in Mexico, at 54,300 hectares in 1993 (Otero 1999, Chap. 5). The situation has since improved slightly, with a higher demand for the fabric, and with a new demand for its seeds, now processed into edible oils and industrialized animal feed. Mexico is also now a large importer of cotton, the commodity being free of tariff under NAFTA.

According to observers of transgenic cotton cultivation, no attention has been paid to the potential problems of genetic flows or environmental concerns due to the insecticide properties of the plant when the crop was allowed to be cultivated (González and Chauvet 1999). As part of its pre-commercial or pre-deregulation status, it is now cultivated under the terms of a research agreement between Monsanto and INIFAP. The aim of the research project is to monitor the implementation of integrated pest management practices, aimed at preventing the resistance of the insects to the toxin produced by the cotton variety (Gálvez and González 1998; Chauvet 2000).

Despite a lot of research and development into flower cultivation since the 1970s (especially in terms of micropropagation and genome preservation), much of the contemporary flower production in Mexico is done under the aegis of transnational corporations, especially from the United States or Holland, and destined for export. The TNCs impose their technological and input packages, including genetic modification to control the color or lifespan of cut flowers. All research and application of biotechnology on flowers (genetic engineering and micropropagation/cloning) is done abroad, and the seed, plant, and all inputs for its cultivation are exported to Mexico for production. Most of the genetically engineered flowers are then shipped to the United States for consumption. Typical of this enclave type of production, only a marginal part of production goes for sale in Mexico (Massieu Trigo 1997).

A study of the impact of the genetically engineered flower industry on the local industry and on labor concluded that, due undoubtedly to the enclave qualities of this sector, genetic engineering had only relatively marginal impacts. However, when surplus production was sold in Mexico, it negatively affected traditional producers. Also, when extra labor was hired, it was often the wives of the male producers, thus revealing significant gender consequences (Massieu Trigo 1997). Moreover, due to the failure of domestic research in this sector to yield any commercial application, flower biotechnology products are only available from abroad, thus raising considerably the entry barriers into this activity. The dependence of this sector on foreign technology should not be underestimated, as the imported genetic material and royalties account for up to 30 percent of the infrastructural and raw material costs of intensive greenhouse flower production (Massieu Trigo 1995).

Another product of genetic engineering that is currently being reviewed for commercialization is Monsanto's herbicide-resistant soy, with field tests in 1998 reaching 12,500 hectares. It should be noted that Mexican regulation does not set a limit on the area of field tests, which in this case makes it hard to distinguish it from commercial production (Gálvez and González 1998, 11). Finally, a number of other agricultural products are under review before being granted a license for field testing.

As should be evident by now, all of the agricultural products for which biotechnology has been used in Mexico so far are export or agro-industrial ones, with a strong presence of transnational capital. This is in line with the trends of increased export and agro-industrial orientation and transnationalization of the Mexican countryside that have been identified earlier. The only exception to this so far is the virus-resistant varieties of potatoes developed by CINVESTAV discussed earlier, which as of 2000 were still being tested in the experimental stations of INIFAP (Chauvet 2000).

A Mexican Life Science Industry? Financial Capital and Biotech in Mexico

The transnational corporations' dominance over commercial transgenic products in Mexico is obvious. As we have seen, the transgenic dairy hormone rBST is Monsanto's, and so are most of the genetically engineered tomato, cotton, and soy planted in Mexico. As for approval of field tests granted by the Ministry of Agriculture through the 1990s, transnational companies also greatly prevail. Of the list of twenty organizations that have received permits, only two are public institutions, CINVESTAV and CIMMYT, one being Mexican and the other one part of the World Bank–based CGIAR group. The rest are transnational firms such as (beyond those already mentioned) DNA Plant Technology, Seminis, and CIICA[2] (three Pulsar subsidiaries), Zeneca, Campbell, Pioneer (now a subsidiary of DuPont), Agrevo, and Upjohn (Gálvez and González 1998).

The national agro-industry is largely absent from this list as it has suffered from the crisis and has been mostly unable to introduce the most advanced technologies. In a recent study of the Mexican biotechnology industry (in Pedraza et al. 1998), many structural problems were reported, such as low proportion of sales revenues going to research (less than 1 percent, while for transnational corporations the proportion is 6 to 15 percent); absence of adequate research installations; lack of qualified personnel; only short-term and less risky projects are initiated; and a traditional dependence on licensing agreements from abroad. Many domestic firms are using techniques of micropropagation (such as CIICA, Tequila Cuervo, and Biogenética Mexicana), but none are using genetic engineering, which is the truly advanced and competitive biotechnology (Massieu Trigo 1995; Briceño 1999; Pedraza et al. 1998). In fact, the only firm of mainly Mexican capital (though it is a transnational corporation) that does advanced agro-biotechnology research is Pulsar. But all of its genetic engineering research is done abroad, through its subsidiaries in the United States and Europe. Even CIICA, the main agricultural Pulsar subsidiary located in Mexico, used to perform genetic transformation in its Chiapas laboratories, but the genetic engineering programs have now been transferred to a subsidiary, DNAP, located in California. The virus-resistant and prolonged shelf life papaya and banana under research at CIICA during the 1998–2000 period were genetically modified in California and then micropropagated for field testing in Mexico (Briceño 1999). Micropropagation is labor intensive, and this explains why it is performed in Mexico, where labor is cheaper.

Pulsar is in fact typical of the new economic power holders in Mexico. It has diversified increasingly over the years, gone through a major process of corporate growth despite the national and world crisis, and most recently consolidated some of its businesses, especially its agro-biotechnology one, through acquisitions on three continents. It has overall sales of more than US$2,800 million (Massieu Trigo and Barajas

2000), and ranked fifth among life science companies in 2000 (RAFI 2000). This life science corporation is distinct, however, from its European and North American counterparts. It is notably more diversified than these once chemical/pharmaceutical companies turned pharmaceutical/agro-biotechnology firms. The subsidiaries of Pulsar, beyond the seed and agro-biotechnology ones, include finance, insurance, telecommunications, construction, and food packing. With the latter (Empaque Ponderosa and Empaq), it has also reached a higher level of vertical integration into agro-industry than the other life science giants.

Pulsar first entered agriculture through tobacco, with the purchase in 1985 of Cigarrera La Moderna, the largest Mexican tobacco producer. This subsidiary was sold in 1997, during a time of restructuring toward agro-biotechnologies and vegetable seeds. From 1994 on, it has acquired a number of seed and seed genetics companies and grouped them under the leadership of Agromod (or Agroindustrias La Moderna, once Empresas La Moderna). In the United States, it acquired Asgrow and DNA Plant Biotechnologies; in Europe, Bruinsma, Petoseed, and Royal Sluis, which were merged with Asgrow in 1995 to form Seminis; in Korea, the Hung-Nong Seed Company and Coong Ang Seed Company; and in India, Nath Sluis. These acquisitions have positioned Pulsar as one of the global leaders in biotechnology and placed the company first in the world for vegetable seed production (Massieu Trigo and Barajas 2000).

Moreover, as a result of this consolidation, the company does not have the problems mentioned above with Mexican biotechnology industry in general. At a global level, it invests 14 percent of its sales revenues in research and development, and is planning to increase this proportion to 18 percent; it has thirty-five agronomic research centers (of which twenty-three are in the United States, with the rest in Latin America and Asia) and fourteen experimental fields. All the recent acquisitions it has made also mean that it now has five hundred qualified scientists worldwide at its disposition. Finally, these acquisitions respond to the same strategies used by other life science giants: to get access to technological capability, varied and local germplasm, and large distribution networks. In sum, Pulsar has followed the typical pattern of restructuring followed by Mexican finance capital: advanced technological restructuring and transnationalization. The transnationalization of the company also means, however, that in its R&D, corporate development, and sales strategies, it responds to the imperatives of the global market, not to the needs of peasants, be they Mexican or not. The potential development of new biotechnologies toward output traits is more likely to orient their products toward niche markets for the high- and to some extent middle-class food markets of the industrialized countries as well as for the elite markets of developing countries.

In sum, with the private biotechnology sector depicted here, dominated by foreign and domestic transnational corporations with goals of global profit maximization rather than national ones, and with an anemic public research structure, an alternative strategy to reorient the use of biotechnologies toward rural and sustainable development is largely foreclosed at the moment.

Conclusions

The process of de-capitalization of the countryside in Mexico and the new model of transnationalized agro-export agriculture is likely to lead to a new process of rural producers segmentation, with the largest companies benefiting most from the state support programs and being most able to survive in these conditions. These are the same companies that are and will be able in the near future to use the genetically modified seeds, plants, and animals, and for the needs of which the transgenic varieties will be designed. This scenario is bolstered by the recent legislative reforms to the intellectual property regime of Mexico and by new seed regulation protecting the rights of private investors, which at this point have been identified with the interests of transnational life science corporations (see Jansen and Roquas, this volume).

A developmentalist scenario would also envision that the development of a private domestic biotechnology sector could lead, with the right incentives orchestrated by a benevolent state, to the orientation of the agro-biotechnologies toward the needs of small or middle-size farmers. But given the global marketing and monopolistic strategies of the life science giants, combined with a regulatory environment that is favorable to these, this option is now forfeited.

One element that could contribute to an alternative scenario is a strong public research program in biotechnology, in order to provide cheap access to biotechnology products and to develop products designed for small producers' needs, while taking full consideration of the environmental implications of such technology. The CINVESTAV virus-resistant potato project is interesting in this regard, but has three general failings: dependence on the goodwill of private corporation for funding; lack of consultation with the supposed beneficiaries of the new potato variety; and lack of funding for technical assistance and for the distribution of the new seeds to these beneficiaries. The reorientation of public research institutions toward participatory technology development practices could change this, as exemplified by the apparent success of the CIAL in a number of other Latin American countries in Honduras (Humphries et al. 2000; Ashby et al. 2000).

Notes

1. While CIMMYT, located close to Mexico City, also has a biotechnology research program using genetic engineering, it is not counted here as part of the Mexican public agricultural research system as it is not funded by the Mexican government, and its research programs and human resources are truly international in outlook and origin, Mexico being only one of its focuses.
2. Centro Internacional de Investigación y Capacitación Agrícola (International Center for Agricultural Research and Training), founded in 1994 by the Pulsar group and located in Tapachula, Chiapas.

References

Aboites, Gilberto, and Francisco Martínez. 1995. "Situación de la Legislación Mexicana en Materia de Biotecnología y Recursos Genéticos." *El Campo Mexicano en el Umbral del Siglo XXI*. Ed. A. E. Rodríguez. Mexico D.F.: Espasa Calpe.

Appendini, Kirsten. 1994. "Transforming Food Policy over a Decade: The Balance for Mexican Corn Farmers in 1993." *Economic Restructuring and Rural Subsistence in Mexico: Corn and the Crisis of the 1980s*. Ed. Cynthia Hewitt de Alcántara. San Diego, Calif.: Center for U.S.-Mexican Studies, UCSD, pp. 145–157.

Ashby, J. A., et al. 2000. "Power to the Poorest" and "The Mature CIAL." *Investing in Farmers as Researchers: Experience with Local Agricultural Research Committees in Latin America*. International Centre for Tropical Agriculture, pp. 23–43 and 64–69.

Barajas, Rosa Elvia. 1991. "Biotechnología y Revolución Verde: Especificidades y Divergencias." *Sociológica* 6:16 (May).

Barkin, David, and Blanca Suarez. 1983. *El Fín del Principio: Las Semillas y la Seguridad Alimentaria*. Mexico City: Centro de Ecodesarrollo.

Bélez, Amada. 1999. Personal interview at SAGAR's Plant Sanitation division. Mexico City, 15 July.

Briceño, Guillermo. 1999. Personal interview at CIICA. Tapachula, Chiapas, 1 July.

Chauvet, Michelle. 1999. "Perspectives for Biotechnological Applications in Mexican Agriculture." *AgBiotechNet* 1 (April).

———. 2000. "Los Cultivos Transgenicos en México." Presented at the XXII International Congress of the Latin American Studies Association, Miami, March 16.

Chauvet, Michelle, José Luis Solleiro, and Rosa Luz González. 1998. "Metodologías para el Análisis del Impacto Socioeconómico de la Biotecnología Agrícola: Lecciones de los Estudios de Caso." *Transformación de las Prioridades en Programas Viables: Actas del Seminario de Política Biotecnología para Latina América.* Ed. John Komen et al. The Hague: Intermediary Biotechnology Service.

Chavez, Marta. 1999. Personal interview at SAGAR's Animal Health. Mexico City, 22 July.

Commandeur, Peter. 1996. "Private-Public Cooperation in Transgenic Virus-Resistant Potatoes: Monsanto, USA—CINVESTAV, Mexico." *Biotechnology and Development Monitor* 28 (Sept.), pp. 14–19.

Gálvez M., Amanda, and Rosa Luz González A. 1998. *Armonización de Reglementaciones en Bioseguridad.* Mexico City: UNAM.

González, Rosa Luz, and Michelle Chauvet. 1999. "Los Cultivos Transgénicos en México." *Este Pais* (March).

Humphries, Sally, et al. 2000. "Searching for Sustainable Land Use Practices in Honduras: Lessons from a Programme of Participatory Research with Hillside Farmers." *Network Paper No. 104.* London: Overseas Development Institute. www.odi.org.uk/agren/publist.html.

James, Clive. 2001. *Global Review of Commercialized Transgenic Crops: 2001.* Ithaca, N.Y.: ISAAA.

Lopéz H., Augustín, and Felipe de J. Orozco M. 1995. "Presente y Futuro de la Propiedad Intelectual de las Variedades Vegetales en México." *El Campo Mexicano en el Umbral del Siglo XXI.* Ed. A. E. Rodríguez. Mexico D.F., Espasa Calpe.

Manjunath, T. M. 2005. "A Decade of Commercialized Transgenic Crops—Analyses of Their Global Adoption, Safety and Benefits." The Sixth Dr. S. Pradhan Memorial Lecture, delivered at Indian Agricultural Research Institute (IARI), New Delhi, on 23 March. Available at http://www.agbioworld.org/word/decade-commercialized.doc (consulted 8 May 2007).

Massieu Trigo, Yolanda. 1995. "La Modernización Biotecnológica de la Agricultura Mexicana: Otro sueño enterrado durante el sexenio salinista." *Cuadernos Agrarios,* 11–12.

———. 1997. *Biotecnología y Empleo en la Floricultura Mexicana.* Mexico D.F.: UAM-Azcapotzalco.

———. 1999. "Cultivos Transgénicos en México." *La Jornada Ecológica* 82 (7 Dec.).

Massieu Trigo, Yolanda, and Rosa Elvia Barajas. 2000. "Empresas La Moderna: Una multinacional mexicana, un nuevo actor social en la agricultura." *Revista Mexicana de Sociología* 62:3, pp. 79–107.

Massieu Trigo, Yolanda, and Rosa Luz Gonzalez. 2000. "Ingenería Genética y Biodiversidad: La Necesidad de la Regulación." Presented at the XXII Congress of the Latin American Studies Association. Miami, 16 March.

OECD. 1997. *Review of Agricultural Policies in Mexico.* National Policies and Agricultural Trade Series. Ed. Gérard Bonnis and Raphael Patrón Sarti. Paris: OECD.

Otero, Gerardo. 1999. *Farewell to the Peasantry? Political Class Formation in Rural Mexico.* Boulder, Colo., and Oxford: Westview Press.

Pedraza, Lorena, et al. 1998. "La Biotecnología en México: Una Reflexion Retrospectiva 1982–1997." *Biotecnología: Nueva Era* 3:3 (Sept.–Dec.).

Quintero, Rodolfo. 1996. "Biotecnología para la Agricultura." In *Posibilidades para el desarollo tecnológico del campo Mexicano, Tomo 1*. Ed. José Luis Solleiro, María del Carmen del Valle, and Ernesto Moreno. Mexico City: UNAM.

———. 1999. Personal interview. Mexico City, 10 Dec.

RAFI (Rural Advancement Foundation International). 2000. *The Seed Giants: Who Owns Whom? Seed Industry Consolidation Update 2000*. Dec. 2000. www.etcgroup.org, accessed February 2001.

Rodriguez C., Dina. 1994. "Tratado de Libre Comerico: Posible Implicaciones en el Desarollo Agrobiotecnológico Nacional." *Apertura Económica y Perspectivas del Sector Agropecuario Mexicano Hacia el Año 2000*. Ed. Emilio Romero Polanco et al. Mexico City: UNAM, pp. 277–291.

Rubio, Blanca. 1998. "El Dominio" Desarticulado' de la Industria Sobre la Agricultura: La Fase Agroexportadora Excluyente." Presented at the Fifth Latin American Congress of Rural Sociology. Chapingo, Mexico: October.

Savidan, Yves. 1999. Personal interview at the Applied Biotechnology Centre of the CIMMYT. Mexico, 30 April.

Serratos H., J. Antonio. 1998. "Evaluación de Variedades Novedosas de Cultivos Agrícolas en su Centro de Origen y Diversidad: El Caso del Maíz en México." *Transformación de las Prioridades en Programas Viables: Actas del Seminario de Política Biotecnología para Latina América*. Ed. John Komen et al. The Hague: Intermediary Biotechnology Service.

Solleiro, J. L. 1995. *Biotechnology and Sustainable Agriculture: The Case of Mexico*. OECD Working Papers 3:5.

———. 1996. "Propiedad Intellectual: Promotor de la innovación o Barrera de Entrada?" *Posibilidades para el desarrollo tecnológico del campo Mexicano, Tomo II*. Ed. José Luis Solleiro, María del Carmen del Valle, and Ernesto Moreno. Mexico City: UNAM, pp. 211–224.

Solleiro, J. L., and Beatriz Coutiño. 1998. *Políticas de Biotecnología y Biodiversidad: Estrategias de Gestión de la Propiedad Intelectual para la Industria de Semillas*. Mexico City: UNAM.

Solleiro, J. L., Hilda Hernández, and Ana Silvia Rocha, eds. 1999. *Marco Regulatorio Nacional de Bioseguridad y Salud*. Mexico City: BIODEM, 1999.

Solleiro, J. L., and Alma Rocha L. 1996. "Cambio Técnico e Innovación en la Agricultura Mexicana." *Comercio Exterior* (Aug.).

SRE. 1991. *Preparación del Programa Nacional de Cooperación en Biotecnología*. Mexico City: SRE, CONACYT, and UNDP.

UNIDO. 1991. Biotechnology Policies and Programs in Developing Countries: Survey and Analysis. New York: United Nations Industrial Development Organization.

CHAPTER 6

Importing Corn, Exporting Labor: The Neoliberal Corn Regime, GMOs, and the Erosion of Mexican Biodiversity

ELIZABETH FITTING

Some of the earliest maize cobs on record have been found by archaeologists in the Tehuacán Valley of southeastern Puebla (MacNeish 1972), the same region where Mexican government tests recently confirmed the presence of genetically modified (GM) corn among traditional cornfields (INE-CONABIO 2002). Imported from the United States to serve as animal feed, as grain for tortillas, or for industrial processing, GM corn made its way to regional markets in Mexico's "cradle of corn," where small-scale Mexican cultivators unknowingly purchased and then planted the grain. This finding helped mobilize Mexican and transnational activist networks and amplify an international debate about the extent to which corn imports from the United States pose a threat to maize biodiversity in the crop's center of origin, domestication, and biological diversity. Beyond the environmental risks of gene flow, however, these imports raise the question as to whether the expansion of neoliberal globalism,[1] particularly free trade agreements coupled with economic restructuring in the global south, grants an unfair advantage to transnational corporations and large-scale northern farmers. The latter often enjoy hefty government subsidies for the production of basic grains (Bartra 2004).

In this chapter, I want to propose, as some activists and academics in the recent maize debates have, that it is not simply the lack of sufficient regulation on transgenic corn imports that poses a risk to in situ conservation of maize landraces, the gene reservoir upon which the development of future corn varieties depends. Rather, the increasing hardships and out-migration of small-scale corn cultivators who are struggling to adapt to economic crisis and neoliberal reforms also jeopardize in situ maize biodiversity. I focus here on the *social* aspect of biodiversity, defined as a dynamic process in which maize landraces are maintained through exchanges between fields and between cultivators (Serratos, Willcox, and Castillo-González 1996, viii). While "modern" corn varieties or "cultivars" refer to those varieties developed by plant breeders, "landraces," in contrast, are those "crop populations that have become adapted to farmers' conditions through natural and artificial selection" (Aguirre Gómez, Bellon, and Smale 1998, 7). In Spanish, *criollos* refers to both landraces and creolized varieties, the latter of which are the outcome of an intentional or accidental mix of landraces with modern, improved varieties. The storage of *criollos* ex situ in seed banks is a crucial means of protecting maize biodiversity, but is often viewed as an insufficient measure on its own. The UN Convention on Biological Diversity, for instance, considers in situ conservation the best approach for safeguarding biodiversity (Glowka, Burhenne-Guilmin, and Synge 1994). Most corn grown in Mexico is white *criollo* corn, popular for making tortillas and other foods because of its high flour content and fine texture, while in the United States, modern varieties of yellow corn make up the bulk of production. Yellow corn is used as animal feed or in the manufacture of corn starch, high-fructose corn syrup, industry-made tortillas, and other products.

This chapter will examine how rural households have adapted to and been affected by the neoliberal corn regime, based on some sixty interviews in the southern Tehuacán Valley town of San José Miahuatlán and a year and a half of fieldwork (2001–2002, with return visits during the summers of 2005 and 2006). I use the term *neoliberal corn regime* to refer to the series of policies introduced since the mid-1980s associated with the ideology of neoliberal globalism, which prioritize market liberalization, trade, agricultural "efficiency," and the reduction of state services over domestic corn production. Policy that tends to favor the interests of the biotech industry also constitutes part of this regime (Otero, Scott, and Gilbreth 1997).

In the southern valley, as in much of the Mexican countryside, maize production has been central to the survival of the rural household. However, in the absence of sufficient or sufficiently remunerative local or regional employment, households increasingly combine corn production with U.S.-bound migration and other off-farm income. This local strategy to adapt to crisis is remaking agriculture and social relations in significant ways: maize has become a more significant share of local agricultural production in the southern valley since the 1980s, but agriculture and agricultural land use have decreased overall, particularly from the 1990s onward. While older residents have switched from other crops to corn production, younger residents increasingly prefer non-agricultural wage labor. Young residents may of course take up corn production as they age, but many claim that they will not simply because there is "no money to be made in the *milpa*" (or cornfield). Before turning to examine the effects of this household strategy, I will first discuss the neoliberal corn regime and the GM corn debates.

The Neoliberal Corn Regime: From Self-sufficiency to Imports

While agricultural and trade policies from the administration of Miguel de la Madrid (1982–1988) to Vicente Fox (2000–2006) have varied, throughout this period a commitment to the neoliberal agenda has deepened, affecting corn production and consumption in Mexico. After the debt crisis of 1982 austerity measures and the restructuring of government agricultural extension services, marketing agencies, and the rural credit system exacerbated the long-standing difficulties faced by Mexican agriculture (Otero 2004).

Previous administrations had pursued the goal of national self-sufficiency in maize and a related commitment to support small to medium producers, while importing corn to meet an increase in demand (Appendini 1992; Austin and Esteva 1987; Otero 1999). Under neoliberal reform, however, the goal of self-sufficiency was replaced by a series of policies that focused on providing urban consumers access to cheap tortillas through grain imports "which enjoyed low, subsidized international prices and could be obtained with cheap credit" (Appendini 1994, 148). After more than 50 years, the tortilla consumer subsidy was also eliminated in late 1998. Rural areas that practiced small-scale, rain-fed agriculture were classified as zones of "low productive potential" and viewed as areas of rural poverty in need of social welfare assistance (de Teresa Ochoa 1996, 190).

In the 1990s, Article 27 of the 1917 Constitution was amended to allow for the rental and sale of *ejidos* (land grants resulting from the agrarian reform process) and the North American Free Trade Agreement (NAFTA) was implemented as part of a strategy to restructure the agricultural sector. Under NAFTA, Mexico was to gradually open its doors to corn imports in exchange for guaranteed access to the market for horticultural products and other labor-intensive crops in Canada and the United States. The assumption was that, according to neoclassical international trade theory, Mexico has a comparative advantage in producing such crops because of its surplus in labor and lower production costs.

Furthermore, this increased reliance on U.S. corn imports was deemed necessary to meet the demands of industry and a growing population, but some researchers have argued that Mexico could meet its own corn needs domestically with state support and the right technology (INIFAP cited in Appendini 1994, 153; Turrent et al. 1996).

One of the most significant effects of NAFTA in Mexico was the increase in corn imports from the United States, the world's largest producer and exporter of the crop. Between 1994 and 2000 imports from the United States went from 14 to 24 percent of the total consumption of corn in Mexico (Bartra 2004). In the year 2000, Mexico was the second largest importer of U.S. corn, and 21 percent of corn grown in the United States was *Bt* corn, a transgenic variety with genes from the soil bacterium *Bacillus thuringiensis.* This variety produces insecticide toxins that kill the European and Southwestern corn borers, neither of which are found in Mexico (Ackerman et al. 2003, 11). Mexico now imports roughly 6 million metric tons of U.S. corn annually, up to 60 percent of which is transgenic (USDA 2006).

NAFTA changed the previous system of import licenses into a duty free, tariff-rate quota system. The import quota for corn was set at 2.5 metric tons for the first year, and was scheduled to increase gradually over a fifteen-year transition period until the quotas and tariffs are eliminated in 2008. As part of the agreement, Mexico removed price support mechanisms that had been in place for forty years. According to economist Alejandro Nadal (2000, 5), "the planned fifteen-year transition period was actually compressed into 30 months. Between January 1994 and August 1996 domestic corn prices fell by 48 percent, thereby converging with the international market some twelve years earlier than provided for under NAFTA, and forcing the Mexican corn producers into rapid adjustment." Corn imports far surpassed the agreed-upon level and prices dropped. The drop in grain prices, however, was not passed on to the consumer through lower tortilla prices. In an effort to stabilize tortilla prices, the government exempted imports from the tariff payment and increased subsidies to the corn flour industry, but tortilla prices soared (Nadal 2000, 5, 38). NAFTA also treats Mexican yellow corn and white corn as equivalent commodities even though the latter is on average 25 percent more expensive on international markets (ibid., 16).

In the southern Tehuacán Valley and the more marginalized areas of rural Mexico, at least two assumptions that informed the design and rhetoric of NAFTA are incorrect. A first assumption was that a national market flooded with imported corn and lower prices would not adversely affect subsistence producers. In the southern valley and elsewhere, however, subsistence producers *are also petty corn sellers* who tend to sell at a disadvantage after the harvest when there is an abundance of low-priced local and imported corn. These same producers are then obliged to purchase corn when their stored supply of local corn has run out and prices are higher (Nadal 2000, 8, 24).

A second line of rhetoric about NAFTA was that it would generate off-farm employment for noncompetitive, displaced rural producers (Nadal 2000, 24). While clothing and poultry factories have expanded in the Tehuacán Valley, many positions are not sufficiently remunerative. The remaking of a flexible labor in the valley has been a gendered process. Many such factory positions are considered low-paying "women's work." To make a better wage, young residents, predominantly men, migrate to Mexican or U.S. cities, supplementing this income when back in the valley with short-term stints in factory, construction, and agricultural work. Connected to this rhetoric, then, is the notion that the Agreement would displace a large number of rural producers. In the southern valley, while migrant remittances and off-farm wages contribute to the reproduction of rural households and corn production, this strategy simultaneously takes young laborers away from corn production for long periods and remakes agricultural practices.

The Mexican Maize Debates

The Mexican campaign against GM corn exposed a regulatory gap between GM crop field trials and the import of GM corn for food, feed, and industrial uses. While scientific field trials of GM corn, conducted in Mexico since 1993, were halted under the de facto moratorium of 1998, GM corn has been imported without adequate regulation or monitoring. The United States does not require its distributors to separate GM corn from other varieties so that once in Mexico, imported corn has been difficult to track or control, due in part to the nature of informal seed exchange between cultivators and gene flow between cornfields. Despite recent improvements to GM regulations in Mexico, some GM corn continues to enter the country without labeling or monitoring.

The debate over GM corn began in the Mexican scientific and government regulatory community in the mid-1990s with the impending commercial release of GM corn in the United States. In Mexico, the General Directorate of Plant Health (DGSV) of the Ministry of Agriculture permitted the first scientific field trials of GM crops in 1988. Field trials of corn were first conducted by research institutes in 1993, followed by industry trials in 1996 (SENASICA 2005). The Directorate was advised by an ad hoc committee consisting of scientists from various disciplines and government agencies, which became the National Agricultural Biosafety Committee (CNBA) in 1992. (This advisory committee is now known as the Specialized Agricultural Subcommittee of the recently created Interministerial Commission on Biosafety or CIBIOGEM.) In late 1998, the Directorate decided to impose a de facto moratorium on GM corn trials because the traits most commonly tested were not of any particular benefit to Mexico (Alvarez-Morales 1999, 91). There were also concerns about the possibility of GM corn mixing with and displacing landraces and wild relatives (Serratos 1996; Serratos et al. 1996).

There is little known about the effects of GM corn on maize biological diversity. Maize was domesticated in central Mexico between five and nine thousand years ago (Matsuoka et al. 2002). Today, Mexico is home to more than forty-one racial complexes of maize, many *criollos*, and its closest wild relative, teosinte (Nadal 2000; Turrent and Serratos 2004). Like modern varieties or cultivars—those varieties developed by plant breeders—GM corn can displace or mix with landraces if widely adopted. However, the potential risks of GM varieties also include the development of pests resistant to transgenic plants or weeds tolerant to herbicides through genetic exchange with GM crop relatives (Serratos 1996, 69).

While the moratorium on field trials was in place, environmentalist groups began to suspect that GM corn was making its way into Mexico in shipments of imported corn from Canada and the United States. In 1999, Greenpeace Mexico tested samples taken from ships in the port of Veracruz carrying U.S. corn and found GM corn among the grain (Greenpeace n.d.). They launched a campaign against the field testing and import of GM corn, as did the Environmental Studies Group (Grupo de Estudios Ambientales or GEA), also based in Mexico City, and the Canadian-based environmental and farmers' rights organization, the ETC group (formerly called RAFI), among others (Greenpeace 2001). They were soon joined by peasant, indigenous rights groups and international NGOs in demanding an end to unlimited, unlabeled, and unmonitored corn imports.

When scientific studies found GM corn in peasants' cornfields and rural markets, the anti–GM corn campaign gained international attention. A study by Dr. Ignacio Chapela and David Quist from the University of California at Berkeley, published in the journal *Nature*, found a gene promoter among Mexican maize fields of the Sierra Norte de Oaxaca (Chapela and Quist 2001). Gene promoters are DNA sequences that make the transgene function with a plant's genome. Their controversial findings led to criticisms of the study's scientific methods, and in an unprecedented turn, *Nature* withdrew its editorial support for the study in April 2002, despite the fact that it had reviewed and accepted the paper. The Mexican Ministry of Agriculture and biotech industry representatives also suggested that if the study's findings were correct, the gene flow between GM varieties and *criollos* was merely part of a natural, and even beneficial, process of hybridization. In contrast, GM critics have argued that the finding of GM corn in Mexico was not benign hybridization, but rather a form of genetic pollution or contamination (Soleri et al. 2005). Because Mexico is the crop's center of origin, this pollution was also considered the first of its kind to occur anywhere in the world.

In order to verify the Berkeley findings, the Mexican National Ecology Institute (INE) and Biodiversity Commission (CONABIO) tested samples from different localities in the states of Oaxaca and Puebla and confirmed the presence of the CaMV 35S promoter used in many commercially planted and sold GM crops.[2] The study also found GM corn among the grain from some of the government's own rural DICONSA supply stores. Since that time, DICONSA began to restrict its purchase of corn to domestic grain only.

Corn falls under the purview of conflicting regulatory categories because it is both grain *and* seed. GM corn entering Mexico as grain destined for food or use in animal feed cannot be restricted under the 2003 modification of the Cartagena Protocol on Biosafety. Mexico was a signatory to the Cartagena Protocol, which was adopted in January 2000 as part of the UN Convention on Biological Diversity (CBD). However, if that same corn enters Mexico as "seed" then it is covered by the Cartagena Protocol as a Living Modified Organism, and the Mexican government has the right to refuse or delay the shipment. The Protocol contains the "precautionary principle" which enables a country to demand that such products be labeled as GMOs or to ban the import of GMOs until such products are proven safe for human health and the environment. The Mexican Senate ratified the Protocol in April 2002 and it became legally binding in the international system in September 2003.

In 2004, Mexico also signed a trilateral agreement with Canada and the United States requiring shippers to label GM food or feed imports *only* when the shipment contains 5 percent or more of GMOs.[3] Shipments which contain less than 5 percent of GMOs are considered equivalent to a non-GMO shipment and do not require identification. Critics argue that up to 5 percent of approximately six million metric tons of imported corn annually is an unacceptably large quantity of GM corn. Additionally, the agreement makes the labeling of such imports available only to distributors and not consumers.

In 2005, Mexico finally passed a law on Biosafety and Genetically Modified Organisms, but the legislation contains contradictions and does little to remedy the problems of the 5 percent rule (Bartra et al. 2005; GEA et al. 2006). The Law establishes that a special regime is needed to protect maize and other plants for which Mexico is the center of origin. However, there is nothing in the Law itself about this regime. Rather, the Law requires that the Ministries of Agriculture and the Environment establish the necessary measures to protect such species with the help of other state agencies. This includes determining which regions of Mexico are centers of origin and diversity of maize, in order to prohibit or approve, on a case-by-case basis, GM field trials or the commercial release of the species in question. Among the criticisms of the Law, however, is that the boundaries of such centers of origin and diversity are imprecise and difficult to define, but even more important, in the case of gene flow, they are permeable (Nadal 2005).

While the defenders of GM corn call for more accurate testing and evaluate GM corn according to a framework of risk, critics question this framework. In the international regulatory community, GMOs are often evaluated according to a framework of risk which prioritizes gene flow and marginalizes non-risk concerns such as ethics, rural livelihoods and food security, quality, and sovereignty (Heller 2002; see also Beck 1986; Cleveland and Soleri 2005; Herrick 2005; McAfee 2003; McAfee, this volume). GM critics, however, evaluate more than the risks of gene flow; they also consider trade liberalization, rural policy, migration, and other problems facing small-scale agricultural producers. Anthropologist Chaia Heller (2002) demonstrates that the French anti-GM movement challenges the discourse of risk as the central framework for evaluating and debating GM crops. Much like the French campaign which portrayed small-scale farmers as artisans central to the survival of the national cuisine and culture, in the Mexican campaign, maize farmers are portrayed as the producers and guardians not just of traditional corn varieties, but of national cultural practices and traditions (Fitting 2006). This anti-GM corn campaign has had some success. When the affected communities of the Sierra Norte de Oaxaca requested an investigation by NAFTA's Commission for Environmental Cooperation of North America (CEC) in 2002, the CEC set up a series of consultations and wrote a report that was submitted to the environment ministers of all three NAFTA signatory countries. The report argues that GM corn is unacceptable in Mexico largely because of *social and cultural reasons*, rather than for known risk factors for maize biodiversity or human health (CEC 2004). While the three governments have largely ignored or rejected its recommendations, the report provided an important forum for critics to link the question of gene flow between transgenic and traditional varieties to the broader issues of import dependency for basic grains, subsidy cuts to the countryside, rural livelihoods, and cultural and culinary practices. The debate about GM corn in Mexico is particularly heated because it is a flashpoint for wider struggles over the future of the peasantry and food security. Maize has long been a symbol of the Mexican nation (Pilcher 1998; Warman 1988), but among critics of NAFTA and GM corn imports, it has come to represent Mexican culture and food sovereignty under attack in an age of neoliberal globalism.

By linking GM corn to these broader issues, the anti-GM corn campaign asks not only about the effects of GM corn on landraces and wild relatives, but how to protect in situ corn biodiversity when small-scale corn producers face impoverishment and out-migration under the neoliberal corn regime (Altieri 2003; Esteva and Marielle 2003; García Barrios and García Barrios 1994). In situ corn biodiversity is a concern despite the fact that there has been an increase in the area devoted to domestic, small-scale corn production following NAFTA, contrary to policy predictions and intentions (Barkin 2002, 2003; Dyer and Yúnez-Naude 2003; Nadal 2000). This increase in the area devoted to corn is considered by some to be evidence that landraces are not threatened or being displaced by modern or GM varieties. While 70 to 80 percent of all Mexican cornfields continue to be grown with *criollos* (Aquino 1998, 245; Turrent 2005), some research suggests that landraces, on their own, account for only 50 percent of the total area of maize production (Ortega-Paczka 1999). There are various factors that affect the loss or abandonment of landraces, including the proximity of farmers to commercial markets or the influence of government modernization programs. Additionally, some landraces have a higher risk of being replaced for other varieties, such as very early or very late maturing varieties (ibid).

Several studies have demonstrated that the area devoted to corn has expanded under NAFTA as part of a "corn subsistence—labor migration" strategy in poorer, largely rain-fed, agricultural regions of south-central Mexico (de Janvry et al. 1997; Nadal 2000). While small-scale corn production offers a rural safety net in times of hardship, it also increasingly depends on remittances and other off-farm income for the purchase of needed agricultural inputs (Hewitt de Alcántara 1994). This is also the case in the Tehuacán Valley, where residents cultivate more irrigated than rain-fed corn. Most types of corn grown in the valley are landraces and creolized varieties (*criollos*) adapted to the local climate and soil conditions. Hybrids are unpopular in the harsh conditions of the southern valley because they do not grow well and when replanted the second generation shows a considerable reduction in yield. But, as happens elsewhere in Mexico, cultivators experiment with new seed when it appears in markets or neighbors' fields (see also Louette 1997). Although the Tehuacán region was one of the locales where GM corn was found,[4] none of the valley producers from San José Miahuatlán with whom I spoke in 2001–2006 had ever heard of *maíz transgénico* (transgenic corn). When small-scale producers from the southern valley see a new kind of corn, they borrow some from their neighbors or buy some at the market and plant it to see how it fares. Thus, when experimenting, cultivators could unknowingly plant GM corn, and the exchange of genetic information between transgenic varieties and landraces may take place even in those regions relatively inhospitable to hybrids, where cultivators prefer *criollos*. GM varieties can be introduced into fields through seed purchases at the market, exchange between neighbors, the curiosity of producers to experiment with new seed, or by pollination between fields unknown to the cultivator.

In contrast to southern valley producers, various indigenous communities of the neighboring state of Oaxaca are politicized around the issue of GM corn and indigenous identity. As previously mentioned, a coalition of indigenous farmers and environmental groups from the Sierra Norte of Oaxaca petitioned the Commission on Environmental Cooperation of NAFTA to investigate the effects of imported GM corn on their communities (Greenpeace 2002). In the campaign against GM corn, indigenous participants often represent themselves as the "people of maize" whose livelihoods, traditions, and cultural beliefs are based on maize agriculture.[5] Indigenous communities and small-scale maize producers who are not politicized or organized around the issue of GM corn, like those of the southern Tehuacán Valley, discuss their livelihoods in terms of a different set of risks and challenges. In San José Miahuatlán, interviewees frequently mentioned the risks and challenges posed by the recent cuts to rural subsidies, the lack of employment opportunities, crop pests, local agricultural labor shortages, and environmental problems such as declining levels of spring water for irrigation. Let us turn now to look more closely at the case of corn production in the southern Tehuacán Valley.

The Tehuacán Valley

The semi-arid valley of Tehuacán is located in south-central Mexico in the state of Puebla. The valley descends from north to southeast continuing toward Teotitlán, Oaxaca. Indigenous peoples from the valley and surrounding sierras look to the rapidly expanding city of Tehuacán, located at the valley's midpoint, for employment in water bottling plants, the poultry industry, and blue jeans factories or *maquiladoras*. The area has a mixed heritage of Popoloca, Mixteca, Chocho, and Mazateca peoples, although Nahuatl became the common language of the valley through migration and Aztec domination prior to the Spanish conquest (Aguirre Beltrán 1986).

In the semi-arid, rural environs of the city, an irrigation system of water springs and underground tunnels and chain wells (*galerías filtrantes*) is central to agricultural production (Enge and Whiteford 1989; Henao 1980). Corn and beans are the most widely grown crops in the region. As in many other areas of Mexico, rain-fed white corn is largely grown for human consumption. In the valley, however, irrigated white corn is also grown for sale as corn on the cob, called *elote*. Other commercial crops include alfalfa, tomato, squash, garlic, melon, flowers, and sugarcane. Besides agriculture, significant rural activities include goat herding, the production of construction materials (bricks, cinder blocks), and artisanal products such as baskets and embroidered fabric for clothing.

The municipality of San José Miahuatlán, with a population of 11,675, is located in the southern end of the valley in Puebla. The head town where I focused my interviews makes up just over half of this population (INEGI 2000). Landholdings tend to be small, two to five hectares, and are private, communal, or land grants (*ejidos*). In comparison to the neighboring sierra, San José is well off, with a major road, electricity, and potable water, and is classified as an indigenous area of "less marginalization." However, San José is one of the poorest municipalities of the valley and has suffered spring water scarcity.

Flexible Agriculture

The household strategy that combines labor out-migration with corn production manifests certain specificities in the southern valley, where maize is cultivated both for grain consumption and for sale as *elote*. While many residents without irrigation water cultivate rain-fed subsistence corn, those with access to water resources try to sell their *elote* through intermediaries to regional markets and Mexico City. The valley has an advantage cultivating *elote* during the fall-winter cycle when other regions suffer from frosts. In the town of San José, *elote* became a more important part of agriculture during the 1980s. Local residents took up *elote* production and became U.S.-bound migrant laborers later than their valley neighbors. Although residents and Ministry of Agriculture officials report that agricultural production overall has been on the decline since the 1980s, it was precisely at that time that *elote* production expanded. In a town faced with both diminishing irrigation water and rising inflation, it is counterintuitive that the irrigation-dependent *elote* would become popular. Irrigated *elote* is more costly to cultivate than rain-fed corn because of higher labor requirements—which translate into the hiring of extra-household workers—and the purchase of expensive inputs like irrigation water. However, this increase is explained by the fact that *elote* not only commands a much higher price than grain, but can be dried and consumed as grain. Although the price for *elote* can vary greatly and cultivators do not always make a profit, *elote* cultivation enables residents to use their corn even when the crop's market price is not profitable. If the market price is too low the *elote* is left to dry for household consumption. For those who feel it is worth the risk or can afford it—those who have the water resources, income, or family members to provide the *milpa* labor—*elote* can generate a modest profit if harvested at the right time of year. Moreover, if prices decline, farmers leave the *elote* to dry in the field, reducing the household's need to buy corn grain for tortillas.

In San José, farmers plant a "six-month" *criollo* (a landrace or creolized variety) of irrigated maize that has a narrow kernel preferred for *elote*, called *chicuase*. After five months they may harvest the crop for *elote*, but if the price is not high or there are no interested buyers, the crop is left to grow for another month (including another round of irrigation), and then dried for grain to be sold in small amounts locally or consumed in the household. The other main corn is a "four-month," rain-fed variety (*nahuitzi*) that can be harvested at three and a half months for local *elote* or grown the full four months to be consumed as grain. *Nahuitzi* is not suitable for sale to Mexico City because it has wide kernels. Another variety used in San José for household consumption is the wide and sweet *macuiltzi*. It can be eaten as *elote* at four and a half months or dried for grain. The decision to harvest the crop as *elote* or grain thus largely depends on the market price for *elote* and whether there are interested buyers or intermediaries.

The overall decline in agriculture coupled by the switch to *elote* production was related to local and regional problems such as soil fatigue, crop pests, population pressures, and a decline in irrigation water. There were also other external problems to contend with, such as the rising input costs of insecticides, fertilizers, yoke or tractor rentals, and cuts in rural subsidies. The majority of farmers interviewed described an increased difficulty in affording agricultural inputs faced with continuing economic crisis and the decline in irrigation water. While state reforms have introduced new agricultural subsidies in order to cushion the abrupt transition to free trade, *Sanjosepeños* argue that little help can be expected from the government. For instance, Procampo grants, a transitional rural subsidy designed to buffer the impact of liberalized trade, tend to be insufficient to make up for the overall effect of inflation and the cuts to price supports and other rural subsidies (see also Myhre 1998; Nadal 2000). The switch to *elote* cultivation was also due in part to the construction of an improved road in the mid-1970s, which connected San José to San Sebastián Zinacatepec, a larger *municipio* of *elote* producers and buyers. *Sanjosepeño* farmers sell to intermediaries—both transporters and wholesalers—who come from nearby towns to take the fresh corn to regional markets and to Mexico City, in particular to the *Central de Abasto* in Iztapalapa (Olivares Muñoz 1995, 39).

While the preference for *elote* cultivation and U.S.-bound migration clearly developed as common household strategies in the 1980s, the decline in agriculture is more difficult to accurately gauge. There are entirely non-agricultural households in San José that are dependent on the purchase of grain for making tortillas. Conversely, there are also former *Sanjosepeños* who commute from the regional city of Tehuacán to San José, now just an hour and a half away by bus, to grow *elote* there. As in other areas of Mexico, some urban residents regularly return to their home towns to cultivate corn on weekends. But a nascent pattern of land use tells an unmistakable story. As more U.S. dollar remittances make their way to San José, the town's inhabited area is expanding and the agricultural area is shrinking. While *elote* makes up a larger share of the overall agricultural production in recent years, the parcels of land on the edges of town, which served as cornfields just seven years ago and were passed within families to their younger members, are now occupied by migrants' cinderblock houses. Let us turn now to look more closely at U.S.-bound migration and its effect on agricultural production.

Flexible Labor

Crisis and neoliberal reform has contributed to an increase in out-migration throughout rural Mexico (Cornelius and Myhre 1998; de Grammont 1996; Delgado Wise 2004). Paradoxically, however, transnational migration constitutes part of a local strategy to remain on the land. Formerly a town of few U.S.-bound migrants, San José is now a community reliant on remittances. Almost two-thirds of my sixty interviewees (2001–2002) either had household members in the United States at the time of the interview or had themselves spent time working in the United States within the last five years. Half of this number had left for the first time after the economic crisis of 1994, and the majority of the remaining interviewees had left previous to this, during the 1980s or early 1990s.

Most households combine both subsistence and commercial agriculture with income from several different sources such as construction, agricultural day labor, transnational (and less frequently national) migration, running a bakery, small store, or corn mill out of one's house, piecework embroidery, *maquiladora* work, and selling goods at the *tianguis* or local market. All but the first three activities are primarily women's work in San José. Although residents have labored in regional agriculture and industry throughout the twentieth century, in recent years, female residents comprise a nascent labor force for valley *maquiladoras*. Recent labor migration is largely undocumented and transnational in the sense that it "is not simply a unidirectional and one-time change in residence from one country to another" (Pries 1999, 3). Rather, at this stage of the process, work activities as well as personal and financial commitments move in both directions across the U.S.-Mexico border (Basch et al. 1995; Cohen 2001; Smith 1998).

In San José, households are transnational, maintained not just beyond the borders of the town but now also the nation-state. Many of the younger migrants (in their teens and twenties) send what they earn beyond their U.S. living expenses to their parents, just as *maquiladora* workers who live with their parents hand over their earnings to the household head, while keeping some spending money for clothes and small items. This pattern changes when a migrant ages and decides to build and establish his or her own house in town, or when a young woman marries or becomes a mother, with a much smaller amount, if any, given to their parents. Most migrants tend to come home once every two or three years for a period of four months or so. Some work for a few years in the United States and then return home for an equal amount of time hoping to repeat the cycle. And there is, of course, a smaller number who never return.

Although undocumented migration can be a dangerous and difficult experience for *Sanjosepeños*, for many it is also a source of pride, a means to start their own household or to contribute to that of their parents. Young migrants remit savings to San José in order to subsidize the corn agriculture of their parents, cover the cost of household maintenance, or build a house of their own. At the same time, the wage labor experience in valley factories and in the United States engenders, or in some cases, strengthens their aversion to agriculture. When back in San José, young migrants tend to work in the valley industries while older migrants work the fields.

Remittances sent to Mexico from the United States reached 20 billion dollars in 2005 (Bank of Mexico 2006). In contrast to an earlier literature that took a critical approach to the effects of rural migration in Mexico as a kind of "addiction" to the U.S. labor market, newer research has painted a more positive picture of the impact of migradollars on the rural and even the national economy, while overlooking the social impact of such migration (e.g., Durand, Parrado, and Massey 1996). Leigh Binford (2003) has argued that this more recent work on migration conflates consumption with investment and makes three related questionable assumptions: that remittance "investment" generates long-term local employment; that there is no saturation point for local investments; and that all economic strata are participating or benefiting equally in migration. In San José, some households are fortunate enough to use wages and remittances for the purchase of items beyond mere social reproduction and consumption, that is, to invest in machinery or a truck. Although the purchase of a truck may create work for several people in town if, say, a construction business is opened, how many such businesses can open in a town of 7,000 before a market saturation point is reached?

Some of the households interviewed have more resources than others: they reported having either access to irrigation water, five hectares of land or more, and/or some piece of income-generating equipment such as a tractor, a truck, or an electric corn grinder (*molino*). Remittances have enabled some households to move up the social and economic ladders. On the other hand, less-well-off migrants reported not earning enough income in the United States to save any money; or spending their money on television sets, wedding parties, or constructing their house instead of "investing" in goods related to employment or agricultural production. Most of the households I interviewed, with or without "resources," as defined here, had migrant members. For the most part, I found that migration in and of itself did not guarantee movement up the social or economic ladder. There is no question, however, that over the past decade or two, remittances have become a fundamental part of household reproduction.

To summarize then, residents have adapted to economic crises and in some cases improved their standard of living through out-migration and corn production. But this increased reliance on migrant remittances does not, for the most part, contribute to the generation of local employment alternatives to migration, nor to the long-term reproduction of corn agriculture. Regardless of how successful or unsuccessful they interpreted their experience in the United States to be, *many migrant interviewees have returned to the United States for work.* The next section explores how migration affects local agricultural practice.

Migrants and the future of local corn

When corn producers in San José were asked whether their children know how to grow corn and select seed, the majority responded in the affirmative. When migrants in their twenties or younger were interviewed, however, they admitted to having limited knowledge about the details of corn production and what specific qualities to look for when selecting seed. One migrant explained that there is a difference between migrants around his age (thirty-six) who send money home to work the land and work the *milpa* themselves when in town, and a younger generation of migrants who do not know how to work the land nor want to return to it. As a successful migrant who built a house and bought four hectares, access to irrigation water, and a tractor with his income, he articulated the benefits of migration in helping his family and the town. But he also expressed doubts about the town's future: "What's going to happen to this town if no one wants to work the land?" His perception of a generational difference among migrants was confirmed by my interviews.

Migrants in their teens and twenties have less experience and less interest in the fields than migrants and residents in their thirties or older. As boys these young migrants may have helped their fathers' farm, but by fifteen or so they are working full-time in valley factories or starting the journey north. Previous generations of migrants—the *bracero* migrants of the 1950s and 1960s, or those migrants who began to work abroad in the late 1980s or early 1990s—also left as young adults but tended to be older by a few years and with more agricultural experience under their belt. Moreover, prior to the mid-1990s migrant labor in the United States *was itself often agricultural,* unlike the work in fisheries and restaurants of most young migrants today. Most are non-agricultural workers despite the fact that since 2004, around 150 *Sanjosepeños* have been recruited annually to work on a contract basis in Californian tomato cultivation. Today's migrants are also returning home less frequently than those who migrated in earlier years because of the expense and heightened border security.

Residents prefer tortillas made from regional, white *criollo* corn, but with the decline in agriculture, where will the maize come from? Both men and women select seed for cultivation, and women are sometimes responsible for the off-field decisions about corn cultivation in the absence of their male relatives, such as hiring laborers. The labor in the cornfield, however, is still considered "men's work." Yet many young male interviewees prefer to migrate north and are accustomed to an hourly wage or at least to thinking in its terms. They describe agriculture as arduous and unreliable. As a result there has been an aging of maize producers. With younger men in the United States or uninterested in agriculture, it is largely older migrants and residents who are responsible for the household's unpaid corn cultivation. Additionally, some agricultural work in San José is undergoing monetization. When sons and husbands are abroad, many households pay day laborers in cash for *elote* production and *sometimes even subsistence corn production*. During the twentieth century, subsistence corn was largely performed by the unpaid male labor of the household or through sharecropping arrangements. In recent years, however, those households without enough available labor rely on paid day workers for this task. Furthermore, returned migrants hired to work the cornfield expect to be paid regular wages rather than through sharecropping arrangements. In this way, remittances and wage labor experience in the United States are contributing to the monetization of corn production (Fitting 2004).

In an interview with two migrant brothers age seventeen and twenty-three, who were back in San José after working most of the year in the U.S. fish packing industry and service sector, they reported wiring money to their parents both for savings and for their parents' use on food, clothes, and agriculture. The eldest had built his own one-level cinderblock house replete with furniture and a television. Most of the year he lives on the American boat where he works. When the brothers are in San José for a few months of the year they eat meals at their parents' house. Like many others, the eldest is in the transition to establishing his own household, a process that may take many years or may never be completed in San José. The younger brother was back in San José indefinitely after a few years up north working at Burger King. They both left school around the age of fourteen to work full-time. As kids they helped with goat herding, worked in *maquiladoras*, and helped their father in the fields. Neither of them is able to describe the details of corn cultivation, such as the timing of irrigations in the crop cycle, landraces best suited for different soils, and so on. Like other migrants their age, they are fluent in *Nahuatl* but hard pressed to come up with the basic terms that cultivators use to differentiate types of maize. They explain agriculture has no future, since "you can't make any money in the *milpa!*"

This preference for non-agricultural work could, of course, turn out to be a generational phase in the household life-cycle, since rural households and subsistence corn production provide a safety net for those who are unemployed, ill, injured, or elderly. Although young men (and women) may learn the details of maize agriculture as they age, they tend to prefer non-agricultural, regional employment when back in San José, or prefer to return to the United States for work because they say this offers a steadier income and a faster return than waiting to sell a harvested crop. They are converting some cornfields into residential neighborhoods. Thus, although transnational labor migration constitutes part of a local strategy to maintain the corn-producing household in San José, current trends—such as young migrants' lack of interest in agriculture and the increasingly difficult economic and environmental conditions faced by corn producers—spell the long-term erosion of corn agriculture and the related displacement of in situ landrace diversity and abundance.

Conclusion

The development of new corn varieties throughout the world relies on the genetic information from landraces found in Mexico and elsewhere. Industrialized agriculture is thus dependent upon seed stored ex situ in germplasm banks and in situ, maintained and developed in peasants' cornfields. Without landraces such agriculture would be unable to adapt to new conditions or pests (Altieri 2003; Fowler and Mooney 1990). This chapter has argued that the neoliberal corn regime has exacerbated the problems faced by maize producers and in situ diversity in several ways. Despite the restrictions on cultivating or testing GM corn in Mexico and recent steps to strengthen GM regulation, the rise in corn imports has introduced transgenic corn into the Tehuacán region among other areas of the country. Additionally, although the rural strategy to adapt to crisis and reform by increasing maize production seems to suggest that the future will bring an abundance of local maize, in places like the southern Tehuacán Valley, overall agricultural production is on the decline due to problems such as increased input costs, inflation, water crisis, population pressures, and agricultural pests. While it remains to be seen whether migrants will take up maize production later in life, the conditions for them to do so are increasingly difficult. This chapter has thus argued that the fate of Mexican corn biodiversity not only depends upon clear and enforceable biosafety regulations but also upon policies that support and enhance sustainable agriculture and small-scale Mexican maize cultivators.

Notes

This chapter is a revised version of an article published in *Agriculture and Human Values* (2006) 23: 15–26. I would like to thank David Barkin, María Mercedes Gómez, Regino Melchor Jiménez Escamilla, Ricardo F. Macip, Gerardo Otero, and Antonio Serratos for their comments during different stages of this project, and my hosts in San José Miahuatlán for their hospitality. All errors, of course, are my own. I also thank the Wenner-Gren Foundation for generously funding my dissertation fieldwork.

1. I use *globalism* here to correspond to the preferred term used in this book and to signal the ideological aspect of the process of globalization. See my brief discussion on the next page and that found in Gerardo Otero's Introduction.

2. These results were presented at two conferences in 2001 and early 2002; at the Organization for Economic Cooperation and Development (OECD) conference in the United States and the In Defense of Maize Forum in Mexico City (Ezcurra et al. 2002; INE-CONABIO 2002). Two research institutions based in Mexico also conducted their own studies. After testing seed at their gene bank, the International Maize and Wheat Improvement Centre (CIMMYT) did not find evidence of transgenes in their collection, while the National Institute of Research on Forestry, Agriculture and Animal Husbandry (INIFAP) found low proportions of transgenes among the sites it tested in Oaxaca (CIMMYT 2001; Turrent and Serratos 2004, 39). In late 2003, a group of NGOs presented their own findings which suggested that the presence of GM corn was not confined to the states of Oaxaca and Puebla, but was much more widespread (CECCAM et al. 2003).

3. The trilateral "Documentation Requirements for Living Modified Organisms for Food or Feed or for Processing," signed by Victor Villalobos, Coordinator of International Affairs of the Ministry of Agriculture (SAGARPA), is available online at http://www.cibiogem.gob.mx/normatividad/Documento%20Trilateral/acuerdo.htm.

4. The INE-CONABIO study identifies the Tehuacán Valley as one of five locations where corn samples showed higher frequencies of transgenic introgression (between 10 and 35 percent) (Ezcurra, Ortiz, and Soberón 2002, 280). Upon closer inspection of the list of these sampled sites, it appears that the localities identified as the Tehuacán Valley are outside the valley proper but located within the larger Tehuacán-Cuicatlán region.

5. For instance, at the Second In Defense of Maize Forum in 2003, a Zapotec participant from Oaxaca, Aldo González Rojas, explained: "Maize is not a business but food for survival, our sustenance and happiness. When we plant it we bless it to ask for a good harvest for all. But we have recently found that native

maize varieties had been contaminated with transgenic seeds. This means that what our indigenous peoples took thousands of years to develop will be destroyed in no time at all by companies that trade in life" (quoted in Ribeiro 2004, 1). In contrast, none of my interviewees in the Tehuacán Valley framed maize in such terms.

References

Ackerman, F., T. Wise, K. Gallagher, L. Ney, and R. Flores. 2003. "Free Trade, Corn, and the Environment: Environmental Impacts of U.S.-Mexico Corn Trade under NAFTA." Global Development and Environment Institute, Working Paper 03–06. Tufts University, Medford, Massachusetts.

Aguirre Beltrán, G. 1986. *Zongolica. Encuentro de dioses y santos patrones.* México: Fondo de cultura económica.

Aguirre Gómez, J. A., M. Bellon, and M. Smale. 1998. *A Regional Analysis of Maize Biological Diversity in Southeastern Guanajuato, Mexico.* Economics Working Paper 98–06. CIMMYT.

Altieri, M. 2003. "Socio-Cultural Aspects of Native Maize Diversity," in *Maize and Biodiversity: The Effects of Transgenic Maize in Mexico, for the Secretariat of the Commission for Environmental Cooperation of North America.* www.cec.org/maize/ (accessed 3 June 2004).

Alvarez-Morales, A. 1999. "Mexico: Ensuring Environmental Safety While Benefiting from Biotechnology," in *Agricultural Biotechnology and the Poor.* CGIAR/NAS.

Appendini, K. 1992. *De la milpa a los Tortibonos: La reestructuración de la política alimentaría en México.* México: El Colegio de México/UNRISD.

———. 1994. "Transforming Food Policy for over a Decade: The Balance for Mexican Corn Farmers in 1993," in *Economic Restructuring and Rural Subsistence in Mexico*, ed. C. Hewitt de Alcántara. San Diego: UNRISD/Center for U.S.-Mexican Studies.

Aquino, P. 1998. "Mexico," in *Maize Seed Industries in Developing Countries*, ed. M. Morris. Boulder, Colo.: Lynne Rienner Publishers.

Austin, J., and G. Esteva, eds. 1987. *Food Policy in Mexico: The Search for Self-Sufficiency.* Ithaca: Cornell University Press.

Bank of Mexico. 2006. "Remittances Reach $20 Billion in 2005," in *El Universal.* 1 February 2006. http://www.eluniversal.com.mx/miami/16816.html.

Barkin, D. 2002. "The Reconstruction of a Modern Mexican Peasantry." *Journal of Peasant Studies* 30(1): 73–90.

———. 2003. "El maíz y la economía." pp. 155–176 in G. Esteva and C. Marielle, eds., *Sin Maíz, No Hay País.* México, DF: CONACULTA, Museo Nacional de Culturas Populares.

Bartra, A. 2004. "Rebellious Cornfields: Toward Food and Labour Self-Sufficiency." pp. 18–36 in *Mexico in Transition: Neoliberal Globalism, the State, and Civil Society*, ed. G. Otero. London: Zed Books.

Bartra, A., et al. 2005. *Transgénicos, ¿quién los necesita?* DF: Centro de Producción Editorial, Grupo Parlamentario del PRD, Cámara de Diputados, Congreso de la Unión, LIX Legislatura.

Basch, L., C. Szanton, and N. G. S. Schiller. 1995. *Nations Unbound: Transnational Projects, Postcolonial Predicaments, and Deterritorialized Nation-States.* New York: Gordon and Breach.

Beck, U. 1986. *Risk Society: Towards a New Modernity.* London: Sage Publications.

Binford, L. 2003. "Migrant Remittances and (Under) Development in Mexico." *Critique of Anthropology* 3(3): 305–336.

CEC. 2004. *Maize and Biodiversity. The Effects of Transgenic Maize in Mexico. Conclusions and Recommendations.* Montreal: Commission for Environmental Cooperation, November 2004. Available online: http://www.cec.org/maiz.

CECCAM et al. 2003. "Contaminación transgénica del maíz en México: Mucho más grave." Press release, Mexico City, 9 October.

Chapela, I., and D. Quist. 2001. "Transgenic DNA Introgressed into Traditional Maize Landraces in Oaxaca, Mexico." *Nature* (414): 541–543.

CIMMYT. 2001. "Initial Tests Find Mexican Landraces in CIMMYT Gene Bank Free of Promoter Associated with Transgenes." http://www.cimmyt.org/whatiscimmyt/init_test.htm.

Cleveland, D., and D. Soleri. 2005. "Rethinking Risk Management Process for Genetically Engineered Crop Varieties in Small-scale, Traditionally Based Agriculture." *Ecology and Society* 10(1): article 9.

Cohen, J. 2001. "Transnational Migration in Rural Oaxaca, Mexico: Dependency, Development and the Household." *American Anthropologist* 103(4): 954–967.

Cornelius, W., and D. Myhre, eds. 1998. *The Transformation of Rural Mexico.* San Diego: Center for U.S.-Mexican Studies.

de Grammont, H. C., ed. 1996. *Neoliberalismo y organización social en el campo mexicano.* México: Plaza y Valdés, UNAM.

de Janvry, A., G. Gordillo, and E. Sadoulet. 1997. *Mexico's Second Agrarian Reform.* San Diego: Center for U.S.-Mexican Studies.

de Teresa Ochoa, A. (1996) "Una radiografía del minifundismo: Población y trabajo en los valles centrales de Oaxaca (1930–1990)," in *La sociedad rural mexicana frente al nuevo milenio,* ed. H. C. de Grammont and H. Tejera. México: UAM-INAH-UNAM Plaza y Valdés.

Delgado Wise, R. 2004. "Labour and Migration Policies under Vicente Fox: Subordination to U.S. Economic and Geopolitical Interests." pp. 138–153 in G. Otero, ed., *Mexico in Transition: Neoliberal Globalism, the State, and Civil Society.* London: Zed Books.

Durand, J., E. Parrado, and D. Massey. 1996. "Migradollars and Development: A Reconsideration of the Mexican Case." *International Migration Review* 30(2): 423–444.

Dyer, G., and A. Yúnez-Naude. 2003. "NAFTA and Conservation of Maize Diversity in Mexico," prepared for the Commission for Environmental Cooperation (CEC).

Enge, K., and S. Whiteford. 1989. *The Keepers of Water and Earth. Mexican Rural Social Organization and Irrigation.* Austin: University of Texas Press.

Esteva, G., and C. Marielle, eds. 2003. *Sin Maíz, No Hay País*. Mexico: CONACULTA and Museo de Culturas Populares.

Ezcurra, E., S. Ortiz, and J. Soberón. 2002. "Evidence of Gene Flow from Transgenic Maize to Local Varieties in Mexico," pp. 277–283 in *LMOs and the Environment: Proceedings of an International Conference*. OECD. Raleigh-Durham, N.C., 27–30 November 2001.

Fitting, E. 2004. "'No hay dinero en la Milpa': El maíz y el hogar transnacional del sur del Valle de Tehuacán." pp. 61–102 in L. Binford, ed., *La economía política de la migración internacional en Puebla y Veracruz*. Puebla: Benemérita Universidad Autónoma de Puebla.

———. 2006. "The Political Uses of Culture: Maize Production and the GM Corn Debates in Mexico." *Focaal: The European Journal of Anthropology* 48.

Fowler, C., and P. Mooney. 1990. *Shattering: Food, Politics, and the Loss of Genetic Diversity*. Tucson: University of Arizona Press.

García Barrios, R., and L. García Barrios. 1994. "The Remnants of Community: Migration, Corn Supply and Social Transformation in the Mixteca Alta of Oaxaca," in C. Hewitt de Alcántara, ed., *Economic Restructuring and Rural Subsistence in Mexico: Corn and the Crisis of the 1980s*. San Diego: Center for U.S.-Mexico Studies/UNRISD.

GEA et al. [Grupo de Estudios Ambientales]. 2006. "Científicos y organizaciones no gubernamentales de México presentan el Manifiesto por la protección de maíz mexicano" (Scientists and NGOs Present the Manifesto for the Protection of Mexican Maize). Press Release, 25 June. Available at www.gea-ac.org.

Glowka, L., F. Burhenne-Guilmin, and H. Synge. 1994. *A Guide to the Convention on Biological Diversity*. Gland, Switzerland: IUCN.

Greenpeace Mexico. n.d. *Expedientes ambientales: Documentos de campaña, maíz transgénico*. México.

———. 2001. *Mexican Environment and Farming Groups Launch Formal Complaint Process against GE Corn Imports*. 11 December Press Release. Accessed 7 September 2003. http://archive.greenpeace.org/pressreleases/geneng/2001dec11.html.

———. 2002. "Mexican Groups Appeal to NAFTA Environmental Commission to Force Action against Genetic Contamination," 23 April Press Release. Accessed 7 September 2003. http://archive.greenpeace.org/pressreleases/geneng/2002apr23.html.

Heller, C. 2002. "From Scientific Risk to *Paysan* Savoir-faire: Peasant Expertise in the French and Global Debate over GM Crops." *Science as Culture* 11(1): 5–37.

Henao, L. E. 1980. *Tehuacán: Campesinado e irrigación*. Mexico: Editorial Edicol.

Herrick, C. 2005. "'Cultures of GM': Discourses of Risk and Labeling of GMOs in the UK and EU." *Area* 37(3): 286–294.

Hewitt de Alcántara, C., ed. 1994. "Introduction" in *Economic Restructuring and Rural Subsistence in Mexico: Corn and the Crisis of the 1980s*. San Diego: Center for U.S.-Mexico Studies/UNRISD.

INE-CONABIO. 2002. "Evidencias de flujo genético desde fuentes de maíz transgénico hacia variedades criollas." Presented by E. Huerta at the En Defensa Del Maíz conference, 23 January 2002, Mexico City.

INEGI. 2000. *Anuario estadístico del estado de Puebla.*

Louette, D. 1997. "Seed Exchange among Farmers and Gene Flow among Maize Varieties in Traditional Agricultural Systems," in *Gene Flow among Maize Landraces, Improved Maize Varieties, and Teosinte.* Eds. A. Serratos, M. Willcox, and F. Castillo-González. Mexico: CIMMYT.

McAfee, K. 2003. "Corn Culture and Dangerous DNA: Real and Imagined Consequences of Maize Transgene Flow in Oaxaca." *Journal of Latin American Geography* 2: 18–42.

MacNeish, R. 1972. "Summary of the Cultural Sequence and Its Implications in the Tehuacan Valley," in *The Prehistory of the Tehuacan Valley.* Vol. 5. Ed. R. MacNeish. Austin: University of Texas Press.

Matsuoka, Y., Y. Vigouroux, M. M. Goodman, G. J. Sanchez, E. Buckler, and J. Doebley. 2002. "A Single Domestication for Maize Shown by Multilocus Microsatellite Genotyping." *Proceedings of the National Academy of Sciences.* USA 99: 6080–6084.

Myhre, D. 1998. "The Achilles' Heel of the Reforms: The Rural Finance System," in *The Transformation of Rural Mexico.* Eds. W. Cornelius and D. Myhre. San Diego: Center for U.S.-Mexican Studies.

Nadal, A. 2000. *The Environmental and Social Impacts of Economic Liberalization on Corn Production in Mexico.* Oxford: WWF/Oxfam.

———. 2005. Presenter at the Forum on "Monsanto's Law," Casa Lamm, Mexico City, 2 May 2005.

Olivares Muñoz, F. 1995. "Estudio de mercado: Producción y comercialización del maíz elotero como hortaliza en la región de Tehuacán, Puebla." M.A. thesis, Dept. de Economía Agrícola, Universidad Autónoma de Chapingo, Mexico.

Ortega-Paczka, R. 1999. "Genetic Erosion in Mexico." Pp. 69–75 in *Proceedings of the Technological Meeting of the FAO World Information and Early Warning System on Plant Genetic Resources.* Prague: Research Institute of Crop Production.

Otero, G. 1999. *Farewell to the Peasantry? Political Class Formation in Rural Mexico.* Boulder, Colo.: Westview Press.

Otero, Gerardo, ed. 2004. *Mexico in Transition: Neoliberal Globalism, the State and Civil Society.* London: Zed Books and Nova Scotia: Fernwood Publishing.

Otero, G., S. Scott, and C. Gilbreth. 1997. "New Technologies, Neoliberalism, and Social Polarization in Mexico's Agriculture." pp. 253–270 in J. Davis, T. Hirschl, and M. Stack, eds., *Cutting Edge: Technology, Information Capitalism and Social Revolution.* N.Y.: Verso.

Pilcher, J. 1998 *¡Que vivan los tamales! Food and the Making of Mexican Identity.* Albuquerque: University of New Mexico Press.

Pries, L. 1999. "New Migration in Transnational Spaces," in *Migration and Transnational Social Spaces.* Ed. L. Pries. Aldershot: Ashgate Publishers.

Ribeiro, S. 2004. The Day the Sun Dies: Contamination and Resistance in Mexico. *Seedling.* July: 5–10. http://www.grain.org/seedling_files/seed-04-07-02.pdf.

SENASICA (Servicio Nacional de Sanidad, Inocuidad y Calidad Agroalimentaria). 2005. Ensayos de productos genéticamente modificados autorizados en México de 1988 al 11 de octubre de 2005. SAGARPA. http://web2.senasica.sagarpa .gob.mx/xportal/inocd/trser/Doc2060/ensayos_OGM_1988_2005.pdf.

Serratos, A. 1996. "Evaluation of Novel Crop Varieties in Their Center of Origin and Diversity: The Case of Maize in Mexico," in *Turning Priorities into Feasible Programs: Proceedings of a Policy Seminar on Agricultural Biotechnology for Latin America.* Ed. J. Komen, C. Falconi, and H. Hernández. The Hague/Mexico: Intermediary Biotechnology Service.

Serratos, J. A., M. Willcox, and F. Castillo-González, eds. 1996. *Flujo genético entre maíz criollo, maíz mejorado y teocintle: Implicaciones para el maíz transgénico.* Mexico: CIMMYT.

Smith, R. C. 1998. "Transnational Localities: Community, Technology and the Politics of Membership within the Context of Mexico-U.S. Migration." pp. 196–238 in *Transnationalism from Below*, special edition of *Comparative Urban and Community Research*. Vol. 6, New Brunswick, N.J.: Transaction Publishers.

Soleri, D., D. Cleveland, F. Aragón, M. Fuentes, H. Ríos, and S. Sweeney. 2005. "Understanding the Potential Impact of Transgenic Crops in Traditional Agriculture: Maize Farmers' Perspectives in Cuba, Guatemala and Mexico." *Environmental Biosafety Research* 4(3): 141–166.

Turrent, A. 2005. "La diversidad genética del maíz y del teocintle de México debe ser protegida contra la contaminación irreversible del maíz transgénico." pp. 51–60 in *Transgénicos: ¿Quién los necesita?* DF: Centro de Producción Editorial.

Turrent, A., and J. A. Serratos. 2004. "Chapter 1: Context and Background on Maize and Its Wild Relatives in Mexico." *Maize and Biodiversity: The Effects of Transgenic Maize in Mexico.* Commissioned for the Secretariat of the Environmental Commission on Environmental Cooperation. http://www.cec.org/maize/resources/chapters.cfm?varlan=english.

Turrent, A., R. Aveldaño Salazar, and R. Moreno Dahme. 1996. "Análisis de las posibilidades técnicas de la autosuficiencia sostenible de maíz en México." *TERRA* 14(4): 445–468.

USDA. 2006. "Adoption of GE Crops in the United States: Corn Varieties." National Agricultural Statistics Service (NASS).

Warman, A. 1988. *La historia de un bastardo: Maíz y capitalismo.* México: Instituto de Investigaciones Sociales, UNAM/Fondo de Cultura Económica.

CHAPTER 7

Political Economy of Agricultural Biotechnology in North America: The Case of rBST in La Laguna, Mexico

GERARDO OTERO, MANUEL POITRAS,
AND GABRIELA PECHLANER

At the turn of the twenty-first century, the commercial use of genetically modified organisms (GMOs) in agriculture was into its second decade. Dramatic advances were made in the late 1980s in recombinant veterinary products for animal farming and in genetically modified (GM) or transgenic crops by the mid-1990s. Yet the future of these advances raises some uncertainties, accompanied as they are by numerous legislative hurdles, consumer hesitation or outright rejection, active resistance by environmental and consumer groups, and even financial troubles facing biotech companies. In addition to these social and economic actors—the state, nongovernmental organizations (NGOs), consumers, biotech firms, and finance capital—there are the direct users of the technology, farmers themselves, who have a significant role to play in determining the future of agricultural biotechnology.

This chapter investigates state policies and farmers' adoption patterns with respect to the controversial recombinant bovine somatotropin (rBST), a milk productivity-enhancing hormone for dairy cows. After a brief discussion of the rBST technology itself, the first section provides a comparative analysis of its regulation around the world, focusing on debates in the North American Free Trade Agreement (NAFTA) region. The second section provides a comparative discussion of the profitability of rBST through the use of La Laguna fieldwork materials in combination with studies on the U.S. experience. La Laguna, the most important dairy-producing region in Mexico, encompasses fifteen municipalities of two bordering states in the north-central part of the country: Coahuila and Durango. Fieldwork with dairy farmers in this region was conducted separately by two of the authors at the end of 1999 and in the summers of 2000 and 2001. The third section presents fieldwork data on the qualitative assessments by La Laguna dairy farmers on rBST's yield and profitability, as well as Monsanto's marketing strategy in the region, in order to derive some insight into the motivations behind adoption. It should be noted that the sample size of our fieldwork material is too limited to achieve results that may be generalized. Rather, our focus has been to draw out the qualitative aspects of rBST use for further study. Lastly, an interpretation of these qualitative data is offered, as well as a research agenda based on the empirical questions that remain for a better understanding of the larger political economy of agricultural biotechnologies in the Americas.

Recombinant Bovine Somatotropin (rBST), also known as Bovine Growth Hormone (rBGH), is a synthetic version of a naturally occurring hormone. It is promoted as a yield-enhancing drug with the potential to considerably increase milk production in lactating cows (by 10 to 20 percent), although the actual percentage varies, as will be discussed. The case of rBST has particular significance, as the drug was one of the first commercial agro-biotechnologies. It is important to note, however, that although rBST is produced through recombinant DNA techniques (genetic engineering), the product itself is not genetically modified, nor does it contain any foreign DNA.[1] This means that most of the environmental and health problems usually associated with genetically modified organisms (GMOs) and transgenic crops, which contain genes foreign to their species, do not apply to rBST. This is true especially in terms of contamination of other organisms. As a productivity-enhancing drug, rBST offers the potential to reduce herd sizes for a given level of milk production. This feature not only benefits countries where increased milk production is needed, but promises to reduce the environmental pressures caused by dairy herds in any milk-producing country. Increased yields imply that fewer cows are needed for the production of the same amount of milk, reducing the amount of manure and decreasing the pressure on resources such as land, water, and feed. Feed efficiency is improved as the additional feed needed to produce the extra milk goes only toward that milk's production and not toward the maintenance needs of the cow (Fetrow 1999). Despite this promise, the technology is not necessarily problem-free, with respect to its health, environmental, or socioeconomic implications, as the controversy over its use has suggested.

Concerns over the use of rBST range from animal health concerns (such as increased mastitis, or udder infections) to human health concerns (such as concerns over consumption of the hormone itself[2] and of the antibiotics necessary to treat increased animal health incidents) and environmental concerns over the related increased intensity of dairy pro-duction. Social concerns are more specific to developing or developed countries. In countries that already have surplus-milk production, such as the United States and Canada, use of the drug raises concerns over declining prices, increased competition, and greater incidence of bank-ruptcies, particularly for smaller family farms. Social concerns particular to developing countries are related to issues of technological dependency, concentration, and inequality, and will be discussed in this chapter with respect to Mexico. It should be noted that many of these implications are not related directly to the genetic aspect of the technology, but rather to its being a productivity-enhancing technology that generally favors large-scale operations.

Regulation of rBST Adoption in North America and Beyond

RBST has been under review in the United States since the early 1980s. Since then, most countries have drawn regulations on the use of rBST, ei-ther allowing or banning its use. The first official approval of rBST hap-pened in 1988 in South Africa, followed by the present Czech and Slovak Republics in 1989 (then Czechoslovakia) and a number of other countries in 1990. Mexico was the sixth country to officially approve the use of rBST in May of 1991, though many accounts in La Laguna claim that it had been in use in that region since 1989. The United States was the twelfth country to approve its use in 1993 (see Table 7.1). Many countries have also banned or deferred judgment on the use of rBST, among which are the members of the European Union, Australia, New Zealand, and Japan. Canada banned the use of rBST in 1999, primarily in response to tests showing an increased occurrence of mastitis in milking cows treated with the drug.

   Except the United States, most countries that have banned the use of rBST on their territory are surplus-dairy-producing countries. This is the case with Australia, New Zealand, the European Union, and Canada. In contrast to these surplus-producing countries, those countries that approved rBST use (particularly early adopters) have dairy-production deficits. Mexico, for instance, one of the earliest adopters of the technol-ogy, is also one of the largest importers of milk in the world. Between 1990 and 1998, it imported anywhere from 8 percent to 32 percent of all milk domestically consumed annually, adding $250 million to its balance-of-payments deficit (SAGAR 1999). As a University of Wisconsin report

Table 7.1. First Countries to Approve the Use of Monsanto's rBST Product, by Date of Approval

	Country	Date of Approval
1	South Africa	November 1988
2	Czech/Slovak Republics	July 1989
3	Zimbabwe	February 1990
4	Namibia	February 1990
5	Brazil	March 1990
6	Mexico	May 1990
7	Bulgaria	July 1990
8	CIS	February 1991
9	Jamaica	April 1992
10	Costa Rica	October 1993
11	European Union*	January 1993
12	USA	November 1993
13	Puerto Rico	March 1994
14	Malaysia	October 1994

*In the European Union, a moratorium was imposed on the use of rBST soon after its official approval.
Source: Based on Krimsky and Wrubel 1996, p. 186.

indicates: "Prior to the NAFTA, Mexico's level of self-sufficiency in milk production was variously reported as being 50 percent–60 percent. USDEC estimates that Mexico's self-sufficiency levels will remain around 70–75 percent for the next few years" (Dobson and Proctor 2002, 14).

While there seems to be a clear correlation between surplus or deficit dairy production and rBST banning or approval, respectively, the United States offers a most important exception to this correlation. This exception seems to highlight different variables at work in governments' banning or approval. The relative economic power among stakeholders appears to be the chief factor. Most notably, it is highly significant that the main firms producing rBST are U.S.-based transnational corporations with powerful lobby groups.

As with other biotechnology products, it is difficult to ascertain the long-term effects of rBST on other biological organisms. Given this scientific uncertainty, there is much less constraint on the policy process, as documented by social and political science literatures (see for example Jasanoff 1986, 1997). This leads to highly controversial regulatory debates, in which contradictory health and safety arguments are used toward the enhancement of political and economic agendas rather than public safety interests. Such tactics are raised from the national to the international forums through free trade agreements, such as the World Trade Organization (WTO). Australia, for instance, has invoked the protection of its foreign export market for dairy products as the reason to decline the use of rBST on its dairy farms, incidentally challenging emerging free-trade rules preventing the use of domestic economic protection measures to limit the import of foreign technologies.

These regulatory debates often center on the opposing principles of "substantial equivalence" (used by regulatory agencies in the United States and promoted in trade bodies such as the WTO) versus the "precautionary principle" (used by countries wanting to bar biotech products and entrenched in the Cartagena Protocol on Biosafety under the UN Convention on Biological Diversity). According to the principle of substantial equivalence, if the product of biotechnology is substantially equivalent to the traditional one, it can be assumed to be safe and should not be forbidden. In contrast, the precautionary principle argues that if long-term effects cannot be ascertained, countries can limit the use and import of such products (see Jansen and Roquas, this volume).

The international power dynamics behind these debates are well in evidence. In 1999, the Codex Alimentarius, the food safety standards organization operating under the Joint FAO/WHO (Food and Agricultural Organization/World Health Organization) Food Standards Programme,[3] refused to certify rBGH as safe. Rather, it "effectively tabled the rBGH issue as a way of saving face for the United States, which would have lost a formal vote" (Ehrenfeld 2002, 4).

Within the United States itself, approval of rBST was the subject of a protracted struggle that took place from the mid-1980s until its final approval in 1993, with implementation legalized in 1994. Contrary to other rapid adopters of the technology, the United States had milk overproduction problems for decades, largely due to the wide array of productivity-enhancing technologies that agrochemical corporations already provided for the sector. In this context, a common perception was that another such technology was simply not needed. In fact, state policy had been concerned with controlling production downward, so as to reduce the number of bankrupting family dairy farmers. Paradoxically, while some branches of the U.S. government contributed funds toward the development of rBST, others were trying to control supply (Otero 1992).

It is not surprising, then, that public opposition to rBST in the United States was considerable from several fronts. Farmers' organizations, particularly in the Midwest and New England, where small- to medium-size family farms prevail, were at the forefront in the struggle against development and approval of rBST. Other organizations, ranging from animal-rights and environmentalist to anti-biotechnology groups, also joined to oppose rBST. In the late 1980s, the public image of rBST was so negative that large milk-processing firms such as Kraft and large supermarket chains declared that they would not use or distribute milk from cows treated with the hormone. This public outcry was so loud and clear that it seemed the technology would be shelved indefinitely. However, millions of dollars had been invested by four major pharmaceutical and chemical companies to develop rBST: Monsanto, Eli Lilly, Upjohn, and American Cyanamid. They were all racing to get the product approved and ready for the market, and were active on the issue of public perceptions. Monsanto, in particular, launched concerted publicity and advertising efforts to counter rBST's bad press (Kleinman and Kloppenburg 1991). Claims in favor of the drug for its environmental benefits were ultimately invoked by U.S. regulatory agencies. That rBST was eventually approved in the United States despite widespread objection is testament to the political clout of these major pharmaceutical or agrochemical companies. Such opposition, however, continues post-approval in the form of dairies that label milk for consumers as containing no artificial growth hormones. Monsanto's lawsuits (e.g., Oakhurst Dairy) against such labeling are a means of circumventing this public opposition.

The case of the United States thus shows that regulation in favor of or against allowing the use of rBST cannot be limited to its connection to deficit or surplus production. A number of additional factors come into play, such as the strength of the environmental, anti-biotech, consumer groups, on one hand, and the strength of industry lobbies in shaping the regulatory process itself, on the other. The power dynamic behind rBST adoption is not limited to the American dairy industry, however. In Canada, the approval process was placed under close public scrutiny when scientists assigned to the review of the hormone resigned in protest over pressure from higher-ranking officials to approve the hormone. This pressure was the alleged result of research funding offered by Monsanto to the regulatory agency (Bueckert 2001; Canada 1999). In Mexico, rBST was swiftly adopted during a meeting between Monsanto and state officials, after a cursory review process based only on company documentation that concluded in favor of approval in 1991 (Poitras 2000).

The political economy of rBST will not be developed further here; however, these issues need to be kept in mind as crucial context for investigations of adoption, dis-adoption, or non-adoption at the farm level.

Profitability and rBST Adoption Patterns by Dairy Farmers

An initial puzzle that motivated this study was that rBST enjoyed a greater adoption rate in La Laguna than in its country of origin. In fact, U.S. dairy farmers have been rather less than enthusiastic about adopting this productivity-enhancing hormone (Barham et al. 2000; Stephanides and Tauer 1999; Tauer and Knoblauch 1997). Furthermore, we had the impression that this technology might not even be profitable, given a number of problems that increase costs and reduce its economic benefits (Butler 1999). In fact, an initial field study by one of the authors revealed that the profitability of the use of the hormone in the La Laguna context is ambiguous at best, thus jeopardizing the usual rational-economic explanations of why these large capitalist dairy operations would favor its adoption. We use the term *capitalist* in this context to clearly differentiate this type of operation, which is capital-intensive and relies on the hiring of wage workers, from small peasant farmers and the collectively organized *ejidos* that arose out of the first years of agrarian reform in Mexico (see Otero 1999, Chaps. 3 and 5).

The La Laguna region is made up of fifteen municipalities from the states Coahuila and Durango. Despite the fact that the region borders the Chihuahuan Desert, La Laguna has become the main dairy-producing region of the country, with some of the most technologically advanced and vertically integrated agro-industrial operations in Mexico. Therefore, if the impression of a lack of profitability of rBST is accurate, then the next question becomes: why have the capitalist dairy farmers of La Laguna so readily adopted the technology? Our initial hypothesis was that, short of a straight economic explanation for adoption, we must resort to a cultural explanation. It could be that a macho culture prevails in La Laguna, so that each dairy farmer is trying to be ahead of the next, at least in terms of milk yields which are publicly available. An alternative hypothesis, also within the cultural realm but adding economic incentives, is that the La Laguna farmers are more susceptible to the marketing campaigns of the large transnational corporations that promote rBST in the region—for example, Monsanto and Eli Lily. This may be a case in which the farm operators of a developing country have become more aggressive technology adopters than U.S. farmers, with much greater faith in its beneficial and "progressive" impacts. Not adopting the new technology would make them look too conservative before their neighbors. Considering that the market is much less regulated than in either the United States or Canada, it is possible that a combination of these two hypotheses goes a long way toward explaining the wholesale adoption of rBST in La Laguna by over 80 percent of capitalist farmers (estimates from all interviewees in La Laguna, including a former Monsanto veterinarian and chief promoter of rBST in the region). This is in sharp contrast with a U.S. overall adoption rate of less than 20 percent.

Many observers from the agricultural-inputs industry as well as some agricultural economists predicted that there would be a fast adoption rate in the United States after its approval. By 1999, however, the adoption rate was already rather disappointing from the point of view of industry. Based on Monsanto's figures, rBST was used on 15–17 percent of the nation's dairy farms. Accounting for size of farms, and an average treatment of 50 percent of the herd, this amounts to only 17 percent of the nation's dairy cows being treated with the hormone (Barham and Foltz 2002, 17). This is well short of predictions that 50 percent of U.S. cows would be treated by the year 2000 (Barham et al. 2000). Table 7.2 below contains empirical figures on adoption rates in the state of Wisconsin, the second largest dairy-producing state after California.

According to most strands of economic theory, technological change under market conditions can usually be correlated to profitability: the entrepreneur or capitalist farmer is enticed by higher rates of profit that new technologies can confer to the enterprise through higher productivity. In the case of rBST, the relationship between profitability and adoption is not that clear. The following analysis drawn from La Laguna economic conditions shows that the relative profitability of using rBST in La Laguna is ambiguous at best (Poitras 2000).

Profitability calculations in such a scenario as La Laguna are limited in their explanatory potential as the gap between such calculated conclusions and what actually occurs on the farm is usually great. RBST offers particular challenges in this regard as numerous factors external to the drug itself affect the cost-benefit ratio. While extensive surveying of Latin American dairy farms about all the variables affecting costs and profits of rBST use would be necessary for an adequate empirical assessment of profitability, the difficulties of achieving this are already amply demonstrated by U.S. studies, some of which will be discussed presently. The limitations of available data with respect to rBST use in a developed

Table 7.2. Percent of Farms Using rBST in Wisconsin, USA, by Size of Milking Herd (%, 1999)

Size Categories	1995	1997	1999
1 to 49 Cows	2.2	3.3	5.3
50 to 99 Cows	10.4	13.9	15.3
100 to 199 Cows	20.8	30.1	34.9
200+ Cows	46.7	48.3	75.0
All Dairy Farmers	6.6	11.8	15.4

Source: Barham et al. 2000.

country such as the United States, however, are multiplied exponentially with respect to a developing country such as Mexico. In the face of such limitations, extrapolations can be made from more limited surveys. Therefore, based on available data and a limited set of interviews with dairy producers, a rough calculation of profitability of rBST use in La Laguna is provided below. The following calculation is intended as a heuristic device from which to guide our further assessments. As an agricultural economist writing on the profitability of rBST on U.S. dairy farms expressed it: "The economics of rBST can be as simple or as complicated as you want to make it" (Butler 1999). The point here is not to offer statistically significant evidence of the profitability of rBST, which would be much beyond the scope of this chapter, but rather to give an illustration of how profitability may be affected on La Laguna dairy farms.

The Profitability of rBST

RBST is applied every two weeks after lactation has started (calculated here with applications starting after the tenth week of lactation), with lactation lasting a total of approximately 305 days per year (but see Ott and Rendleman 2000 for a calculation with fewer lactation days). This results in about seventeen treatments of rBST, at a cost of MX$55 per treatment. Interview respondents who used rBST claimed that with the recommended application of the drug, which is injected, the yield of milk per cow increased on average between 2 and 3 liters per cow per day, thus well below the 3.5 to 5.5 liter/cow/day that Elanco, the rBST distributor, cited on its pamphlet (Poitras 2000). According to interview results as well as data on the cost structure of dairy farming in La Laguna from the

Table 7.3. Annual Marginal Profitability of the Use of the Regular Application Program of rBST on an Average La Laguna Dairy Farm (1999)

Additional Annual Revenues, per Cow			Additional Annual Costs, per Cow		
	2.5	Additional lt./cow/day	–	935	MX$55 per shot of rBST, applied every two weeks during treatment, averaging 17 treatments per lactation period.
X	3.20	MX$/lt (the regular price paid by LALA in 1999)	–	620	MX$ (for the extra feed given to the cow over 241 days, increased by an average of 7.5% according to interview data, calculated on annual average feed costs of $12,537 per cow on a specialized La Laguna dairy farm—FIRA, 1998) ($34 per day*241*7.5%)
X	241	Days of treatment (based on 305-day lactation with use starting in 10th week, equaling 242 days of treatment)			
=	1,928	MX$	=	–1,555	MX$

=	373	MX$ (profit per cow per year)

Fideicomisos Instituidos en Relación con la Agricultura (FIRA), a subsidiary of the Bank of Mexico, the use of the hormone leads to additional revenues from the increased milk yield of MX$1,928 per cow per year (FIRA 1998). The costs are increased by MX$1,555 per cow per year, thus yielding profits of MX$373 per cow per year (Table 7.3).

Using an exchange rate of 9.50 pesos per U.S. dollar, the profit is US$39.26, or about $40 U.S. per cow per year. In addition to the profit from the regular application of the hormone, many producers also use the hormone in ways that might have other marginal positive impacts on the profitability of production. For example, rBST can be used to help sick cows recover, or to increase productivity of cows in their last lactation or with reproductive or physiological problems (i.e., cows that are bound for the slaughterhouse). Since the hormone not only increases production but can also maintain lactation to an affordable yield level for up to a total of 1,000 days, it can grant a longer productive and profitable life to those animals.

Therefore, on the surface, this quick calculation makes rBST use in Mexico appear moderately profitable. There are a number of additional costs from rBST that are not taken into this account, however, and farm-level data are crucial for actual assessments of profitability. In 2000, average milk production per cow in Mexico was still "only a fraction of that of the U.S." (Dobson and Proctor 2002, 16). But there are different milk production systems in Mexico, ranging from small operations based on dual purpose (milk and meat) or seasonal (milking beef cows during rainy season), to larger intensive production farms (Chauvet and Ochoa 1996, 2), such as in the La Laguna region. Milk production per cow on these commercial farms "is reportedly near levels found on farms in the southwestern U.S." (Dobson and Proctor 2002, 16). Therefore, intensity of production for commercial farms in La Laguna mimics that in the United States and, with qualifications, data from U.S. studies can be used to supplement our basic understanding.

Even in the United States, ex-post studies of rBST profitability are limited. Those studies that do exist, however, report only marginal improvements in profitability resulting from its use. Using New York Dairy Farm Business Summary data, Tauer and Knoblauch (1997) used a temporal comparison (1993–1994) and found rBST use to generate $27 more per cow net farm income, while Stephanides and Tauer (1999) compared 1994 to 1995 users and non-users and found its use to result in a loss of $39 per cow, though neither result was statistically significant (cited in Ott and Rendleman 2000, 5). Continuing with the same New York data source, Tauer (2001) more recently compared four years of data (1994 to 1997) and concluded that the estimated profit impact of rBST was generally positive, but again, statistically zero (Tauer 2001, 1). Ott and Rendleman (2000) used United States Department of Agriculture (USDA) survey data of dairy producers to calculate the economic impacts of rBST in the U.S. Assuming sixteen treatments, and incorporating such detailed cost variables affected by rBST application as extra feed, labor, veterinary, and medical costs, returns to management, and milk hauling charges, they conclude that rBST increased profits by $126 per cow.

Two points are consistently raised with all studies. The first is the overwhelming difficulty in assessing profitability at the farm level. Nearly all studies pay some tribute to the challenge of measuring the determinants of farm profitability "since there are numerous causation factors, many of which cannot be quantified" (Tauer 2001, 6). Complicating the already existing difficulties of calculating profitability are the normal market fluctuations in such factors as the price of feed and the price paid for milk, which can drastically alter rBST's profitability. The second point is that despite all the variables complicating assessments of profitability of rBST, all studies have consistently found a statistically significant increase in yields at the farm level. It needs to be kept in mind that nationally, numerous technologies and improvements in genetics, nutrition, and management practices have already produced two decades of milk productivity growth (Barham and Foltz 2002, 17). Consequently, Barham and Foltz (2002) calculate that only 3.5 percent of these two decades of productivity increase can be attributed to rBST adoption (2002, 17). Similarly, Barham, Foltz, Jackson-Smith, and Moon (2004) used panel data from Wisconsin dairy farmers and found that while rBST users had the highest productivity levels overall, non-adopters had the largest growth in productivity, suggesting "that in terms of productivity increases among Wisconsin dairy farmers, rBST adoption represents only a small portion of the action" (2004, 67).

It should be added that due to the particularities of rBST, profitability may be more individualized. Ott and Rendleman (2000), for example, conclude that while rBST use may be profitable on some cows, it may not be profitable on all cows; consequently, "producers should evaluate each cow or productive group of cows before committing to rBST" (Ott and Rendleman 2000, 7). The authors establish an optimum application rate of approximately 73 percent of the herd (Ott and Rendleman 2000, 8). While statistically zero, trends in profitability were more than once found to be generally positive (Barham et al. 2004). Consequently, Tauer (2001) argues:

> These results are for the treatment average. It may be that there are farms profiting from the use of rBST. The implication then, is that some farmers may be losing money using rBST. But since the output impact of rBST is unambiguously positive, it may be difficult for individual farmers to determine if that output impact is translating into profit. (Tauer 2001, 8)

Tauer further notes that profitability assessments are not based on individual cow records, and therefore "precludes analysis on rBST use tactics, which may be complex and unique by farm" (2001, 4). Similarly, Butler argues that while rBST use has "probably very little impact on the competitive position of adopters vis a vis non-adopters" (1999, 8), it is very difficult for producers—who do not have the time or the technologies to monitor responses for individual cows—to calculate the profits from rBST: "You guess at individual profitability on each cow—while costs continue to increase" (dairy producer speaking about rBST, cited in Butler 1999, 7).

While conclusive data with respect to rBST's profitability in the United States is difficult, the above discussion has already provided sufficient grounds to seriously question it. U.S. farmers themselves have found sufficient reason to doubt the technology. RBST adoption in the United States has not only plateaued, but dis-adoption has become increasingly prevalent. Barham and Foltz found that 25 to 40 percent of those who have tried rBST no longer use it (2002, 17). Dis-adoption rates in the United States vary by state, ranging from a low of 7.5 percent in Wisconsin in 2001, to a high of 20.7 percent in Stearns County, Minnesota, in 2000 (Barham and Foltz 2002, 16). Barham et al. (2004) found that, of those who dis-adopted, 82 percent cited that "rBST was not cost effective" as their reason for discontinuation.

In the context of Mexico and of La Laguna, weak profitability results should be considered with even more circumspection, as many factors make profitability less stable than in the United States. Further, the existing studies fail to deal with a host of other indirect, but nevertheless crucial, issues associated with the use of the hormone. For one, the same literature that predicted a rapid diffusion of the hormone in the United States also claimed this technology to be scale neutral. From this assumption it could be inferred that its diffusion should be more or less rapid and easy, and could benefit a host of small producers in developing countries. Yet, while use of the drug itself can indeed be scale neutral, everything associated with its use is not, as is now supported by studies finding positive relationships between farm size and rBST adoption (Tauer 2001; Barham et al. 2004). In reality, herd size is often related to the sophistication of the technologies and infrastructure of the farms and of the resources available, both in advanced and developing countries. RBST is beneficial only in optimal production conditions, and if some elements are missing in the optimal management scheme, the hormone can be more destructive than beneficial (Poitras 2000).

Although not significant to the U.S. context, adequate feed can be a challenge for many producers in developing countries. According to an editorial in the *Biotechnology and Development Monitor* (1996), "the single major cause of poor livestock productivity is poor animal nutrition and rBST is unable to bypass this reality." While feed is a most basic issue, the whole management of the farm is important. In a region like La Laguna, many factors make it difficult to achieve optimal farm and cow handling management: low availability of sufficiently skilled labor; the heat, which in itself makes the region non-optimal for the use of the hormone; periodic or even chronic droughts, which lead to poor-quality fodder and/or more expensive and more complicated logistics for fodder and feed provisioning; macro-economic instability, which can drastically increase the costs of imported inputs necessary for such optimal management; and so on.

As noted, the heat is another issue of particular significance to La Laguna. According to industry literature, the only consequence of less than optimal treatment of the cows is that it can lead to less efficiency of the hormone. Some interviewees, however, decided to stop using the hormone in the summer, in periods of heat, or when no fodder of good quality was available, as they had too many premature deaths or too many problems of fertility, which they associated with the application of the hormone (Poitras 2000). Even if this perception is incorrect and the health of the cows is not really negatively impacted by the application of the hormone in times of heat, the small margin of profitability that the use of the hormone confers means that marginal losses can easily be incurred in less than optimal conditions.

There is another issue of particular significance to La Laguna. Even strong supporters of rBST agree that more intensive production leads to a shorter productive life span for the cows (almost half compared to more conventional modes of operation) and a decreased level of fertility (as most of the energy of the cow goes toward milk production, with little left for reproduction). Fertility is inversely proportional to production. Therefore, if the hormone increases milk production, fertility necessarily declines.

> You get lots of litres of milk out, as with using the hormone, but fertility declines and vice-versa. This is a natural mechanism of the animal due to the stress brought about by exploitation or overexploitation. Therefore, the animal's life does indeed get decreased with rBST use: the animal is used up sooner and becomes more susceptible to illness due to the stress of overexploitation. (Emilio Rodríguez Camacho, agricultural-inputs supplier in La Laguna, personal communication, 2004)

Similarly, increased mastitis, lameness, and other problems compromising animal health are not direct consequences of the hormone itself, but are due to a number of factors, including the strain on the animal due to increased milk production. Considering that the milk-producing life of a cow is already relatively short, the combined decrease in fertility and in the lifespan of the cow can have a significant impact on the costs of using rBST. These two problems combined make self-sufficiency in the replacement of "wasted" or used-up cows impossible, and consequently require calves to be purchased, primarily from abroad (mostly Canada and the United States). The cost of imported calves has steadily increased over the years, from some $1,200 (in U.S. dollars) in 1998, $1,600 in 1999, $1,750 in 2001 (cost also varies with genetic quality), to about $2,000 in 2004 (Rodríguez Camacho, personal communication, 2004). Therefore, in addition to the less quantifiable variables of necessary associated technologies and external factors such as heat, the increased cost of replacement cows—particularly in countries where these cows must be imported—comprises a very tangible factor that remains outside of most calculations of rBST profitability, and that likely has very particular effects for developing countries.

Finally, a longer-term perspective would consider the environmental and social impacts in La Laguna of the intensive dairy production toward which the use of rBST is oriented. Dairy production in Mexico is shifting northward, increasing the share of dairy production occurring in the three major milk-producing states by 1 percent between 1994 and 2000 (Dobson and Proctor 2002, 17). The La Laguna region extends between Coahuila and the bordering state of Durango, which had production increases of 3 percent and 1 percent, respectively (Dobson and Proctor 2002, 18). This northward shift represents a shift in production to larger intensive-production dairy farms, as opposed to the smaller, semi-confined, dual purpose or seasonal type production (Dobson and Proctor 2002, 16).

Some La Laguna businessmen (interviewed by Otero) argue that given the environmental impacts, dairy production should not be practiced at all in this region. Most particularly, with milk being the single most important animal product in the region, such intensive production means that the scarce water resources of La Laguna have been increasingly directed toward feed. The proportion of irrigated land used for the production of feed crops rose from 12 percent in 1970 to 53 percent in 1998 (see Figure 7.1). The main feed crop, alfalfa, accounted for between 45 and 65 percent of all feed crops cultivated in La Laguna between 1970 and 1998 (SAGAR-Laguna 1980–1998), and used 22 percent of all underground water consumed in the region. This otherwise very advantageous and efficient fodder has a much higher water-to-product conversion rate than any other major crops cultivated in the region. It has a ratio of liters of water used to one kilogram of fresh product of 279, in comparison to 150 for maize or 134 for sorghum (Aguilar Valdés and Luévano González 1999).

Increases in the intensity of production and expansion of the dairy industry in the region have thus led to a situation today where underground water is used at a radically unsustainable rate. While decreasing food dependency in Mexico is highly desirable, it has to be asked whether such intensive dairy production is socially and environmentally viable in the long term. In the municipality of Matamoros, for instance, with one of the highest concentrations of dairy farms, underground resources are disappearing at a rate of two meters a year (Aguilar Valdés and Luévano González 1999). This has production cost implications (with deeper and more expensive wells having to be dug, and more electricity required for pumps); animal health implications (when the water becomes poisoned due to mineral imbalances in the soil); agronomic implications (due to acidification of the land); and, finally, human costs, with arsenic poisoning not uncommon in several towns of La Laguna. Further, as dairy farming is not labor intensive, these costs cannot be balanced with respect to its prospects for employment, which further compromises its social justification.

With greater certainty than in the United States, all of the factors discussed above suggest that the profitability of using rBST in La Laguna is tenuous at best. Furthermore, environmental and social sustainability of the La Laguna farm economy at large is being questioned. Based on the most intensive production technologies developed in Canada and the United States, such as rBST, dairy production in La Laguna has a series of potentially negative social, health, and environmental effects. One of the questions that remain is then: why, in the face of such low or tenuous profitability, do many La Laguna dairy farmers continue to use rBST? One possible answer might be that dairy farming allows relatively low rates of profit, so that any apparent means to improve them will be tried by many farmers. RBST does increase production noticeably, acting on the most powerful of all criteria by which farmers assess each other: yields. Moreover, farmers report being constantly subjected to the pressure of the distributing companies, some of them in regular contact with the farmers for the sale of other veterinary products. These more cultural or marketing elements should not be underestimated, as those who reject the use of rBST are often subjected to subtle social pressure: they may be regarded as backward, conservative, or worse, not "man" enough to take the risk of a new technology. The next section will draw from interviews conducted

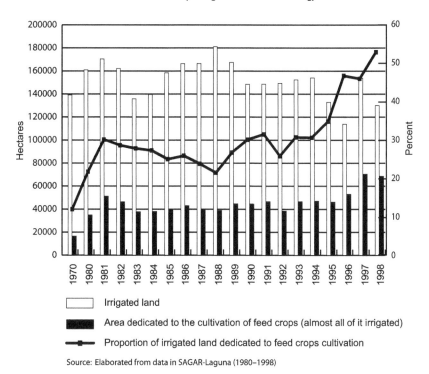

Irrigated land

Area dedicated to the cultivation of feed crops (almost all of it irrigated)

Proportion of irrigated land dedicated to feed crops cultivation

Source: Elaborated from data in SAGAR-Laguna (1980–1998)

Figure 7.1. Dairy Farming and Irrigated Land Use

in La Laguna in the summer of 2001 to explore some of these more inter-
subjective reasons underlying the decision to adopt rBST or not.

Narratives of rBST Adoption in La Laguna

The previous section presented an economic calculation and extrapolations from U.S. data to estimate rBST's profitability in La Laguna. It
should be clear that while this calculation method is not completely sat-
isfactory, it offers sufficient grounds to question the economic explanation for the wholesale adoption of rBST by capitalist dairy farmers in La
Laguna. Therefore, in an attempt to more fully understand the logic of
rBST adoption in La Laguna, it is necessary to investigate its cultural
underpinnings. This is done by drawing on narratives of adoption as
presented by dairy farmers themselves. While the number of interviews
conducted for this section was limited to a dozen, farms were selected in
a targeted manner in order to arrive at a preliminary understanding of

types of farming practices in the region. Through this approach a typology of four farming styles, or management approaches, was identified.

It should be noted that this approach to data gathering is not free of methodological problems, either. For instance, in the first two interviews, both informants started out with very firm statements about the great benefits of using rBST, but then reversed their positions toward the end of the interview. Initially, they both seemed convinced that it had been a success. This opinion change indicates that the very questions raised during the field interview may have had an impact on the interviewee's perception—for example, by calling attention to issues or factors that may not have been considered in previous cost calculations and adoption decisions. This appears somewhat indicative of the difficulty of assessing the true profitability of rBST, beyond its success in terms of yields. While this type of methodological problem requires caution, as sociologists we must contend with the "reflexive" nature of our discipline, and exercise as much "epistemological vigilance" as possible in assessing the truths that we find *and* help formulate.

The types of farming practices identified and their motivations are characterized below through the narratives of selected exemplar farmers.

Martín[4] began by establishing the technical conditions under which use of rBST is warranted. Cows must be within a specific range of weight, which they have come to classify as anywhere between an index of 3 to 3.5. If the cow is between 3.5 and 4, then it is regarded as too fat; if it is under 2.5, then it is too thin. In either case, using rBST would result in problems for the cow. Some guidelines given by industry promoters did not really work. For instance, he was told to use the hormone with fat cows, but he thinks it does not work in these cases.

Yields and Acceptable Revenue

Martín says that at a larger farm that belongs to the same owners as the farm he operates, they are producing 36–37 liters per cow per day, partly due to using some weather control with curtains. They also use rBST and have a herd of 7,000 cows. His farm has 1,500 milking cows, 270 dry cows, 350 in gestation, and 320 calves. According to Martín, farmers regard a total cost per cow of between 48 and 52 percent of revenues, with an average yield of 30 liters per day, as acceptable.

One problem related to the use of rBST is the fact that cows now last only 2.5 lactation periods, when they used to last 4 to 5 before the use of the hormone. With the high replacement cost that this entails, some farm operators began to doubt the profitability of using the drug. Martín thinks that, in general, veterinarians in the region do not like to use the drug, for it also generates more health problems. Instead, they recommend using the drug much more selectively with cows that exhibit some specific problems of lactation. A fat cow, for instance, cannot be made to produce more milk by using the drug. Also, there used to be higher fertility before using the drug. In 2000 at the larger farm, out of 7,500 cows in production, 3,400 were "wasted" or sacrificed at the slaughterhouse because they had ceased to be productive. Because the reproduction within the farm could not handle such a high level of replacement, they had to purchase 1,700 cows from the United States. It cost $1,750.00 to replace each cow. In this larger farm, where Martín worked previously, use of rBST also resulted in plenty of abortions. When some cows were withdrawn from the treatment, their yields dropped abruptly from 31 liters per day to only 24. After two years without using rBST, the problems continued, but the yield also stayed at 24 liters per day.

Martín then narrated a tale of another farmer, several years ago, whose cows produced 18 liters per day when those of others produced 21–22 liters. But he says that those were the times when this farmer was making the most profits, not least from selling his surplus cows: he was breeding more cows than needed for replacement.

Upon further reflection, then, Martín thinks that the initial attraction of using the drug was increased milk yield. Well managed, he believes it may work. And yet, he considers that 30 to 32 liters could be achieved without using rBST. Martín then concludes that he would get rid of the drug ("yo la quitaría"), because the cows become addicted to it. The decision to adopt was made by him and the farm owner once they saw that others who adopted it had increased their yields from 28 to 31 liters. Many others followed suit.

Armando has been in dairy farming for twenty-two years. He has been applying rBST (Monsanto's product) for four years, and he says that the farm owner made the decision to adopt based on his relatives' experience; they had adopted it previously. One thing that has been demonstrated also in Armando's practice is that if you stop applying the drug, yields drop dramatically, and they will never go up again without the drug. In his operation, the average yield per cow was 27.57 liters/day. The average daily feed intake per cow cost MX$39. At a quota milk price of MX$3.30 per liter, daily revenues are: MX$90.98 per cow. But what about the other costs such as infrastructure, electricity, water, veterinarian, labor, etc.? How are these costs affected by rBST use? Farmers' economic calculations, at least during interviews, seem to be limited to the most immediate and direct costs involved in using rBST.

During three years, Ramón changed rBST product once a year, from Lactopropin (Monsanto's trademark) to Boostin (a Korean trademark), and back. He thought that the former was the better product, offering greater stability in increasing milk yield. Some farmers use the hormone only when a given cow is about to dry up, to extend its productive (i.e., milk-producing) life. Yields for him are 31 liters on average for the year, but there are months with averages of 34 liters and others with averages of 28. Births have declined from eight to ten per cow several years previously, to fewer than four in 2001.

Ramón confirms the pattern seen by other farmers in the two brands of rBST: Lactopropin is smoother in its boosting effect, while Boostin has a very sharp effect, going quickly from, say, 29 to 32 liters per day. But the latter, if you stop its use, also drops the yields very quickly. This causes considerable imbalance in the cattle, increasing the incidence of mastitis and requiring more management.

During summer, stress conditions increase for cows, and some farmers reduce or eliminate the use of rBST so as not to increase the stress. Other farmers, though, continue to use it if they think that their cows are in good physical shape to withstand the extra stress. Part of the problem in this season is that cows ingest less solid material, which results in a less healthy condition.

For his part, Felix uses the services of a veterinarian that records and monitors individual cows' yields, births, and other variables for him and seventeen other farmers. This veterinarian has produced comparative statistics of the eighteen dairy farms he keeps track of, and Felix came out on top as the most profitable of all farmers. Some of the factors included in the calculation are: milking times per day (two or three), use of rBST or not, use of other feed additives to enhance milk production or not (some of these may stress cows, like Elanco's Rumencin®), type of feed, whether feed is purchased or produced on the farm, etc.

Felix's success is based on a low-cost approach, a focus on trying to increase the cows' comfort levels, and using the best genetic materials available. In the past, during the five years when he was using rBST, Felix was not able to replace wasted cows, let alone have surplus calves for sale. Then he decided to go back to the old methods. He has more than five hundred milking cows. His average yield is a modest 23 liters per cow per day. He uses the very best semen available. The highest price is about US$28 per dose, but semen prices have gone down. With this approach, he has always had a surplus of cows that he can sell as a result of adequate fertility and low mortality rates. In his view, those who milk their cows three times a day, use rBST, and feed more are engaged with a logic of increasing milk production, but their operations are not necessarily the most profitable. The benefits of Felix's approach are lower costs, less waste, better reproduction, and production of surplus cows.

Mentality toward Technological Innovation

With regard to technological innovations and achieving yield increases, Martín said that local farmers' mentality is captured in the saying *Pa'delante lo que quieras; pa'tras, ni pa'agarrar impulso* ("Forward, as much as you want; backward, not even to launch yourself"). As captured in this local saying, farmers will adopt any technological innovation that will improve their yields. In Armando's view, there are at least the following factors determining yield, in order of priority or importance: genetics, feed, comfort, and management, including technology such as heat deflection, an inseminator, and the semen applicator. In this region, he says, farmers all go for the full risk, "for the kill" (*todos se tiran a matar*). In other words, if one operator or farm owner decides to adopt the hormone, they will use it on their entire milking herd. In his case, he had to buy 60 new cows for replacement, out of 1,358 in lactation.

Aggressive technological innovation may be the dominant mentality or cultural approach to management in the region. According to other interviewees, however, before using the hormone there were fewer problems. As a result, some farmers apply the hormone on a selective basis, so the precise extent to which it is adopted within farms is an empirical question for future research.

Alfredo, for instance, applies rBST to only 25 to 30 percent of his cows. He has been using Boostin for two years. In his view, and contrary to Armando's opinion, few farms apply the hormone to the entire herd. He uses it because he thinks that each cow yields more profits that way. He uses rBST as a means to extract more milk from "waste" cows. These are the cows that have to be sent to the slaughterhouse anyway, because they are already in their last lactation period, and the hormone allows lengthening that last milking period. He has had problems with low fertility, however. This is compounded by the fact that, when a cow's production declines, it may have to be sent to the slaughterhouse, also losing its calf.

Alfredo's technology innovation approach seems to be on the conservative side. He says, for instance, that rBST is applied in his farm only to cows whose yield is less than 30 liters per day. He also uses semen of medium or so-so (*regular*) quality, and lower-priced semen for second or third pregnancies.

Felix, the farmer with the low-cost/high-comfort approach to management, presents an interesting contrast with Alfredo. He tends to use most of the advanced technologies available, but does so selectively to increase the cows' comfort levels. In the case of rBST, he used it eight years before, for three years. He wanted to produce lots of milk, reaching an average of 30 liters per day per cow, and milking three times a day. His current strategy, though, is for low costs, even if milk yields are reduced. There used to be too many wasted cows before. He is interested in expanding his herd size, but with lower input costs, and milking only twice a day to reduce labor costs. "They think I'm staying behind, but for me this turns out more money," he said with a big smile on his face.

Felix had been using his current system for the past five years. The idea for the new approach came from reading a study, which concluded that operators with greater milk yields were not necessarily the most profitable. For him, an indicator of which are the most profitable dairy farming operations is whether they have surplus cows for sale or not: those who sell calves while preserving a stable or growing herd size are the most profitable.

For his part, Ramón regards cows as mere machines that are to be exploited to the maximum. Milking three times a day is now the norm, and he would like to move into four times per day. He uses baths to reduce heat stress during the summer days. When it comes to sending a wasted cow to the slaughterhouse, for him it does not represent a loss, but eliminating a problem. As may be seen, then, Felix and Ramón represent two distinct approaches to profitability: While the former emphasizes lower costs, at the expense of lower yields, Ramón is the all-out high inputs, intensive farmer. Other farmers seem to have a less consistent, more ad hoc approach. Which of these approaches is ultimately most profitable is an empirical question, as is the question of their environmental sustainability.

Relations with LALA and other Milk Processors

The dominant milk processing plant in the region is LALA (for "La Laguna," the region's name). Initially started as a cooperative by La Laguna dairy farmers, LALA is now a vertically integrated corporation with 23 companies. It is ranked 58 out of Mexico's 500 most important companies, and controls 26 percent of the Mexican milk market (Farrar 2000, 1).

In 1998, LALA introduced a quota system directly linked to the number of shares that each farmer owns in it. Other processors also have similar quotas to LALA's. Virtually all the capitalist farmers in the region own shares in one or the other processor. Prices paid to farmers vary depending on the proportion of shares they own to the amount of milk they deliver, with prizes for quality. If a farmer goes over his quota, LALA will pay a lower price for this surplus milk. As of 2001, the quota price was Mx$3.30; surplus-milk price was only Mx$2.60, or it could be as low as Mx$2.40 if one is penalized for low quality also. Therefore, despite Mexico's need to increase dairy production, LALA and the other processors do *not* promote increasing milk yields, primarily due to limits in processing capacity and problems in commercialization. This information and analysis put a different perspective on state enthusiasm to approve the technology. Nonetheless, there were farmers who thought that they had to produce more milk, even if it were outside of their quota, but the price disincentive is substantial. This quota system makes the high-input, high-yield technological innovation mentality all the more puzzling.

Monsanto's Marketing Strategy in La Laguna

Many issues with respect to Monsanto's marketing in the region were raised during our interviews, and suggest a need for further investigation into the particularities of Monsanto's strategy in comparison with that in the United States.

For Felix, who had used rBST for five years, its price is now too high. He started with Monsanto's product. This company used to send him veterinarians, nutritionists, and gifts as incentives to adopt. Felix told us the story of rBST introduction in La Laguna, the price wars that ensued, and Monsanto's bid for predominance. His rendering describes it as following three stages: First, Monsanto started out as the only company offering the hormone in 1992. It launched a very robust promotion, which had a strong effect in the form of widespread adoption. One example of the promotion is that Monsanto would give each farmer a nice vertical refrigerator with glass doors for each 1,000 doses of the hormone that were purchased. In time, Felix hoarded three refrigerators. Monsanto sold cheap and provided technical advice. At this time, the daily cost of each dose amounted to that of one liter of milk. Second, toward 1993, Elanco (the veterinary subsidiary of the U.S. transnational Eli Lilly) entered the rBST market with the product called "Optiflex-Elanco®," setting off a price war. Subsequently, it appeared that Monsanto decreased either the quality or content of its product, so that its effectiveness declined. Third, Monsanto purchased Elanco's rights to sell in Mexico and stopped selling its own product, marketing Elanco's product instead. Soon thereafter, in the third stage, the price doubled to the equivalent of two liters of milk per cow daily, and it stopped being profitable to use rBST.

Monsanto had other strategies in the region. Gonzalo, a former president of the Regional Union of Dairy Farmers of La Laguna (Unión Regional de Ganaderos de La Laguna, URGL), has never used rBST on his cows. "They have used us as guinea pigs here in Mexico," he says. However, he was contacted by the U.S. embassy and they sent people from Monsanto to sell their product. Elanco flew Gonzalo to the United States three times on promotional trips. It was not yet authorized in the United States, so his question to a U.S. official was: "Why do they not allow it in your own country yet?" Then he added: "If you had two glasses of milk to give to your child, one with rBST and the other without it, which would you choose?" The official's answer was that he would choose the one without the recombinant hormone.

Gonzalo's conclusion as a dairy farmer was that he could not stay behind in technology adoption, but there are laws in nature which one cannot violate. In his view, many La Laguna farmers have now reduced the use of rBST, after seeing the imbalances caused in the cows: they become more sensitive, eat more, become ill more often, and die sooner. He foresaw that there would be a future price for pure milk (i.e., rBST-free). Although he refuses to criticize its use, because he "knows" that farmers do fine with it, he has refrained from using rBST in his own cows. He notes that treated cows today are more infertile, though they may produce 10 to 20 percent more milk, depending on management. He used it in a clinical experiment with five cows but it did not work to his satisfaction. Some farmers use it off and on, or on selected cattle.

From our exploratory interviews, then, we can identify at least four cultures or managerial approaches among the La Laguna capitalist farmers: (1) that with a focus on yields, but whose ad hoc, haphazard, or inconsistent adoption of new technologies or management practices leads to medium to low profitability; (2) the conservative approach of farmers who may adopt rBST on a selective basis, on small portions of their herds, combined with other medium- to low-quality genetic technologies (e.g., semen), whose profitability is also low to medium; (3) the all-out, input-intensive, high-turnover-rate approach, in which cows are seen as mere machines, with high profitability; and, finally, (4) the approach of former adopters of rBST who stopped its use, but focus on other high-quality genetic technologies, high comfort levels for cows, and tight management practices, with high profitability even with lower yields. Further, these interviews place emphasis on a number of issues that appear to have particular significance for rBST use in La Laguna, such as the potential of selective use; the extra costs of sickness and fertility; and the marketing strategies of pharmaceutical and agrochemical companies in the region.

Conclusions: Toward a Research Agenda

As indicated at the start of this chapter, the adoption of new biotechnologies cannot be considered in isolation from their political-economic context. Two major problems affect most developing countries when it comes to new technologies: (1) the lack of infrastructure and resources to offer an independent safety, profitability, and wider social benefits assessment of new imported technologies; and (2) strong pressures to introduce any new technology, especially one that can improve the yields of basic food commodities for which domestic production is insufficient. Since the mid-1980s, but particularly since the start of NAFTA in 1994, many Mexican dairy farmers have been concerned with the commercial opening of the borders. As of 2003, NAFTA has led to a fresh milk market completely free of tariffs. U.S. producers are perceived by La Laguna dairy farmers to have an advantage because of the availability of cheap credits and much stronger and consistent government support. The opening of markets therefore leads to fears of bankruptcy and renewed efforts to stay ahead of competition and gain competitive status in relation to their North American counterparts. RBST is often perceived as one of the main available tools for such pursuit (Poitras 2000).

If self-sufficiency in dairy production is a serious goal for Mexico, then state officials must question the benefits of having allowed production-enhancing technologies whose environmental *and* economic sustainability remain unproven. Such liberality may have further stretched the limits of a region with considerable problems of environmental sustainability without the traditional trade-offs of profitability and social sustainability.

The importance of the dairy industry in La Laguna cannot be overstated. The "specialized" (intensive dairy production) system, located in the central and northern regions of Mexico, comprises 17 percent of Mexico's milking herd but supplies 59 percent of the national milk production (Chauvet and Ochoa 1996, 2). Nonetheless, many local observers outside of the industry are extremely concerned about the benefits that it brings to the region. A better regional picture will emerge once we are able to complement our current knowledge with further intensive interviews and a questionnaire administered to a representative sample. These data could then be used to articulate detailed comparative studies with the U.S. situation, such as those already conducted in Wisconsin and New York. California and Arizona likely offer the most comparable U.S. farm structures to that in La Laguna.

While our quantitative assessment can only be used as a heuristic device, its supplementation with data from the U.S. experience and qualitative data from interviews with La Laguna dairy farmers suggests that the profitability of rBST use in La Laguna is seriously in question. A few important issues—selective adoption, scale economies, animal health and fertility—were found to complicate profitability assessments, and potentially could have other social implications.

One complication for profitability calculations is that some farmers do not apply the drug to entire herds, but use it selectively. Those farmers who are most aware of the needs of their cattle and the benefits and limitations of rBST are most likely to extract a profit from it. It needs to be kept in mind that despite increasing awareness of the intricacies of the drug, adoption in the United States—where education levels, farm management systems, and data collection and information dispersal are more prevalent—has declined. It is even more difficult to gain the benefits of the drug in regions where these supports are lacking or insufficient.

Secondly, while touted as scale-neutral, rBST is anything but, particularly in developing countries, where adequate feed can be a hurdle to productivity. Consequently, rBST's profits are most likely realized in the specialized or intensive farms, and are likely unattainable from the extensive, dual-purpose, and other small farm operations. In these cases, a transition to intensive management in itself would be likely to increase milk production, but at a loss to the small producers. Therefore, while the drug might foster more efficiency and competitiveness of the specialized and technologically developed dairy sector of Mexico, thus helping it face the world market, the hundreds of other producers excluded from such development face the archetype of choices well known to the excluded in Latin America: poverty and unemployment; low-paying maquiladora employment; and emigration north of the border (Bartra 2004; Delgado Wise 2004).

Therefore, while rBST leads to increased milk yields, it does not necessarily help farmers prosper, and only a few appear to significantly benefit from its use. Neither does it bring more wealth regionally, as what wealth it produces is captured by the source of the technology: Its marginal profitability combined with clearly increased yields points to the fact that the pharmaceutical transnational corporations producing and distributing rBST are extracting all the "rent," or relative surplus value, associated with the use of the hormone (Butler 1999).

If rBST is only moderately or tenuously profitable in La Laguna, as we propose, then a next step would be the empirical testing of the hypotheses suggested here for its adoption: Is it that La Laguna adopters are too alienated into a logic of milk yield increases at any cost? Is it an ingrained macho culture that drives these farmers to compete with each other primarily based on whose cows have the highest milk yields? Or are the marketing strategies of the oligopolistic transnational corporations, which take advantage of a rather permissive regulatory environment, the main explanatory factor? Our preliminary investigation provides some support for all of these explanations based on cultural factors and TNCs' marketing strategies. Furthermore, the fact that many farmers adjusted their perspective of rBST during the course of the interviews indicated that the difficulty of ascertaining rBST's profitability is an additional factor in making the decision to adopt. A full empirical survey needs to be conducted to investigate the relative weight that these factors have in explaining rBST adoption in La Laguna. If marketing strategies are found to strongly determine adoption, in spite of questionable or marginal profitability, then state policy would have to be revised substantially to more closely regulate TNC activities. It would have to be geared to promote more socially and environmentally sustainable technologies.

Notes

We gratefully acknowledge the kind and effective research assistance provided by Emilio Rodríguez Camacho, a dairy-farming-inputs producer and distributor who knows most of the La Laguna dairy farmers and veterinarians. Given his clear understanding of our research goals, his help in targeting farmers with a variety of management styles was critical. Furthermore, Emilio drove Otero and Poitras to many farms for interviews.

1. The synthetic version of the hormone is produced by first introducing the BST-producing gene in a bacterium, which is then fermented for its reproduction. The hormone is then purified from the bacteria, a process which removes the DNA or the genetic information. The genetic modification is thus made on the bacteria, not on the hormone as such.

2. RBST triggers an increased production of IGF-1 (insulin-like growth factor 1), which acts on the mammary cells, increasing their activity and thus milk

production. Concerns over the human health impacts of rBST are linked to this compound. For example, a certain role is thought to exist for IGF-1 in the development of prostate, colon, and breast cancers (Larsen 1998).

3. The Codex Alimentarius website reports that the main purposes of this program are protecting the health of consumers and ensuring fair trade practices in the food trade, and promoting coordination of all food standards work undertaken by international governmental and nongovernmental organizations. http://www.codexalimentarius.net/web/index_en.jsp

4. All names mentioned in the text are pseudonyms to preserve the anonymity of interviewees.

References

Aguilar Valdés, A., and A. Luévano González, eds. 1999. *Impacto Social y Económico de la Ganadería Lechera en la Región Lagunera*, 6th edition, Torreon, Coahuila, Grupo Industrial Lala, S.A. de C.V.

Barham, Bradford, and Jeremy Foltz. 2002. "rBST Adoption in the United States: That Was the Juggernaut . . . That Wasn't." *Choices*. Summer 2002. pp: 15–19.

Barham, Bradford L., Jeremy D. Foltz, Douglas Jackson-Smith, and Sunung Moon. 2004. "The Dynamics of Agricultural Biotechnology Adoption: Lessons from Series rBST Use in Wisconsin, 1994–2001." *American Journal of Agricultural Economics* 86(1): 61–72.

Barham, Bradford L., Douglas Jackson-Smith, and Sunung Moon. 2000. "The Adoption of rBST on Wisconsin Dairy Farms." *AgBioForum* 3 (2–3): 181–187. http://www.agbioforum.org.

Bartra, Armando. 2004. "Rebellious Cornfields: Toward Food and Labour Self-sufficiency." pp. 18–36 in Otero 2004.

Biotechnology and Development Monitor. 1996. "Editorial: A 'Magic Bullet' of Biotechnology Examined." *Biotechnology and Development Monitor*, No. 27. Available online: http://www.biotech-monitor.nl/2702.htm

Bueckert, D. 2001. "Anti-rBGH Scientists in Canada Fight Government Gag Order: Whistle-blowing Scientists Seek Right to Speak to Media." Canadian Press, Friday, 3 August.

Butler, L. J. 1999. "The Profitability of rBST on U.S. Dairy Farms." *AgBioForum* 2(2), Spring. http://www.agbioforum.org.

Canada. 1999. "RBST and the Drug Approval Process." Interim Report of the Standing Senate Committee on Agriculture and Forestry, Ottawa, Canada, March.

Chauvet, M. and R. F. Ochoa. 1996. "An Appraisal of the Use of rBST in Mexico." *Biotechnology and Development Monitor*, No. 27. Available online: http://www.biotech-monitor.nl/2703.htm.

Delgado Wise, Raúl. 2004. "Labour and Migration Policies under Vicente Fox: Subordination to U.S. Economic and Geopolitical Interests." pp. 138–153 in Otero 2004.

Dobson, W. D., and R. Proctor. 2002. *How Mexico's Dairy Industry Has Evolved under the NAFTA—Implications for U.S. Dairy Exporters and U.S. Investors in Mexico's Dairy-Food Businesses.* Babcock Institute Discussion Paper No. 2002-1. Madison, Wisc.: Babcock Institute for International Dairy Research and Development.

Ehrenfeld, D. 2002. "The Cow Tipping Point." *Harper's Magazine*, Vol. 205 (1829). Excerpted from "Unethical Contexts for Ethical Questions." Yale University Presentation, December 2001.

Elanco, 1999, *Lactotropina = + Leche x Vaca*, Mexico, Promotional pamphlet distributed by Elanco Animal Health, a division of Eli Lilly y Cía. de México, S.A. de C.V.

Estrada, J. M., et al. N.d. "El estres calórico y el uso de Posilac®, Somatotropina Bovina, en ganado lechero," mimeo, St. Louis, Mo., Monsanto Agricultural Company, Animal Sciences Division.

Farrar, F. 2000. "LALA Expands, Adds Shop, Upgrades Fleet." Primedia Business Magazines and Media Inc. Available online, LookSmart's FindArticles: http://www.findarticles.com.

Fetrow, J. 1999. "Economics of Recombinant Bovine Somatotropin on U.S. Dairy Farms." *AgBioForum* 2(2). http://www.agbioforum.org.

FIRA. 1998. "Diagnóstico de la Ganadería de Leche," mimeo, Torreón (Coahuila), Subdirrección Regional Norte, Residencia Estatal Comarca Lagunera.

Jasanoff, S. 1986. *Risk Management and Political Culture: A Comparative Study of Science in the Policy Context.* New York: Russell Sage Foundation.

———. 1997. "American Exceptionalism and the Political Acknowledgement of Risk." *Comparative Science and Technology Policy*, ed. S. Jasanoff. International Library of Comparative Public Policy. Vol. 5: 391–411.

Kleinman, D., and J. Kloppenburg Jr. 1991. "Aiming for the Discursive High Ground: Monsanto and the Biotechnology Controversy." *Sociological Forum* 6(3):427–447.

Krimsky, S., and R. P. Wrubel. 1996. *Agricultural Biotechnology and the Environment: Science, Policy and Social Issues.* Urbana: University of Illinois Press. Chap. 9: "Animal Growth Hormones: The Case of Bovine Somatotropin," pp. 166–190.

Larsen, H. 1998. "Milk and the Cancer Connection." NotMilk.com website. First published in *International Health News*, Issue 76, April 1998. Available online at: http://www.notmilk.com/drlarsen.html

Otero, G. 1992. "The Differential Impact of Biotechnology: The Mexico–United States Contrast." pp. 117–126 in *Biotechnology: A Hope or a Threat.* Iftikhar Ahmed, ed. London: Macmillan.

Otero, Gerardo. 1999. *Farewell to the Peasantry? Political Class Formation in Rural Mexico.* Boulder and London: Westview Press.

Otero, Gerardo, ed. 2004. *Mexico in Transition: Neoliberal Globalism, the State and Civil Society.* London: Zed Books.

Ott, S. L., and C. M. Rendleman. 2000. Economic Impacts Associated with Bovine Somatotropin (BST) Use Based on Survey of U.S. Dairy Herds. *AgBioForum* 3(2&3), 173–180. Available online: http://www.agbioforum.org.

Poitras, M. 2000. "Political-Economic Processes of the Introduction of Agro-Biotechnologies in the Mexican Countryside: The Case of rBST in La Laguna Dairy Farming." *Sociedades Rurales, Producción y Medio Ambiente* 1(1).

SAGAR. 1999. *Situación Actual y Perspectivas de la Producción de Leche de Ganado Bovino en México 1990–2000*. México: SAGAR (Secretaría de Agriculura, Ganadería y Desarollo Rural), Centro de Estadísticas Agropecuarias.

SAGAR-Laguna. 1980–1998. *Anuario Estadístico de la Producción Agropecuaria*, Ciudad Lerdo (Durango), SAGAR, Delegación en la Region Lagunera Durango-Coahuila, Subdelegación de Planeación y Desarollo Rural.

Stephanides, Z., and L. W. Tauer. 1999. "The Empirical Impact of Bovine Somatotropin on a Group of New York Dairy Farms." *American Journal of Agricultural Economics* 81(1): 95–102.

Tauer, L. W. 2001. "The Estimated Profit Impact of Recombinant Bovine Somatotropin on New York Dairy Farms for the Years 1994 through 1997." *AgBioForum* 4(2), 115–123. Available online: http://www.agbioforum.org.

Tauer, L. W., and W. A. Knoblauch. 1997. "The Empirical Impact of Bovine Somatotropin on New York Dairy Farms." *Journal of Dairy Science* 80: 1092–1097.

CHAPTER 8

Genetically Modified Soybeans and the Crisis of Argentina's Agriculture Model

MIGUEL TEUBAL

One of the most dramatic consequences of the current Argentine crisis is the suffering of millions of hungry people who have experienced a widespread decline in their living conditions. In 2001, at the height of the crisis, over half the population—20 million persons, according to official figures—was living below the poverty line. Around one in four was suffering extreme poverty, meaning that their income was insufficient to cover their basic food needs, something unprecedented in the social and economic history of Argentina. This panorama was reinforced by media coverage of children starving, either because their parents could not care for them or because of a lack of assistance from the state.

This situation is nothing short of scandalous if we consider that Argentina has an enormous potential for producing sufficient food in quantity and quality to feed several times the country's total population, and that in recent decades there has been a sustained and dramatic increase in agricultural production—mainly of cereals and oilseeds. The country's estimated annual production of cereals and oilseeds is over 70 million tons, almost two tons per capita, with a total of 90 million tons of agricultural produce in general. But these figures come with a caveat: almost half of the country's grain production is soy, virtually all of it genetically modified and for export.

Historically one of the "breadbaskets of the world," Argentina was one of the few Third World countries self-sufficient in food, as well as an important net exporter of grain and other commodities to the world economy. How then can it be explained that a country like Argentina has become submerged in hunger and misery? How could this increase in agricultural production and overall supply of food be accompanied by a significant increase in hunger and poverty? Or, even more dramatically, why is it that one hundred children a day die of hunger in one of the "breadbaskets of the world"? What was the role played in all this by the dramatic boom in the production of genetically modified soybeans?

A useful point of departure for considering these problems is the approach outlined by Amartya Sen in his analysis of famines occurring throughout the world. According to Sen, extreme food shortages and famines occur when the population loses its access to food *entitlements*. Entitlements are what allow the population to obtain certain goods, in this case the food necessary to cover basic food (nutritional) needs. Sen criticizes the FAO and other international organizations for focusing almost exclusively on the promotion of agricultural output. He argues that famine occurs not necessarily because of a lack of food but rather because large sections of society do not have access to it. A famine is a situation in which many people have no food to eat. This is not the same as saying that the community lacks sufficient food resources (see Sen 1981; Sen 1997 [1984], Chaps. 18, 19, 20; Teubal and Rodríguez 2002, Chap. 8).

In Sen's terms, Argentina suffered the collapse of her food *entitlement* relations, that is, the capacity large sectors of society had for covering their basic (food) nutritional needs. In this sense, the "Argentine case" may be read as a model case study of hunger in the presence of an abundant supply of food resources. Unlike other countries, Argentina experienced the collapse of her *food security*—the capacity to supply all her inhabitants with good quality food—although this was not necessarily comparable to the famines occurring in the 1970s, for example, in Bangladesh or Ethiopia (Sen 1981).

Nevertheless, as is pointed out by Sen, "access to food," or more generally to food *entitlements*, depends on the institutional arrangements that exist in a particular society at a specific historical moment. Therefore, what first comes to mind in considering the case of Argentina is the need to analyze the food situation as an integral part of the current crisis, that is to say, in the context of the *structural adjustments* and agricultural model implemented in the country since the 1990s. Indeed, the fundamental causes can be sought in what in this book is termed *neoliberal globalism* (Otero 2004; see Chap. 1, this volume): the ideology that promotes privatization processes, extreme deregulation—particularly of the labor market—and opening up to the world economy, all of which were applied even more drastically in Argentina than elsewhere. This had a significant effect on the levels of poverty and unemployment, as well as on real wages and incomes of the poorest sectors of society.

In this chapter, I intend to show how the issue of access to food is closely linked not only to the *neoliberal globalism* introduced in Argentina but also specifically to the *agricultural and agroindustrial model* that was implemented and that formed part of it. This model is promoted and dominated by large transnational corporations and the technology they control: supermarkets at the distribution end of the food chain, large corporations that control the food industry, a highly concentrated financial sector, and the seed and agrochemical industries, among other sectors (see Teubal and Rodríguez 2002).

I focus on the neoliberal reforms of the 1990s that, combined with aspects of the revolution in biotechnology, transformed the agrifood system and agriculture within it. I also intend to demonstrate how factors inherent in these transformations significantly affected the population's access to food.

One aspect of the *industrial agriculture* model (Vallianatos 2001, 2003) implemented in the country was the widespread introduction of genetically modified crops, particularly genetically modified soybeans. Indeed, Argentina now produces more genetically modified produce than any other Third World country, precisely because of the boom in the production and export of genetically modified soy. In this chapter I intend to show the adverse consequences of this process on Argentina's *food sovereignty*.

From a Breadbasket of the World to the Soybeans Republic?

In this section I present data that illustrate the boom in overall agricultural output in recent decades as well as the growth in importance of soybeans, which, from the mid-1990s onward, increasingly are genetically modified. I then link these trends with the structural adjustments applied both in the agricultural sector and in the economy as a whole in the 1990s. At the same time, we highlight how this transformation helped undermine Argentina's food self-sufficiency and led to the exclusion of numerous farmers.

Argentina in the twentieth century, together with countries like Australia, Canada, and even the United States, was an important supplier of meat and grain to the world economy. At the same time, these exports—beef, wheat, corn, sunflowers, etc.—were also basic foods consumed in the domestic market. Production was concentrated mostly in the Pampa, while other regions focused on traditional industrial crops for domestic consumption: sugarcane (in the provinces of Tucuman and Salta, in Northwestern Argentina [NWA]), cotton, *yerba mate* (in Chaco and Misiones respectively in Northeastern Argentina [NEA]), etc. Furthermore, apples and pears produced in the Rio Negro valley in Patagonia, as well as wine in the Cuyo region (Mendoza and San Juan), were products that began to be exported. Argentina thus produced almost all the food consumed by her population, with the exception of some tropical products, such as coffee and palm hearts.

This production potential was based in large measure on small and medium farmers, who formed an important part of Argentine agriculture, more important than in other Latin American countries. Using census data categories of farm sizes, "medium and large multi-family farms" (basically the *latifundio* sector of large estates) occupied over half of Argentina's farmland and production in 1960 (in Brazil, Chile, Ecuador, and Guatemala this proportion was higher). But "rural family farms," or "small and medium producers," occupied 45 percent of agriculture land and accounted for 47 percent of total output, a larger proportion in both cases than in other Latin American countries. Moreover, the *minifundios* (small holders), farmers that form part of the peasant economy, occupied only 3 percent of the land, compared to 17 percent in Ecuador and 14 percent in Guatemala (Feder 1975 [1971], Table 18, p. 102). These data show the relative importance in Argentina of U.S.-type family farmers and the relative insignificance of the traditional subsistence peasant, except outside the Pampa.

In the 1970s, new grain and oilseed varieties were introduced in the Pampa, permitting farmers to harvest two crops a year instead of alternating crop and cattle production. This was possible because new varieties allowed the sowing of a secondary crop that began to be dominant in combination with wheat production. The boom in soybean production began at this time, since the introduction of a Mexican germplasm in wheat gave rise to wheat-soybeans dual-cropping. According to de Obschatko and Piñeiro, "wheat-soybeans dual-cropping spread rapidly in the *pampas*, especially in the sub-region that typically produced corn. This new dual-cropping partially supplanted corn and sorghum as well as livestock production, which had traditionally formed part of a mixed production system" (de Obschatko and Piñeiro 1986, 11). The essence of the new *agriculturalization* of the Argentine countryside was based on soybeans and the accompanying technological package that it required. It expanded largely at the expense of the production of cattle and traditional grains.

In many respects this process represented the belated application of the Green Revolution to Argentine agriculture (see Chap. 2, this volume).

From that point on, Argentina took on the characteristics of a "newly agricultural country," to use the term coined by Harriet Friedmann (1993, 45), by analogy with the "newly industrialized countries" (NIC) in South East Asia. Soybean production has increased every year since (see Table 8.1). Although Friedmann refers fundamentally to Brazil, her vision is equally applicable to Argentina. The two countries, together with the United States, have become the largest soybean exporters in the world economy (see Chaps. 9 and 10, this volume).

Introduction and Boom of Genetically Modified Soybeans

In the mid-1990s, another innovation was introduced in Argentine agriculture. In 1997, farmers began to plant the genetically modified soybean seed known as Roundup Ready® or RR, by Monsanto's trademark. Roundup® is the commercial brand of glyphosate, the herbicide to which the genetically modified soybean seed is resistant. A new technological package combines this seed with the glyphosate, used in ever larger quantities in a new *no-tillage* system which does not require plowing the land. This new genetically modified seed and the Roundup® glyphosate were designed by Monsanto, although it was transferred under license to Asgrow, which later was bought by Nidera. Monsanto also produces the glyphosate necessary to eliminate the weeds that are left behind due to the *no-tillage* system of production.

In this framework, the soybean complex takes on a great significance. A handful of large transnational seed companies, such as Monsanto and Novartis, induce farmers to incorporate a technological package that they control by providing the seeds that are resistant to glyphosate. Thus, not only do they provide the seeds but also the accompanying technological package, including the agrochemicals that producers are obliged to purchase when they plant the genetically modified crop. Farmers become increasingly dependent, not only on agroindustry and supplies of agrochemical products, but also—to a much greater extent than before—on the companies that provide the transgenic seeds.

As a result of the widespread use of this technology, glyphosate became the most-used phytosanitary product, with total sales growing from 1.3 million liters in 1991, to 8.2 million in 1995, to over 30 million in 1997. Sales amounted to $263 million in 2000, representing 42 percent of the entire agrochemical market (Teubal and Rodríguez 2002). In 2003, they generated sales for an estimated $350 million (Domínguez and

Sabatino 2003). The wheat-soybeans combination, together with corn, which has also seen a genetically modified strain (*Bt* corn represented 30 percent of production at the beginning of the decade), is now among the most "dynamic" crops of Argentine agriculture.

The Soybeans Boom and Agricultural Production

Tables 8.1–8.4 illustrate the soybeans boom in Argentina in the period 1980/1981 to 2002/2003, in its relation with other grains and oilseeds. It is worth noting that, since 1997, when production of the genetically modified soybean variety was introduced, the proportion of total soybean production that was genetically modified increased substantially, reaching almost 100 percent in 2004. This is also a product that is almost entirely exported.

Soybean production has enjoyed an uninterrupted boom since the early 1970s. According to the Secretaría de Agricultura, Ganadería, Pesca y Alimentación (Secretariat of Agriculture, Livestock, Fisheries and Food) (SAGPyA), the 1980/1981 harvest amounted to 3.7 million tons, rising to 10.8 million tons in 1990/1991 and to around 35 million tons in the 2002/2003 harvest, representing almost half the total production of cereals and oilseeds. Wheat production also increased in this period, reaching a high of almost 16 million tons in 1996/1997, a figure only slightly surpassed in 2000/2001. Durum wheat production, which is generally higher quality, also peaked in 1997/1998. By contrast, corn production fell substantially, from almost 13 million tons in 1980/1981 to 7.7 million tons in 1990/1991. But it later rebounded, reaching a record of more than 19 million tons in 1997/1998. This bumper harvest was not repeated subsequently in the 1998/1999–2002/2003 period.

The country's other main traditional oilseed is sunflower, of which Argentina is the world's largest exporter. Production of this crop also rises systematically over the period under consideration, peaking toward the end of the 1990s. Rice production, partially tied to Brazilian demand, follows the patterns of wheat and corn.

In comparison to other crops, the advance of genetically modified soybean production is striking. In 1997/1998, when the *no-tillage* genetically modified soybeans production began, output amounted to 18,732,172 tons. As mentioned above, today's production is almost 35 million tons, all of which is genetically modified. Over this same period, production of sunflowers dropped by 1,885,880 tons, corn by 4,316,127 tons, and rice by 293,505 tons (see Table 8.1).

Table 8.1. Argentina: Production of Principal Cereals and Oilseeds (tons)

	Rice	Corn	Sunflowers	Wheat	Durum Wheat	Soy	Others+*	Total*
1980/81	286,300	12,900,000	1,260,000	7,780,000	194,700	3,770,000	9,387,000	35,578,000
1990/91	347,600	7,684,800	4,033,400	10,992,400	44,200	10,862,000	4,274,700	38,239,100
1996/97	1,205,140	15,536,820	5,450,000	15,913,600	193,103	11,004,890	3,816,346	53,119,899
1997/98	1,011,135	19,360,656	5,599,880	14,800,230	286,590	18,732,172	6,066,606	65,857,269
1998/99	1,658,200	13,504,100	7,125,140	12,443,000	157,600	20,000,000	4,717,000	59,605,040
1999/00	903,410	16,780,650	6,069,655	15,302,560	176,100	20,135,800	5,005,991	64,374,166
2000/01	873,183	15,359,397	3,179,043	15,959,352	187,270	26,880,852	4,920,525	67,359,622
2001/02	709,295	14,712,079	3,843,579	15,291,660	136,160	30,000,000	4,549,860	69,242,633
2002/03	717,630	15,044,529	3,714,000	12,301,442	97,600	34,818,552	4,106,616	70,800,369

*90/91 does not include Rape and Safflower; 96/97 does not include Rape.
"Others" includes Grain Sorghum, Canary Grass, Oats, Beer Malt, Flax, Groundnuts, Safflower, Rape, Rye, Feed Barley, and Millet.
Source: Prepared using data from SAGPyA.

These data reflect a clear trend toward monocropping of genetically modified soy. They are backed up with data showing the area sown with cereals and oilseeds. As can be observed in Table 8.2, the area sown with soybeans rises from almost 2 million hectares in 1980/1981, to almost 5 million in 1990/1991. In 1997/1998, when genetically modified soybeans were first introduced, the crop occupied more than 7 million hectares. In 2002/2003, the area under cultivation with soy—now almost 100 percent genetically modified—was over 12.6 million hectares. In general, the area dedicated to other grains has tended to diminish. While 3,751,630 hectares of corn were planted in 1997/1998, the figure for 2002/2003 was just over 3 million. At the same time, the area dedicated to "other" crops has dropped substantially, compared with 1980/1981 and 1997/1998. A similar trend can be observed with sunflowers and rice, in relation to 1997/1998 (see Table 8.2).

These data clearly reflect a trend toward "monocropping" in Argentine agriculture. As can be seen in Table 8.3, the proportion of total grain and oilseed production accounted for by soybeans has risen from just over 10.6 percent in 1980/1981, to about half of total production in 2002/2003. On the other hand, the proportion of the total production accounted for by wheat, corn, and "other" crops drops over the same period from 29.2 percent, 18.8 percent, and 35.5 percent to 23 percent, 11.3 percent, and 10.4 percent, respectively. Although there was a minor boom in sunflower production at the end of the 1990s, this represented only a slightly higher proportion of the total production in 2002/2003 than it did in 1980/1981 (see Table 8.3).

The tendency toward monocropping is also evident in the proportion of land sowed with different cereals and oilseeds. The importance of soy, as a proportion of total area planted with cereals and oilseeds, increases systematically, from 9.1 percent in 1980/1981 to 46 percent in 2002/2003. Over the same period, the importance of all other crops—except sunflowers and rice—declines as a share of total area sown, in particular since the mid-1990s, when genetically modified crops were introduced (see Table 8.4).

Notice a certain increase in soybean yields per hectare until 1997/1998. But this phenomenon is not observed in the following years, when yields level off. Meanwhile, in about the same period, improvements in productivity of other crops, such as corn and wheat, also diminish. In other words, there is no evidence that the technological breakthrough that takes place with the introduction of genetically modified soybeans results in substantial increases in productivity with a significant impact on

Table 8.2. Argentina: Area Sowed with Principal Grains and Oilseeds (Hectares)

	Rice	Corn	Sunflowers	Wheat	Durum wheat	Soy	Others+*	Total*
1980/81	84,800	4,000,000	1,390,000	6,196,000	90,400	1,925,000	7,535,000	21,221,200
1990/91	98,000	2,160,100	2,372,350	6,178,400	19,800	4,966,600	4,263,100	20,058,350
1996/97	226,573	4,153,400	3,119,750	7,366,850	83,250	6,669,500	3,951,185	25,570,508
1997/98	247,500	3,751,630	3,511,400	5,918,665	81,615	7,176,250	4,087,530	24,774,590
1998/99	290,850	3,270,250	4,243,800	5,453,250	73,700	8,400,000	3,887,785	25,619,635
1999/00	200,700	3,651,900	3,587,000	6,300,000	69,800	8,790,500	3,544,305	26,144,205
2000/01	153,732	3,494,523	1,976,120	6,496,600	67,800	10,664,330	3,443,585	26,296,690
2001/02	126,435	3,061,661	2,050,365	7,108,900	47,650	11,639,240	3,068,687	27,102,938
2002/03	135,170	3,084,374	2,378,000	6,300,210	42,800	12,606,845	2,853,641	27,401,040

*90/91 does not include Rape and Safflower; 96/97 does not include Rape.
"Others" includes Grain Sorghum, Canary Grass, Oats, Beer Malt, Flax, Groundnuts, Safflower, Rape, Rye, Feed Barley, and Millet.
Source: Prepared using data from SAGPyA.

Table 8.3. Argentina: Production of Principal Grains and Oilseeds (percentage of total)

	Rice	Corn	Sunflowers	Wheat	Durum wheat	Soy	Others+*	Total*
1980/81	0.8%	36.3%	3.5%	21.9%	0.5%	10.6%	26.4%	100.0%
1990/91	0.9%	20.1%	10.5%	28.7%	0.1%	28.4%	11.2%	100.0%
1996/97	2.3%	29.2%	10.3%	30.0%	0.4%	20.7%	7.2%	100.0%
1997/98	1.5%	29.4%	8.5%	22.5%	0.4%	28.4%	9.2%	100.0%
1998/99	2.8%	22.7%	12.0%	20.9%	0.3%	33.6%	7.9%	100.0%
1999/00	1.4%	26.1%	9.4%	23.8%	0.3%	31.3%	7.8%	100.0%
2000/01	1.3%	22.8%	4.7%	23.7%	0.3%	39.9%	7.3%	100.0%
2001/02	1.0%	21.2%	5.6%	22.1%	0.2%	43.3%	6.6%	100.0%
2002/03	1.0%	21.2%	5.2%	17.4%	0.1%	49.2%	5.8%	100.0%

*90/91 does not include Rape and Safflower; 96/97 does not include Rape.
"Others" includes Grain Sorghum, Canary Grass, Oats, Beer Malt, Flax, Groundnuts, Safflower, Rape, Rye, Feed Barley, and Millet.
Source: Prepared using data from Table 8.1.

Table 8.4. Argentina: Area Sowed with Principal Grains and Oilseeds (Percentage of Total)

	Rice	Corn	Sunflowers	Wheat	Durum wheat	Soy	Others+*	Total*
1980/81	0.4%	18.8%	6.6%	29.2%	0.4%	9.1%	35.5%	100.0%
1990/91	0.5%	10.8%	11.8%	30.8%	0.1%	24.8%	21.3%	100.0%
1996/97	0.9%	16.2%	12.2%	28.8%	0.3%	26.1%	15.5%	100.0%
1997/98	1.0%	15.1%	14.2%	23.9%	0.3%	29.0%	16.5%	100.0%
1998/99	1.1%	12.8%	16.6%	21.3%	0.3%	32.8%	15.2%	100.0%
1999/00	0.8%	14.0%	13.7%	24.1%	0.3%	33.6%	13.6%	100.0%
2000/01	0.6%	13.3%	7.5%	24.7%	0.3%	40.6%	13.1%	100.0%
2001/02	0.5%	11.3%	7.6%	26.2%	0.2%	42.9%	11.3%	100.0%
2002/03	0.5%	11.3%	8.7%	23.0%	0.2%	46.0%	10.4%	100.0%

*90/91 does not include Rape and Safflower; 96/97 does not include Rape.
"Others" includes Grain Sorghum, Canary Grass, Oats, Beer Malt, Flax, Groundnuts, Safflower, Rape, Rye, Feed Barley, and Millet.
Source: Prepared using data from SAGPyA and Table 8.2.

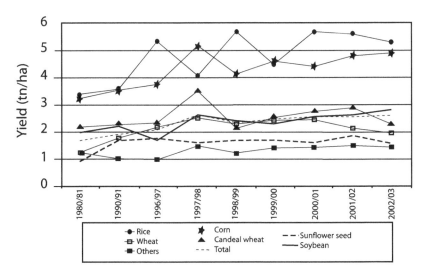

Figure 8.1. Cereal and Oilseed Yields

economic growth. This puts into perspective arguments generally presented with regards to the impact of genetically modified seeds on the overall growth of agriculture in general (see Figure 8.1).

The 2002 National Farm Census confirms the increasing importance of this agriculture model based on the production of genetically modified soybeans in the period between the two last agriculture censuses (1988–2002). Data confirm the increase in the area sown with oilseeds (soybeans and sunflowers) from 5,430,710 to 9,018,447 hectares, representing a growth of around 66 percent. This extraordinary growth is due almost entirely to soy, since, according to SAGPyA data, the area sown with sunflowers in 2002 was similar to that of 1986/1987.

The area under oilseed cultivation increased in many regions: 60.4 percent in the pampas, 86.5 percent in the Northeast, and 138 percent in the Northwest. In the latter region, increases in planted area were at the expense of land devoted to traditional industrial crops, which dropped by 30 percent and 17 percent in the Northeast and Northwest respectively.

The last census also registers a reduction of 3.4 percent in total and incorporated into farming in the 1988–2002 period. Thus, the expansion of soybean production substituted for other farm products: sweet potatoes and sugarcane in Tucuman, dairy farms in Santa Fe and Cordoba, cotton in Chaco, fruit in the pampas, etc. There was also a notable decline in livestock production nationwide, affecting cattle, sheep, and pigs.

To sum up: the boom in genetically modified soybeans came at the expense of a series of traditional farming activities such as milk production, cattle farming, industrial crops, fruit farming, etc. Another cost to be considered was the advance of the farming frontier, which displaced native shrubland and the "yunga," especially in the provinces of Chaco, Santiago del Estero, and Salta.

Although soybeans have spread across the length and breadth of the country, the most important regions in the new map of soybean production are Santa Fe, Cordoba, and Buenos Aires. In the province of Cordoba, soybean production leaped 62 percent since 1988, but this growth was accompanied by a drop of 17 percent in cattle herds and dairy farms throughout the province. In Santa Fe a similar process was experienced. Soybean production there increased by 59 percent, also at the expense of milk production. Between 1988 and 2003, the number of dairy farms throughout Argentina dropped from 30,141 to 15,000 (www.sagpya .mecon.gov.ar).

A study pertaining to the province of Cordoba indicates that

out of the six million hectares dedicated to agriculture over the past 30 years, more than three-and-a-half million have been turned over from traditional meat and milk production to crop farming. This increase would not be problematic if the new area had been adequately exploited, with rotation of directly sowed crops, the use of fertilizers to provide necessary nutrients according to the volume of the harvest, and the adequate control of pests, weeds and diseases. But agricultural expansion has mainly consisted of soybeans monocropping, with a very limited use of fertilizers. This has led to the physical and chemical deterioration of the soil . . . Monocropping of any grain is detrimental for the sustainability of any productive system . . . The specific case of soybeans is worsened by the fact that it produces little stubble, thus favoring erosion. (Salinas et al. 2003)

These overall trends have been corroborated by a series of studies on particular provinces and regions, as well as concrete case studies showing the displacement of crops or livestock production by soy. These studies also point out the importance of the soybean boom as a leading example of how the agricultural model is based on the activities of large companies and producers, which are devoted to supplying foreign markets and generating foreign currency, to the detriment of production of basic consumer foods produced by small- and medium-sized farmers and peasants.

The decline of rice production is a case worth considering. Since the appearance of genetically modified soybeans, the area dedicated to this prominent component of the national diet dropped 47 percent (see Table 8.2). It is surely no coincidence that in Entre Rios, where this rice crop was widespread, soybean production has increased by 101 percent. Where once there was rice, now there is genetically modified soy.

The case of cotton has serious overtones. "The cotton harvest will be the worst of the past 80 years" (*Clarín*, 8/3/03).

> After years of exporting cotton, this year we will import an estimated 35,000 tons of fiber, for a value of 80 million dollars. . . . The area once destined to the cultivation of what used to be known as "white gold" has disappeared. It has been occupied basically by soy, the country's star crop. . . . Mainly large farmers are those who made the switch. The phenomenon is not as significant in the case of small producers and peasants, who have lower costs at harvest time because of the use of family labor. (*La Nación*, 1/3/03, quotes from Domínguez and Sabatino 2003.)

To further illustrate this case, according to census data in the cotton-producing provinces of Chaco and Formosa, soybean cultivation increased by 120 percent and 361 percent respectively. "The soybeans boom replaced cotton monocropping in Chaco, where only 15 of the 64 licensed gins are in service and over one thousand harvesters are idle because of a lack of production" (INFOSIC, 24/5/03). While the textile industry is today recovering, the cotton it uses is no longer produced locally, as the area under cultivation has fallen 40 percent in Chaco and 78 percent in Formosa. As a result, cotton to a large extent is being imported.

In other places, such as Bandera in Santiago del Estero, 200,000 hectares were planted with soy, making Santiago del Estero the fourth largest soy-producing province in Argentina. In 1995/1996—before an RR soybean was introduced—the area under oilseed cultivation in that province amounted to 94,500 hectares. The latest National Farm Census (NFC) records 315,000 hectares cultivated with oilseeds, mainly soy. In the province of Catamarca, producers are harvesting soybeans twice a year. It is believed that this method could spread throughout the irrigated area of Northwestern Argentina (Backwell and Stefanoni 2003).

In sum, the *soyification* process is not limited to the pampas; it includes all the regions of Argentine agriculture. According to Walter Pengue (2000), an expert in Vegetable Genetic Improvement at the University of Buenos Aires, "other crops and productive processes are being replaced.

If this could be changed the following year, it would not be a problem, but in reality entire areas of shrubland, fruit and dairy farms are being ripped up and eliminated, in order to plant soy, all of which affects the diversity of production." According to Pengue, "we are witnessing the beginnings of shortages of some of the basic components of the Argentine diet, because we are installing a model of monoproduction and soybeans are becoming ubiquitous" (cited in Backwell and Stefanoni 2003).

The genetically modified soybean "revolution" is supplanting crops traditionally destined for the domestic market. Some of these did not generate foreign currency to the same extent as soy, but they did ensure a varied, sufficient, and accessible supply of foodstuffs for the local population. In short, there has been a change in the geography of Argentine agriculture, previously oriented much more to satisfying the needs of the domestic market (by means of a diversified supply). Traditional crops have been displaced by the monoculture of soybeans for export. This process is to have numerous social and economic consequences.

Neoliberal Globalism and the Agrifood System: The Impact on Agriculture

Neoliberal structural adjustments implemented in the economy in the 1990s had numerous effects on the agrifood system as a whole, and especially on agriculture. A key element that significantly influenced agriculture was the Deregulation Decree of 1991 that eliminated, at a single stroke, the various agriculture boards that had regulated agriculture activities since the 1930s. Argentine agriculture suddenly became one of the most deregulated in the world, subject more than any other to the vagaries of the world economy (see Teubal and Rodríguez 2002, Chap. 7). Here lie some of the reasons for the lack of active policies tending to regulate the production of basic foodstuffs in the country, as well as policies that could have formed the basis of support for medium farmers and peasants.

Privatization and deregulation processes, as well as an almost indiscriminate opening up of markets to foreign competition, were measures applied throughout the economy, presumably so as to achieve "a greater integration into the world economy." They had significant impacts on the patterns and varieties of agricultural activity, on production prices, access to credit, overall profitability, and the living conditions of the majority of the groups that make up agriculture. Agriculture was also affected by transformations in the agrifood system as a whole, of which it is a part—that is, changes in industrial processing of agricultural produce, in the marketing and distribution of food oriented to domestic and export markets, etc.

Thus, structural adjustments influenced transformations in the agrifood system as a whole, which in turn affected the agricultural sector. On the one hand, there was concentration and centralization of capital in food processing agroindustry and in the wholesale and retail distribution of food (the spread of supermarkets). On the other, a reduced group of companies achieved exclusive control over the supply of seeds and other inputs to farmers. These trends went along with an increase in foreign ownership of many firms, especially toward the end of the nineties. Together with increased vertical integration in many agro-industrial complexes there were significant changes in the relations within the agrifood system as a whole. Increased vertical integration was accompanied by contract farming and other "agro-industrial" articulations and forms of organization. Large non-agricultural companies of the agrifood system acquired greater power in relation to small and medium farmers, who tended to lose their power (capacity) for autonomous decision making (Teubal and Rodríguez 2002).

The liberalization, opening up to foreign competition, and deregulation of the economy offered large companies a favorable framework for expanding their control over key segments of the agrifood system as a whole. They thus obtained a dominant position in storing, processing, and marketing of agricultural produce and in the production and supply of seeds and other agriculture inputs. These processes led to the consolidation of oligopsonies and/or oligopolies in various agroindustrial complexes: in the dairy complex, seven companies, notably SanCor and Mastellone, which control 80 percent of the market; six companies—Molinos Río de la Plata, Cargill, and Aceitera General Deheza, among others—account for 90 percent of refined sunflower oil sales. Fargo, Bimbo, and La Veneciana account for 85 percent of bread product sales, while the French company Danone (Bagley), Nabisco from the U.S. (Terrabusi, Mayco, Capri, and Canale) and Arcor have cornered 80 percent of the biscuit market ("Cash," economic supplement of *Página 12*, 6/6/2004). In other cases, the concentration and centralization of capital was instrumental in cornering highly profitable business areas, such as the Hilton quota,[1] where five companies (Swift Armour, Quickfood, Friar, Gorina, and Finexcor) dominate 55 percent of the market. The joint share of the leading ten companies in the meat sector is as high as 77 percent of the export market. In the case of transnational corporations dedicated to grain exports, seven companies (Cargill, Bunge, Nidera, Vincentín, Dreyfus, Pecom-Agra, and AGD) account for 60 percent of the volume of exports. In line with these trends, there was also a noticeable increase in the concentration of the market for farm inputs with a large dependence on the seeds and "technological package" provided by Monsanto in soybeans and corn.

Meanwhile, the marketing of food also became increasingly concentrated, with the boom in the expansion of supermarkets (both wholesale and retail). This process introduced a new dynamic in the agrifood system, since hyper- and supermarkets "as powerful clients of the food industry changed the previous rules of marketing and the participants' relative negotiating power" (Gutman 1999, 36). In vegetable oils dominated by soy, there was also an increased concentration of capital. The oilseed complex is the country's second most important after cereal production, but it has become the main export complex. Nevertheless, the processing of oilseed products does not generate much employment, even less than the tobacco sector (Teubal and Rodríguez 2002, 70–71). One aspect related to the soybean complex is the concentration of soybean oil and flour exports. While the top five and top eight exporters in 1990 accounted for 50 percent and 67 percent, respectively, of the volume of soybean flour exports, by 2002, the equivalent figures were 79 percent and 92 percent. A similar situation occurred with soybean oil exports: the five and eight leading exporters of this item accounted for 53 percent and 72 percent of exports respectively in 1990, while the equivalent figures for 2002 were 80 percent and 92 percent (CIARA 2003).

Increased concentration and centralization of capital in the food industry, in the retail and wholesale distribution of food, and in the supply of seeds had significant impacts on agriculture. A trend toward greater vertical integration can be observed within the complexes that make up the agrifood system, together with the spread of "contract farming." Farmers—mostly small and medium sized, as well as peasants—have tended to lose their relative autonomy, as well as their capacity to negotiate more favorable terms with regard to prices, credits, and the conditions under which they market their output.

We may conclude that the profound changes wrought on the agroindustrial complexes have restricted farmers' decision-making capacity with regard to production, inputs used, and the choice of production techniques. In addition, producers have experienced a weakening of their power to negotiate prices with agroindustry and brokers and, in the case of genetically modified seeds, with the seed suppliers. All these factors can be compounded by a lack of predictability caused by the volatility of international agricultural prices, which have been directly passed on to farmers since the market was deregulated and opened up to world markets.

Increased concentration of capital in the various complexes has been accompanied by the elimination of all regulations that established minimum or support prices, as a result of the Deregulation Decree of 1991. This has allowed the main decision makers (*polo integrador*) in each agro-industrial complex to increase their profitability by reducing agricultural prices and imposing conditions with regard to quality, presentation, and transport of produce, as well as the farmers' choice of crop variety and inputs, etc. The situation has worsened since genetically modified crops were introduced (see Teubal and Rodríguez 2002, Chaps. 6 and 7).

Social Differentiation and the Agricultural Crisis: The Disappearance of Medium and Small Farmers and Peasants

The major issues facing agriculture and the problem posed by genetically modified soybeans become evident when we consider who were the winners and who the losers over recent years. The great majority—more than 90 percent—of those who participate in agriculture (but not those who own the majority of the land) are small and medium farmers, peasants, and rural laborers. A consequence of the crisis that hit the sector has been the disappearance of a great many agricultural activities, with the bankruptcy and disappearance of numerous cooperatives, businesses, and industries related to the sector, as well as the decline in the living conditions of rural families and the environmental degradation that has occurred under the new model. These problems were particularly evident in the extra-pampa regional economies.

Some commentators have referred in journalistic terms to the "paradox" of the sector, referring fundamentally to the pampas: "How can we explain . . . that in the last decade . . . the harvest and export of grain has almost doubled, with increased investments and the incorporation of technology, but at the same time small and medium-scale producers have been impoverished, experiencing expulsion and concentration" (Seifert 2001).

This is the somber situation in which the agricultural sector finds itself, particularly medium and small farmers and peasants, who are subject more than at any other period to the vagaries of world prices, with little access to credit and handicapped by a series of factors. The wholesale expulsion of farmers from agriculture and, in many cases, their transformation into rentiers who no longer work their own land, has given rise to a "farming without farmers," another emblematic aspect of the country's new agriculture model.

The last NFC of 2002 recorded 318,000 farm units which occupied a total area of 171 million hectares. Compared to the previous census of 1988, this represents a drop of 24.5 percent in the number of registered farms and a 3.4 percent decline in the area under agricultural production (177 million hectares in 1988). According to the INDEC (2003), while the number of farms fell, their average size increased 28 percent, reaching 538 hectares, basically reflecting the disappearance of small farms. It is worth noting that the average size of Argentine farm units is much larger than in the United States, Europe, or elsewhere in the world. In fact, the average size of farms in Argentina reached 470 hectares in 1990, while the equivalent figure for the United States was 182 hectares, in the same year, and just 16.5 hectares in Europe (see Basualdo and Teubal 1998, on the basis of USDA data).

A private survey carried out in much of the Pampa recorded a 31 percent drop in the number of farms in the 1992–1997 period (survey by Mora and Araujo 1997). According to Giberti, "the decline in the number of farms in the U.S. or Europe, for example, has not been anything as rapid" (Giberti 2001, 128). This reflects the clear bias against small and medium farmers inherent in government policies. At the beginning of the nineties, during the Menem presidency, a high-ranking official of the Agriculture and Livestock Secretariat declared that 200,000 small and medium farms would have to disappear, because they were "inefficient."

An Experimental Census in Pergamino carried out by the INDEC in 1999 indicated that the number of farms in that region dropped 24.2 percent in the 1988–1999 period. Moreover, this reduction was greater in the case of land-holdings of up to 5 hectares—which dropped by 38 percent— or of holdings of between 5.1 and 10 hectares—which registered a decline of 44.1 percent. Meanwhile, it should be noted that, as part of the same process of reduction in the number of small landholdings, there was an increase in large farms. The Experimental Census recorded an 18.3 percent increase in the number of holdings of between 500 and 1,000 hectares, and a 38.7 percent rise in those of between 1,000 and 2,500 hectares.

Similar trends were observed in various specific sectors. In Cordoba, the number of dairy farms dropped from 10,102 in 1988, to 7,926 in 1993, a 21.5 percent decline, equivalent to the disappearance of around 435 dairy farms a year. In Santa Fe, the number of dairy farms also fell, from 15,262 to 5,664, between 1975 and 1992 (Blousson 1997). This trend, which saw dairy farms disappearing in the 1990s, accelerated toward the end of the decade, when genetically modified soybeans were introduced. Between 1996 and 2002, for example, the number of operating dairy farms fell from 22,000 to 13,000, a reduction of 40 percent (www.sagpya.mecon. gov.ar). In 2003, the dairy industry declined by another 12.4 percent, and it began to be necessary to import milk from abroad. This decline in the number of dairy farms was due mostly due to their loss of profitability vis-à-vis the production of soy.

A study of the sugarcane sector demonstrates the fall in the number of producers during the 1990s. The study shows that between 1988 and 1996 in Tucuman the number of sugarcane farms fell 25 percent, which meant the disappearance of around 2,500 farms (Giarracca et al. 1997). The 2002 census shows a 40 percent decline in the number of farms in Tucuman, compared to the previous census of 1988.

In the Alto Valle district of Rio Negro, the last agricultural census in 1988 registered 8,000 farms. According to a private study, there were 6,000 farms in 1993, while a recent study by the Federal Investment Board recorded 3,629 (Scaletta 2001).

While small and medium farms declined, the other side of this coin has been the consolidation of large holdings, as mentioned above in the case of Pergamino. The advance of large farm holdings in recent years sets Argentina apart from the majority of the OECD countries. The average size of farm holdings on the Argentine pampas is much larger than in Europe, or countries like the United States, Australia, and Canada, or, for example, in southern Brazil (which has a similar geography and climate).

The expansion of these large landholdings has occurred in conjunction with the appearance and expansion of new methods of production in the pampas: contract farming and "sowing pools" (*pool de siembra*). There are various indications that, both in the pampas and in the country's rural hinterland, stock companies are becoming increasingly dominant at the expense of individual physical firms and family farms. There has also been a boom in large enterprises that own no land, in the form of investment funds or sowing groups that contract large amounts of land. (The former are a specific legally defined entity and are often large in size; the latter are circumstantial groupings, governed by private agreements.) These companies handle large volumes of production, operating under temporary contracts. According to Giberti, there were 79 companies of this type in the pampas in 1997, with total holdings of 600,000 hectares, of which 200,000 corresponded to investment funds. These ventures seek high returns in the short term, due to the increase in tax pressure on landholdings, which means that simply owning land is no longer enough in itself to act as a store of value safe from inflation (Giberti 2001). The participation of the "megaproducers" (Soros or Benetton) also took on increased importance, with the expansion of business ventures formed by groups of investors, operated by agricultural technicians, and administered by private consultancies. These produce on a large scale, using land owned by third parties (investment funds or sowing pools).

Although the disappearance of small and medium farmers is part of a general process of concentration, the arrival of genetically modified soybeans has created a production system that has significantly accelerated the process. The adoption of new technologies, such as glyphosate-resistant soybeans, has eliminated the need for labor to remove the weeds that compete with the crop, since these are eliminated by applying glyphosate. As a result, the introduction of this technological package, apart from increasing the farmer's dependence on agricultural inputs, modifies the sowing process. It establishes the need for new machinery and modifies the previous process of plowing the land. This means an increase in the sales of direct sowing seed drills. The "saving" in labor costs during sowing ranges from 28 percent to 37 percent, in the case of genetically modified soybeans (Domínguez and Sabatino 2003; Teubal and Rodríguez 2002).

In the regional economies, the years immediately after 1991 were also very tough on smallholders and peasants, who were severely affected by the new conditions. On top of the structural problems of low prices and a lack of resources was added the disappearance of all the regulatory mechanisms that served as a framework for negotiations with the large processors or brokers. In some cases, the producers attempted to "flee toward the future," taking out loans in the hope of adapting to the new conditions. Others financed their farm businesses with parallel economic activities. The smallholders in the pampas and the peasants of the North sought other activities in a country where unemployment was mounting on a daily basis (especially in the hinterland, due to the impact of the restructuring and privatization of large companies in the petroleum and steel sectors). Many farmers and peasants attempted a "multi-occupational" approach, one of various family strategies to generate income, or sought help from governmental assistance programs. Nevertheless, conflicts arose almost immediately in these regions, although without the demands being articulated at the national level. It should be remembered that the majority of those who occupied the Plaza de Mayo in the agrarian protest of 1993 came from these regions (Giarracca, Aparicio, and Gras 2001).

At the same time, the cost of services rose due to the privatization process, and the cost of capital increased, despite the implementation of convertibility and resulting economic stability. Convertibility and the resulting overvalued exchange rate induced imports of certain foods from abroad, making soybean production more profitable still in relative terms. The survival and propagation of small and medium farm holdings gradually became more difficult.

The smallholders of the pampas incurred the heaviest debts: since they owned their land, they had the necessary collateral for mortgages and, since they had little operating capital, they resorted to bank loans to finance a supposed overhaul of their production (the incorporation of technology or land to increase the scale of production). So long as international grain and oilseed prices remained high, until 1995, many believed that they could achieve the "professional farm model" predicated by officials and journalists for Argentine agriculture.

At the time, specialists, politicians, journalists, and some academics were predicting a veritable technological revolution. This would apparently transform the country into an intensive producer, but required a scale of production that would exclude half of its farming operations. The supposed advantage of this model was that it would increase exports and reduce food prices. In the most concentrated sectors of food production, however, while agricultural prices remained steady or dropped, food prices rose in relative terms, or failed to drop as far as expected, due to an increase in margins between wholesale and retail prices (see Teubal 1998; Teubal and Rodríguez 2002). This phenomenon, which affected food consumers in general, also affected farmers and rural workers in their role as consumers.

With the decline in international prices of agricultural commodities, the interest on the debts taken out by farmers became unpayable, and together with the deterioration of the national economy, this led them to lose hope in social assistance or the possibility of "escaping into the future." They thus began to participate in social protest measures (see Teubal and Rodríguez 2001; Giarracca 2001; Giarracca and Teubal 2004). Devaluation at the beginning of 2002 and the increase in international soybean prices in subsequent years ushered in a period of relative prosperity for farmers that persists today.

Aspects of the Agrifood or Industrial Agriculture Model:
Final Considerations

The *agroindustrial* or *agrifood* model implemented in Argentina in recent years stems from the North American model of agriculture and agro-industrial development (see Chap. 1, this volume). At a global level, it gathered steam during the so-called "Green Revolution" in the latter half of the twentieth century (see Otero and Pechlaner, Chap. 2, this volume). This model is based on control by large transnational agroindustries over fundamental sectors of the agrifood system—that is, over key aspects of agricultural production, industrial processing, and the retail distribution of agricultural products. It is currently associated with the biotechnological revolution and genetic engineering, as manifested in the widespread introduction of genetically modified seeds. It is a model that encounters stiff resistance internationally, due to the damage it is accused of causing on the environment, in social and economic spheres, and even to human health. Opponents of GMOs point to the possible consequences of their use on biodiversity, the environment, health, and food (see Poitras, Chap. 11, this volume).

This chapter presents an analysis of some of the socioeconomic impacts related to the implementation of this model in Argentina. It is worth noting the impact that the new agricultural model has had in limiting the autonomy of agricultural producers and increasingly pushing them out of existence. In the same way, the loss of food self-sufficiency in the production of basic, mass-consumption food (staple foods) should also be highlighted. For all these reasons it comes as no surprise then that criticism of this model, as well as of the practices and policies of the large agrifood transnationals (agribusiness), constitutes one of the central planks of the world anti-globalization movement, which proposes transcending neoliberalism and that an "other world is possible."

Farmers in Argentina have been at the forefront of the agricultural "modernization" process, for which this model served as a framework. One reason was perhaps the relative abundance of resources in Argentina and the uncritical acceptance of the modernization criteria in the scientific-technological domain. But another factor was that the model was associated with practices that substantially reduced production costs (i.e., no-tillage sowing), which were promoted by the sector's corporate organizations, including the Argentine Agrarian Federation (Federación Agraria Argentina). Indeed, as we have described in this chapter, Argentina became one of the world's main producers and exporters of genetically modified soy. It is also one of the countries that experienced the most rapid growth of *supermarketism* and concentration in agroindustry, displacing countless small and medium, urban, and rural producers, and destroying numerous jobs.

At the same time, there is a growing worldwide opposition to seeds and food with genetically modified components, which may eventually limit Argentina's export capacity, for example to the European Union. In this light, it would probably be a mistake to continue promoting policies in favor of genetically modified seeds, and not consider far more promising alternatives that are appropriate to a more socially sustainable model for the country. These include organic and/or small-scale models of production that could be carried out by medium and small farmers and peasants.

One of the main arguments used by large companies in favor of the adoption of genetically modified seeds rests on the assumption that these represent a solution to the problem of hunger both in Argentina and in the world as a whole. This is the same argument used in the debate on genetically modified organisms (GMO) in general. There are good reasons to believe that this argument is just as illusory as the similar claims that were made at the time for the Green Revolution (see Chaps. 1 and 2, this volume).

There is no doubt that the Argentine agricultural sector has undergone a significant transformation in recent decades, with increased grain production, particularly of oilseeds. Census data illustrate these trends: the expansion of soybeans has been impressive, particularly genetically modified soy, since the mid-nineties. Coincidentally, this is the same period in which hunger and poverty in Argentina increased significantly. Is there a causal relationship between the two trends? The defenders of genetically modified soy, no-tillage sowing, and the use of glyphosate are at pains to deny it.

This chapter concentrated on analyzing the recent transformation of Argentina's agrifood system and some of the consequences of the widespread introduction of genetically modified soybeans as part of the implementation of a new agrifood model. This is a model that has had a significant impact on the capacity of various agricultural sectors to make autonomous decisions with regard to their form of production, their living conditions in general, and the food sovereignty of the country as a whole. All this invites various considerations on the subject.

For one thing, there has been a decline in Argentina's production of traditional, basic, mass-consumption foods that is confirmed by a series of census, sectorial, and national studies. There has been a growing specialization in genetically modified soybeans and an accompanying loss of agricultural and food diversity, vital for the provision of a sufficiently varied and nutritious food supply for the population as a whole. This is true both nationally and at the farm level. At a global level the trend is toward specialization and monocropping; at the farm level, there has been a loss of both biological and productive diversity, and farmers have lost the possibility of producing for their own consumption, which was also a source of food at this level. Together, these elements suggest the country is losing its food sovereignty.

Then there is the disappearance of myriad farms, which has contributed to the increase in unemployment in its various forms. But the development of the agrifood model also affected the prices and quality of mass-consumption foods. Throughout the nineties, food prices increased faster than the general price index. This was evident in 2002, at the height of the crisis, when the increase in the price of milk and other basic foodstuffs was more than double the increase in the general price index. It is evident that this is a phenomenon that affects the access of the population to a decent diet. Some basic foodstuffs, such as milk, are becoming scarce. In summary, the process of concentration, together with the orientation toward exports, based on genetically modified soy, significantly affected domestic food prices (see Teubal and Rodríguez 2002).

Finally, Argentina has become, more than ever, a monoproducer and monoexporter of transgenic soybeans. This represents a great risk for the sustainability of the overall economy. It affects the resources available for tackling social needs, once the proportion of the fiscal surplus that is allocated to foreign debt servicing has been resolved.

In conclusion, the trends outlined in the current chapter have certainly had an impact on the prices of basic foodstuffs and thus on the income and real wages of the population. Together with unemployment, they have had a significant negative impact on access to food, heightening hunger and poverty. In summary, we may conclude that the agrifood model implemented in Argentina contributed significantly to the hunger and misery that can be observed in our society.

The various phenomena highlighted in the current study have tended to steadily reduce the nation's food sovereignty. We mean by *food sovereignty*

> the right of a people to define its own policies and sustainable strategies for food production, distribution and consumption, guaranteeing the entire population's right to food. These rights are based on small- and medium-scale production, respect for local cultures and the diversity of

forms adopted by agricultural workers, fishermen and indigenous peoples for agricultural production, marketing and the management of rural areas, in which women play a fundamental role. (CMA:cad 2002)

These trends in the industrial agriculture and agroindustrial model to which we make reference in this chapter have tended to take us further away from the essential criteria involved in food sovereignty. Surely acting in favor of measures tending toward greater food sovereignty could be the path toward the eradication of hunger and malnutrition that would also guarantee a lasting and sustainable food security for society at large.

The social unrest that has characterized Argentina in recent years has created new forms of social organization for food production and distribution. Community-run gardens, community purchasing, and street and free markets all form part of this phenomenon. Grass roots organizations have also set up a series of production projects—for example, bakeries established by *piketero* (unemployed picketer protestors) organizations to satisfy the needs of their members. Similarly, there is increasing interest in organizing food supplies to schools and the neediest sectors of society, as well as in using the output of small agricultural producers to supply the poorer sectors of society. What is needed is precisely an overall restructuring of the agrifood system as a whole, a reorganization of agriculture, agroindustry, and food distribution, putting them at the service of the community as a whole, and especially its most deprived sectors.

Note

1. The Hilton quota offers the possibility of placing 28,000 tons of beef in the European market. These were high-quality cuts, worth approximately seven dollars a kilo in 2002.

References

Backwell, Benjamín, and Pablo Stefanoni. 2003. "El negocio del hambre en la Argentina. ¿Soja solidaria o apartheid alimentario?" *Le Monde Diplomatique—Cono Sur Edition*, February.

Basualdo, Eduardo, and Miguel Teubal. 1998. "Economías a escala y régimen de propiedad en la región pampeana argentina." Presentation at the XXI International Congress of the Latin American Studies Association (LASA), Chicago, 24–26 September 1998.

Blousson, R. 1997. "El desafío de la lechería." SAGyP, Buenos Aires.

CIARA (Cámara de la Industria Aceitera de la República Argentina). 2003. www .ciara.com.ar.

CMA:cad. 2002. Final Declaration of the Second Regional Consultation of Non-Governmental and Civil Society Organizations in Preparation for the World Food Summit: 5 Years Later. La Habana.

de Obschatko, Edith, and Martín Piñeiro. 1986. *Agricultura pampeana: Cambio tecnológico y sector privado.* Buenos Aires: CISEA.

Domínguez, Diego, and Pablo Sabatino. 2003. "Lo que la soja se llevó." Presentation at the Third Interdisciplinary Symposium on Agrarian and Agroindustrial Studies, PIEAA, Faculty of Economic Sciences, UBA, 5–7 November.

Feder, Ernest. 1975 [1971]. *Violencia y despojo del campesino: El latifundismo en América Latina.* Mexico, Spain, Argentina: Siglo XXI.

Friedmann, Harriet. 1993. "The Political Economy of Food: A Global Crisis." *New Left Review* No. 197 (January–February): 29–57.

Giarracca, Norma. 2001. "El movimiento de Mujeres Agropecuarias en Lucha: Protesta agraria y género durante el último lustro en Argentina." pp. 129–151 in Norma Giarracca, comp., *¿Una nueva ruralidad en América Latina?* Buenos Aires: CLACSO.

Giarracca, Norma, Carla Gras, Paula Gutiérrez, and Karina Bidaseca. 1997. *Conflicto y negociación en el contrato agroindustrial: La actividad cañera desregulada en Tucumán.* Research report, GER, Buenos Aires (photocopy).

Giarracca, Norma, Susana Aparicio, and Carla Gras. 2001. "Multiocupación y pluriactividad en el agro argentino: El caso de los cañeros tucumanos." *Desarrollo Económico. Revista de Ciencias Sociales* 41(162): 305–320. July–September.

Giarracca, Norma, and Miguel Teubal. 2004. "'Que se vayan todos': Neoliberal Collapse and Social Protest in Argentina." pp. 67–90 in Jolle Demmers, Alex E. Fernández Jilberto, and Barbara Hogenboom, eds., *Good Governance in the Era of Global Neoliberalism: Conflict and Depolitisation in Latin America, Eastern Europe, Asia and Africa.* London and New York: Routledge.

Giberti, Horacio. 2001. "Sector Agropecuario. Oscuro panorama. ¿Y el futuro?" *Realidad Económica* (Buenos Aires) No. 177 (January–February): 121–138.

Gutman, Graciela E. 1999. "Desregulación, apertura comercial y reestructuración industrial. La industria láctea en argentina en la década de los noventa." In Daniel Azpiazu, comp., *La desregulación de los mercados.* Buenos Aires: Grupo Editorial Norma.

INDEC. 2003. Resultados Provisionales. Censo Nacional Agropecuario 2002. Press release, Buenos Aires, 25 March.

INFOSIC. www.infosicargentina.com.

Mora and Araujo. 1997. Unpublished survey.

Otero, Gerardo, ed. 2004. *Mexico in Transition: Neoliberal Globalism, the State and Civil Society.* London: Zed Books.

Pengüe, Walter. 2000. *Cultivos Transgénicos. ¿Hacia dónde vamos?* Buenos Aires, Lugar Editorial.

SAGPyA (Secretaría de Agricultura Ganadería Pesca y Alimentación). www .sagpya.mecon.gov.ar.

Salinas, Aquiles, Eduardo Martellotto, Juan Pablo Giubergia, Pedro Salas, and Edgar Lovera. 2003. "¿La intensificación de la Agricultura se está haciendo en forma sustentable?" Economy, Statistics and Information Technology Department, INTA. Marcos Juárez, Córdoba, *Ámbito Financiero*, 19 March.

Scaletta, C. 2001. "Chacareros descontentos y final abierto en el Valle," *Página 12*, 26 May.

Seifert, R. 2001. "La paradoja de la productividad," *La Nación*, 26/05/01.

Sen, Amartya. 1981. *Poverty and Famines: An Essay on Entitlement and Deprivation.* Oxford: Clarendon Press.

———. 1997 [1984]. *Resources, Values and Development.* Cambridge, Mass., and London: Harvard University Press.

Teubal, Miguel. 1998. "Transformaciones en el sistema agroalimentario: El impacto sobre los precios relativos." pp. 75–87 in *Informe de Coyuntura*, Centro de Estudios Bonaerenses (La Plata) Year 8, No. 77, November–December.

Teubal, Miguel, and Javier Rodríguez. 2001. "Neoliberalismo y crisis agraria." pp. 65–116 in Norma Giarracca et al., *La protesta social en la Argentina. Transformaciones económicas y crisis social en el interior del país.* Madrid/Buenos Aires: Alianza Editorial.

———. 2002. *Agro y alimentos en la globalización. Una perspectiva crítica.* Buenos Aires: Ediciones La Colmena.

Vallianatos, E. G. 2001. "All of Africa's Gods Are Weeping." *Race & Class* (Institute of Race Relations, London) 43(1): 45–57.

———. 2003. "American Cataclysm." *Race & Class* (Institute of Race Relations, London) 44(3): 40–57.

CHAPTER 9

Brazilian Biotechnology Governance: Consensus and Conflict over Genetically Modified Crops

WENDY E. JEPSON, CHRISTIAN BRANNSTROM,
AND RENATO STANCATO DE SOUZA

The Brazilian case of biotechnology governance represents a key test in understanding the future global distribution of genetically modified (GM) crops. In Brazil, a legal moratorium against commercial planting of GM crops was in force from 1998 to 2003; however, contraband or "Maradona" soybeans occupied significant areas of cropland, especially in southern Brazil. A policy consensus for biosecurity and regulation of GM crops developed under the center-right coalition of President Fernando Henrique Cardoso (1995–2002). The "Cardoso consensus" allowed experimentation of GM crops, encouraged Brazilian researchers to develop biotechnology expertise, and permitted consumer and farmer groups time to respond to these new initiatives. Opposition to the Cardoso consensus grew from several sources, including Brazilian state governors and important civil society actors, such as environmental and consumer activists and farmers. Opposition has shifted the debate from a technical problem to political, judicial, and environmental terms. A new consensus is taking shape under the new administration of President Luiz Inácio "Lula" da Silva. Temporarily, farmers are allowed to grow GM soybeans from saved (and illegal) seed, coupled with tighter regulation and centralized decision making in the executive branch. Despite federal action, governors, farmers' groups, consumer-rights organizations, and environmental activists will shape the geography of GM production in Brazil, probably maintaining significant GM-free sectors and regions. From this analysis, a sweeping conquest by GM crops of Brazilian agriculture seems unlikely. Even though the federal government allows commercial planting of GM crops, it is not inevitable that GM technology will saturate all production systems in all areas of Brazil.

We begin our discussion of biotechnology governance by focusing on Brazilian agriculture, because Brazil is a major world producer of soybeans, and soybeans are the most cultivated GM crop globally. The chapter then outlines the terms of the consensus on transgenics under the Cardoso administration and the legal moratorium that prevented commercial farming of GM crops. The next section analyzes sub-national politics in opposition to the Cardoso consensus, focusing on the southern state of Rio Grande do Sul, a major soybean producer that developed a set of policies against GM crops. Next, we address opposition to the national consensus by activist and farmer groups. The 1 January 2003 inauguration of President Luiz Inácio Lula da Silva, leader of the leftist Workers' Party (Partido dos Trabalhadores; PT), has contributed definitively to the GM debate. We review the period of ambiguity about GM policy and outline what may be called the "Lula consensus": circumventing the moratorium on GM soybean planting; requiring GMO labeling on consumer products; and reorganizing the national biosafety framework.

Brazil's Agricultural Geography: Between Conventional and "Maradona" Soybeans

Understanding Brazilian soybean production is essential to grasping the significance of the Brazilian biotechnology debate. Brazil has been a major producer of soybeans since the mid-1970s, and accounts for more than 20 percent of world production (Soskin 1988; Warnken 1999). Brazil's 2002 soybean harvest was approximately 42 million metric tons on 15.7 million hectares (USDA 2003, 28–29; Cardoso and Ferreira 2003), and the harvest increased to 56 million metric tons in 2003. For the 2003 harvest, 67.8 percent of Brazil's soybean production is in three states: Rio Grande do Sul produced 8.8 million tons (15.7 percent), Paraná produced 12 million tons (21.4 percent), and Mato Grosso produced 15 million tons (26.7 percent) (USDA 2003, 28–29). Brazil's average soybean yield, 2,708 kilograms per hectare, is slightly higher than U.S. and Argentine average yields (Aliski 2002; Cardoso 2001; USDA 2002).

Brazilian soybean exports depend upon European and East Asian markets. In 2001 European countries purchased nearly two-thirds of Brazil's raw soybean exports and more than three-quarters of its soymeal exports. Holland and France are Brazil's main importers of soybean oilcake (28 and 19 percent of Brazil's soybean oilcake exports, respectively) (USDA 2004, 13). Holland is a particularly significant importer, purchasing 21 and 18 percent, respectively, of Brazil's raw soybean and soymeal exports, while China imports 31 percent of Brazil's raw soybean exports (USDA 2002; USDA 2004, 12). Overall, Brazil's soybean export revenues were predicted to reach US$8 billion, or 12 percent of the country's export revenues, in 2003, a year-on-year increase of US$2 billion (Salvador 2003e).

Herbicide-resistant soybeans are the most popular GM crop planted worldwide. In 2003, the global area of GM crops was 67.7 million hectares, mainly in the United States (42.8 million hectares) and Argentina (13.9 million hectares), one of Brazil's southern neighbors. In 2003 Brazil cultivated approximately 3 million hectares of transgenic soybeans, according to the ISAAA (ISAAA 2004). Soybeans represented 61 percent of world GM crops in area, while maize (23 percent) and cotton (11 percent) were well behind at 15.5 million hectares and 7.4 million hectares, respectively (ISAAA 2004).

Contraband herbicide-resistant soybean seeds, imported illegally from Argentina, are known in Brazil as "Maradona" soybeans, referring to Diego Armando Maradona, a famous Argentine soccer player. The name is as much a reference to the Argentine origin of the GM seeds as to their adulterated or altered state, and their "addiction" to herbicides; Maradona was expelled from the 1994 World Cup for a positive drug test. Maradona soybeans have established a significant presence in southern Brazil and have generated wild speculation on their presence nationally. The most credible source on the extent of Maradona soybeans in Brazil is Luiz Antonio Barreto de Castro, head of EMBRAPA's genetic resources unit, who suggested that two-thirds of Rio Grande do Sul's soybean crop, and nearly 10 percent of Brazil's soybean area, is transgenic (Cardoso 2002). However, it is important to note that there is a tendency to overestimate GM crop production, as inflated numbers serve the interests of both detractors and supporters of GM technology. According to various sources in agribusiness, Maradona soybeans in Rio Grande do Sul account for between 70 and 95 percent of total soybean area (USDA 2003, 38; USDA 2002, 39; Luccas and Stancato 2002; Hahn 2001; Gazeta Mercantil 2002). In 2002 GM soybeans appeared in at least 56 separate fields in the state of Paraná (Angelo 2002) and, in neighboring Santa Catarina state, 10.7 tons of GM soybeans were incinerated later the same year (Meurer and Kiefer 2002). The USDA estimated that 15 percent of Paraná's 2004 soybean crop is transgenic (USDA 2004, 25). Maradona soybeans have even entered settlements of the Landless Workers Movement (Movimento dos Trabalhadores Rurais Sem Terra; MST), where MST farmers allege that the contraband seeds are relatively cheap and reduce production costs (Cardoso 2003).

Table 9.1. GM Crop Experiments According to Trait (to January 2001)

Resistance	Cotton	Maize	Soybeans	Sugarcane	Other crop
Herbicide resistance	39	377	51	11	4
Insect resistance	49	359	12	2	2
Disease resistance	0	0	0	4	4
Herbicide and disease resistance	0	25	2	1	0
TOTAL	88	761	65	18	10

Source: CTNBio 2001.

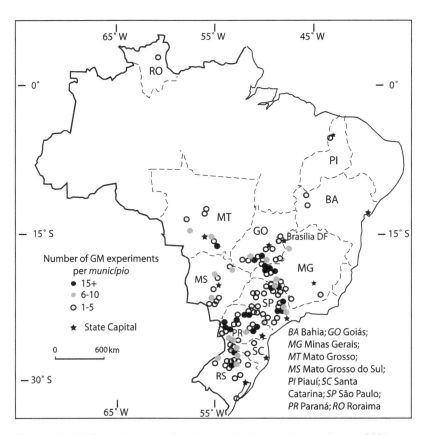

Figure 9.1. GM Experimentation Sites in Brazilian Municipalities to January 2001. (Note: Only states with GM experiments are indicated.)

However, GM soybeans are not the only transgenic crop in Brazil. Between 1997 and January 2001, the Brazilian government approved 942 experiments on GM crops, including maize (761), cotton (88), soybeans (65), and sugar cane (18) (CTNBio 2001). Experiments have been conducted by national agricultural research institutes and transnational corporations, such as Cargill, Novartis, and Monsanto, primarily for herbicide and pesticide resistance in soybeans and maize (Table 9.1) (CTNBio 2001). Experiments on GM crops have been concentrated in Brazil's core agricultural regions, including the states of Paraná, São Paulo, Minas Gerais, and Goiás (Figure 9.1).

Creating the Cardoso Consensus, 1995–2002

A policy consensus under the center-right coalition of President Fernando Henrique Cardoso (1995–2002) comprised federal policies and judicial rulings, which limited the release of genetically modified organisms into the environment and restricted commercial production of transgenic crops in Brazil (Jepson 2002). There are three key components of this consensus. First, the federal government allowed and regulated transgenic research and experimentation through the National Biosafety Technical Commission (Comissão Técnica National de Biossegurança; CTNBio). Second, the government also permitted commercial products, if properly labeled, to contain GM material. Finally, the Brazilian judiciary maintained a broad moratorium against commercial planting of GM crops, with the exception of GM soybeans, which were exempted from the ban in 2003.

Federal Regulation and Judicial Challenges

The Brazilian government passed the Biosafety Law (1995) to control the use of genetic engineering techniques and environmental release of GMOs (Brazil 1995a, 1995b). The 1995 Biosafety Law, which authorized the executive branch to form CTNBio, allowed the commission to regulate experimentation, registration, use, transportation, storage, commercialization, liberation, and waste removal of genetically modified materials. CTNBio, which included biotechnology specialists, government officials, and private biotechnology sector representatives, was authorized to develop technical norms and procedures to deal with biosafety hazards, outline a code of ethics, and assess GM risk for levels of biosecurity. The commission also conferred biosafety quality certificates for GM experimentation and outlined protocols for GM food labeling.

Prohibition of commercial GM cultivation did not originate in CTN-Bio, but from a lawsuit initiated in 1998 by a coalition of nongovernmental organizations (NGOs) opposed to transgenic biotechnology. Although Monsanto won approval from CTNBio to commercialize its herbicide-resistant transgenic soybean (Roundup Ready® soybean; RRS) in 1998, Brazil's nongovernmental Institute of Consumer Defense (Instituto Brasileiro de Defesa do Consumidor; IDEC) and Greenpeace filed suit in federal court to stop this action. The specific technical issue was whether CTNBio had authority to approve commercial transgenic crops without requesting a report demonstrating that transgenic organisms posed no threat to the environment. The government argued that the soybean did not require an environmental impact statement because the GM variety was biochemically identical to conventional soybeans. However, IDEC argued that RRS was substantially different than conventional soybeans. Therefore, under the 1988 Brazilian Constitution, only the environment ministry could issue a report that explicitly permitted the release of this transgenic material into the environment. In June 2000, after almost two years of appeals by Monsanto, federal judge Antonio Souza Prudente declared that CTNBio's authority to waive environmental impact statements was unconstitutional (USDA 2001a; USDA 2001b; Luccas and Stancato 2002; Soares 2003).

In February 2002 another federal judge, Selene Maria de Almeida, ruled to suspend a previous ruling by Judge Prudente, which had prohibited commercial planting of GM crops without prior environmental impact statement and report. But Judge Almeida's colleague, Antonio Ezequiel, delayed the case by asking to examine the documents again, effectively suspending the ruling. Judge Almeida had argued that CTNBio's decision was based on technical study and that CTNBio had proved that GM soybeans presented no risk to human or animal health. Judge Almeida had ruled on CTNBio's legitimacy to authorize GM planting (Gallucci 2002).

The suit was stalled for the rest of 2002. Following the inauguration of the Lula administration, the attorney general's office requested suspension of Judge Almeida's ruling. This request, filed in February 2003, was formalized after Marina Silva, the environment minister who is a well-known opponent of GM crops, sent an official note to the attorney general. However, Judge Almeida rejected the appeal, alleging that the ruling process had already begun (Salvador 2003b).

In August 2003, Judge Almeida released a new ruling that permits planting of GM crops. Although her ruling lacks the approval of two other judges who must also decide on the issue, its legal basis was a law allowing judges to suspend appeals in cases that cause "irreparable damage" to litigants. Judicial delay not only harmed Monsanto, it was argued, but also "paralyzed" Brazil's agricultural and biotechnological development (Escobar 2003c). Immediately afterward, IDEC vowed to appeal, Brazil's attorney general argued that the merits of the case had yet to be decided, and the environment minister stressed that an environmental impact report was still required before GM crops could be planted commercially (Campanili 2003; Formenti and Gallucci 2003; Gallucci 2003).

The NGO coalition engaged in another legal dispute against GM crops in January 2002. The coalition challenged the experimentation on transgenic crops containing the genes from *Bacillus thuringiensis* (*Bt*), a soil bacterium used in organic agriculture as a biopesticide. A legal ruling held that insect-resistant GM crops (Table 9.1) must be reclassified as pesticides and must have a "Special Temporary Register" issued by the environment, health, and agriculture ministries. Because no standard register exists for GM crops, all of Monsanto's experiments with *Bt* maize and cotton have been paralyzed. Brazil's national agricultural research institute (Empresa Brasileira de Pesquisa Agropecuária; EMBRAPA) has stopped its experiments in insect resistance in beans and papaya. Overall, approximately half of all GM experiments approved by January 2001 were suspended until 2003, when IBAMA authorized EMBRAPA to conduct field experiments on GM papaya, beans, and potatoes. In May 2004, IBAMA authorized field experiments with *Bt* maize for Dow Agrosciences (Agência Estado 2004).

Understanding the Cardoso Consensus

The Cardoso consensus on biosafety limited the question of transgenic organisms to a narrow set of technical issues and appeased multiple constituencies among Brazilian agricultural and scientific sectors. Permission to experiment with transgenic organisms was seen to support the scientific community, prioritize Brazilian scientific competitiveness and agricultural expertise, and permit national enterprises to develop their own transgenic seed packages. By limiting the scope of biotechnology governance through CTNBio and allowing GM experimentation, the Cardoso consensus provided the technical framework in which the scientific community could engage in a multitude of GM-related activities without excessive oversight. For example, in January 2001, CTNBio authorized 942 experiments, but the staff included only twenty professionals capable of experimental oversight (USDA 2001a). It is important to note that the policy consensus under Cardoso's administration was not directed at all toward concerned environmental activists and scientists; in fact, the bureaucratization and narrowing of GM policy to technical issues excluded voices of dissent.

The moratorium on commercial planting, another part of the national consensus, also assisted the country's scientific community by delaying the entrance of foreign firms' transgenic technology into the Brazilian biotechnology sector. This delay allowed Brazilian public and private research institutions to develop ties with Monsanto and other agrochemical transnational corporations, leverage technology transfers, and commercialize competing technologies. EMBRAPA used its control over Brazilian tropical soybean cultivars, guaranteed by the 1997 seed varieties law (Lei de Cultivares) and Brazil's intellectual property rights legislation, to license its own seed cultivars to Monsanto. The transnational corporation reciprocated by licensing GM herbicide-resistant technology to EMBRAPA. Under the seed varieties law and EMBRAPA's licensing agreement, Brazilian farmers will retain the right to save seeds for their own use in subsequent seasons, thereby prohibiting Monsanto from introducing the "terminator gene" or "genetic use restriction technology" (EMBRAPA 2000; Leite 1997). In addition, EMBRAPA grants exclusive licensing to national seed companies (EMBRAPA 2000). The effectiveness of Cardoso's consensus was most evident in March 2004 when EMBRAPA announced that it developed eleven herbicide-resistant soybeans that were pending approval for commercial planting (USDA 2004, 26).

The moratorium on commercial planting also allowed the agricultural sector to create a comparative advantage in conventional crops, develop its own GM technology, and capitalize on the demand for non-GM foods throughout the world, especially in Europe and Asia. Some politicians and farmers argue that Brazil should develop a national comparative advantage in conventional soybean products, as it would benefit Brazilian traders US$20 per ton compared to Argentina's GM soy meal (USDA 2001b, 4). Another market-oriented option, promoted by the powerful agro-industrialist and governor of Mato Grosso, Blairo Maggi, was to develop a dual structure that would permit the production of both transgenic soybeans and conventional soybeans (*Produtor Rural* 1999). The national consensus permitted commodity producers and distributors to interpret varying market signals on conventional commodities, develop their own competing technologies, establish GM-free supply contracts, and develop internal regulatory regimes to protect a large portion of the conventional commodity supply from genetic contamination.

Contesting the Consensus: Sub-National Politics

While scientific agencies dominated the Cardoso consensus, state (sub-national) governments challenged their dominance and broadened the scope of the debate to include environmental impact and consumer health. Challenges mounted by the state government of Rio Grande do Sul, long governed by the leftist PT, offer an important example of resistance to the national consensus and the active pursuit of establishing a GM-free zone (Palaez and Schmidt 2004). In the late 1990s, Rio Grande do Sul embarked on a dual strategy to regulate transgenic technology and contest the center-right Cardoso administration's national biotechnology policies. First, led by agriculture secretary José Hermeto Hoffmann, state officials mounted legal cases against private biotechnology companies conducting experiments in their state and used the constitutional challenges against CTNBio's exclusive jurisdiction over biosafety governance to reclaim authority over environmental regulation. Second, state officials employed market strategies to secure GM-free contracts for export of the state's conventional soybeans to European markets. The widely reported incidence of Maradona soybeans, however, suggests that their policies have not been completely successful. Nevertheless, the Rio Grande do Sul case is instructive because it raised the political tenor of the debate over the control or regulation of biotechnology and influenced other state governments to challenge biotechnology regulation. In late 2003 and early 2004, Paraná and Acre states took steps to restrict GM crop cultivation and commercialization in order to create "GM-free" zones.

Legal Challenges

Rio Grande do Sul challenged CTNBio's exclusive jurisdiction over biosafety. In 1991, shortly before the development of the Convention on Biological Diversity, which began an international effort to establish norms for biological safety, the state legislature passed the Law of Biotechnology and Genetic Engineering (No. 9,453), which prohibited commercial production, required the registration of all transgenic experimentations with the state government, and levied heavy fines for noncompliance. In March 1999, after CTNBio had been established by the federal government, the state government passed the State Decree of Biotechnology (No. 39,314), which gave the agriculture secretary authority to confiscate and destroy irregular or unauthorized GM experiments. Within the agriculture secretariat, the Department of Crop Production (Departamento de Produção Vegetal; DPV) supervised enforcement of the state's bio-safety laws. The state regulation required companies or institutions that already had CTNBio approval to register experiments concurrently with the state authorities. State-level registration established an independent, or perhaps more stringent, supervisory mechanism to oversee biosafety controls.

Rio Grande do Sul's PT government also exploited the GM contro-versy to accelerate devolution and wrest environmental regulation from federal control. Using the 1988 federal constitution to back up the state's claim for stringent enforcement of state biosafety laws, agriculture sec-retary Hoffmann suspended experiments of transgenic crops approved by CTNBio. Citing irregularities in the containment of GM material, Hoffman first targeted an improperly registered transgenic rice crop at an experimental station near the state capital, Porto Alegre (Fonseca 2000). The state's challenge demonstrated to CTNBio that the experiment did not comply with federal biosecurity norms. Based on this evidence, CTNBio eventually withdrew its approval and ordered it destroyed. Later that year, Rio Grande do Sul's attorney general attempted to block the rice harvest resulting from another CTNBio-approved experiment owned by Aventis CropScience do Brasil; however, in this instance, the state's challenge failed. Aventis, on appeal, won the right to continue ex-perimentation, and harvested an estimated 400 kilograms of transgenic rice (Zero Hora 2000).

Legislative overtures in the state and federal governments attempted to limit Rio Grande do Sul's jurisdiction over the release of genetically modified organisms and strengthen CTNBio's regulatory authority. At the state level, political opposition in Rio Grande do Sul's legislature at-tempted unsuccessfully to abdicate the state's power to construct its own biosafety laws and centralize regulation. Some members of the national legislature also have opposed Rio Grande do Sul's state-level regulation and advocated a constitutional amendment that would prevent Brazilian establishment of biosafety laws and regulations independent of CTNBio (USDA 2001a, 3).

Exploiting GM-Free Markets

The PT state government buttressed legal initiatives with attempts to deepen the state's competitive advantage in conventional soybeans. First, the state government encouraged commercial ties between producers and importing European food industries by actively seeking out new contracts and buyers for the state's conventional crops. In April 1999, for example, agriculture secretary Hoffman met with the president of Carrefour, a French food-retail conglomerate, to entice the company to import 300 million metric tons of conventional soybeans. Second, the state government developed new capacities to monitor GM crop production in order to prevent Maradona soybeans from seeping into the conventional supply. In May 1999 Hoffman requested that European anti-GM organizations supply laboratory equipment and kits to identify GM material. These materials increased the state government's technical capacity to monitor illegal transgenic crops (DPV 2000). Rio Grande do Sul also applied rigorous supervision over all phases of CTNBio-authorized experiments in the state. Moreover, state officials pursued a local education program on the problems of transgenic release into the environment to discourage planting of Maradona soybean seeds. Finally, the state has begun to set up a certification system for conventional crops to guarantee the commodity chain for European markets.

As part of its enforcement strategy, the DPV mounted a toll-free telephone number to report suspected cases of illegal transgenic material. The GM tip line directed state officials to suspected locations, where they tested materials with the imported GMO detection kits donated by European NGOs. Enforcement efforts are difficult to evaluate, but during the first planting season (1999) covered by the state enforcement, the DPV confiscated 174,300 kilograms of illegal transgenic soybean seeds seized in fourteen municipalities and destroyed 300 hectares of planted transgenic soybeans in two municipalities (Figure 9.2) (DPV 2000). In November 2001 state government officials also seized 450 sacks of illegal transgenic soybean seeds identified during an enforcement sweep. Extra-official enforcement has also occurred in Rio Grande do Sul. In February 2002 anti-GM activists mobilized approximately 1,500 farmers and workers, who identified Maradona soybeans using imported GMO detection kits, and promptly filed a complaint with judicial authorities (Ogliari 2002).

Contesting the Consensus: Consumers, Farmers, and Activists

Defiant state politicians were not alone in opposing the Cardoso consensus. Two important voices included consumer and farmer groups. From

Figure 9.2. Illegal GM Seeds and Fields in Rio Grande do Sul, Brazil, 1999–2000

very different perspectives, both groups expanded the narrow terms of the GM debate to include human health, consumer rights, and environmental impact.

Consumer Groups

IDEC, the Brazilian consumer-defense NGO, joined forces with international NGOs to challenge the legitimacy of government biosafety policy, question CTNBio's scientific authority, and publicize the European-based anti-GM message in Brazil. One of IDEC's most important campaigns is the legal challenge, filed with Greenpeace, against the commercialization of GM soybean production that resulted in the legal moratorium, which lasted until 2003. IDEC is opposed to the Cardoso administration's Biosafety Law because of weak risk assessment and regulation, and to labeling requirements, which allegedly lack technical basis and do not affect most food products (Luccas and Stancato 2002). IDEC also drew upon international resources to compensate poor oversight by federal agencies and CTNBio. In February 2000 IDEC signed a cooperative agreement with the British Consumers' Association to access European laboratories for GM detection tests on food products (Stancato de Souza 2001, 29). In June 2000 IDEC released GM residue test results for forty-two different domestic and imported food products. Their initial study revealed that 28.5 percent, or twelve products, contained residual GM content. Later that year IDEC identified four popular brands containing GM products (IDEC 2000; IDEC 2001).

IDEC's highly publicized consumer campaign adopted the European anti-GM discourse concerned with human health and environmental impact, directly challenging the depoliticized technical approach characteristic of the Cardoso consensus. IDEC used their GM test results to denounce GM contamination in letters to the agriculture, justice, and health ministries alleging violation of consumer protection codes, biosafety laws, and food safety regulations (IDEC 2001). Response to IDEC's lobbying was mixed. The federal consumer health institution only requested voluntary removal, while the São Paulo state public health authority ordered the removal of GM products. Brazil's food industry association challenged São Paulo's legal authority to regulate consumer health matters, which they considered a federal responsibility. IDEC maintained that state health officers had the same authority as federal agents to remove products under the consumer protection law (IDEC 2000).

Consumer groups exacerbated inter-ministerial divisions that were smoothed over by the government's consensus. In 2000, Cardoso's agriculture minister Pratini de Moraes defended CTNBio and claimed (but later retracted) that IDEC was bankrolled by Monsanto's competitors. Health minister José Serra took a more cautious stance. He spoke against CTNBio's policy, defended the legal moratorium on transgenic commercialization, advocated for more health tests and environmental impact assessments of the genetically modified material, and supported product labeling (*O Globo* 2000). The inter-ministerial dispute came to a head soon afterward, in July 2000, when health, agriculture, justice, environment, and science-technology ministers signed a document harmonizing the government's stance on biotechnology. According to insiders, environment minister José Sarney Filho, who until then had been voicing doubt over commercial GM release, was forced into submission (Stancato de Souza 2001; Sato and Weber 2000). However, six months later, Sarney

Filho reasserted his skepticism by forming another biosafety commission, focusing on environmental impact, within his ministry (MMA 2001).

Consumer and environmental groups also have targeted the food processing industry, including Perdigão, a large Brazilian meatpacker. In July 2002 Greenpeace accused Perdigão of using GM soybeans in the manufacturing of some of its products. In a press release, Perdigão responded that it had rigorous controls on its raw materials to avoid the entrance of GM soybeans. On 27 July 2002, Perdigão signed a contract with a laboratory on the campus of the Federal University of Viçosa, Minas Gerais (Figure 9.1), to support its policy against the use of GM soybeans.

IDEC forms an important part of a transnational advocacy network (Keck and Sikkink 1998) formed in opposition to transgenics. This network has established a major presence in Brazil's anti-GM movement. Although a full discussion of this network would require a separate chapter, some characteristics may be summarized here. The campaign "Por um Brasil Livre de Transgênicos" is supported by nine NGOs, including IDEC, Action Aid Brasil, and Greenpeace. The network reports on issues relating to transgenic food and agriculture in a regular news bulletin, including such items as results of a survey IDEC commissioned from Brazil's major polling firm. The results were that 37 percent of people surveyed knew about GMOs, and 71 percent of those who knew about GMOs said they preferred conventional to GM food products (AS-PTA 2003).

In summary, IDEC has expanded the Brazilian GM debate in two distinct ways. First, IDEC exploited information gaps and asymmetries about the very nature of transgenic crops to expand the narrow regulatory scope that CTNBio defined for GM release into the environment. IDEC advocated state-level regulatory control over health and consumer safety as a challenge to complicity and consensus within CTNBio's rulings on biotechnology governance. This challenge also has called into question the authority of federal agencies in regulating GMOs. Although U.S. officials have described the IDEC and Greenpeace strategy as having delayed "the normal path of approval of GMOs in Brazil by exploiting the current 'gaps' in the federal legislation" (USDA 2001a, 6), the NGOs simply challenged the Cardoso administration's inherently unstable technocratic consensus. The achievements of IDEC's campaigns may be understood by the success with which they exploited international anti-GM advocacy networks and used globalized campaign strategies and resources to expand the debate beyond the scientific elite of CTNBio. IDEC's relationship with British consumer groups and global environmental NGOs provided the support necessary to challenge CTNBio's hegemony over the science of biotechnology.

Farmer Groups, Large and Small

Brazilian farmers are not unanimous in their support of GM crops. Certainly, the alleged high incidence of Maradona soybeans in Rio Grande do Sul is evidence that many farmers have opted to experiment with GM soybeans; indeed, farmers report the thrill of spraying herbicide on their illegal RRS fields without damage to their soybean crop. By implication, they argued that the state should allow GM crops rather than continue fighting the inevitable adoption of transgenic technologies. One of the authors of this chapter (Stancato) toured Brazil's main soybean-producing regions in early 2004. He did not observe resistance to GM crops among farmers; indeed, many eagerly awaited legalization of GM soybeans. They were convinced of the economic advantages of GM technology, primarily because soybean monoculture over several years has encouraged the development of weeds resistant to traditional herbicides. In addition, the head of a large soybean trading firm claimed that legalization of GM soybeans in Brazil would encourage better adapted GM varieties than the ones smuggled from Argentina, thus enhancing Brazil's competitiveness (Gazeta Mercantil 2002). Similarly, Monsanto's Brazilian director, Rick Greubel, argued that contraband GM soybeans affected several parties negatively: "Brazil loses credibility in the international market; regulatory agencies become discredited; certified seed companies lose sales; farmers lose access to varieties developed for their regions; and consumers become misinformed" (Escobar 2003b).

Although reports of Maradona soybeans suggest that some farmers are eager to adopt GM technology, interviews of farmer organizations and news reports suggest that eagerness is over-exaggerated, and that not all Brazilian farmers are ready to plant transgenic soybeans. Large cooperatives in southern Brazil play a key role in organizing the commodity chain for conventional and organic soybeans, which provide alternatives to GM crop production (Hisano and Altoé, this volume). Another example of farmers opposing Maradona soybeans comes from western Bahia, northeastern Brazil, where the main producer organization, known as AIBA (Associação de Agricultores e Irrigantes do Oeste da Bahia), is opposed to adopting GM crops. AIBA represents the large, capital-intensive farmers in a region that produces 4 percent of Brazil's soybean crop. In a special publication, AIBA confirmed its commitment to "non-transgenic production" (AIBA 2001). AIBA's environment director affirmed that the organization was not interested in GM crops; rather, its members were more concerned with other factors of production, such as electricity for irrigation and transport infrastructure. More significantly, these farmers were concerned that GM crops would restrict supply of inputs to a single multinational corporation, thus limiting the range of choices that farmers have as consumers of farming inputs (Lopes 2001, 2002).

Similarly, the Fundação ABC, the advanced research arm of three powerful agricultural cooperatives in the state of Paraná, southern Brazil, is hardly enthusiastic about GM crops. For example, the cooperatives sampled soybeans for GM presence since the 2001 harvest. The Fundação ABC's 1,100 members, nationally recognized as technology-savvy farmers who cultivate 185,000 hectares of cropland, believe that "present transgenic technologies offer few advantages" because conventional technologies are sufficient to control present pest and weed problems (Valentini 2001, 2002).

Keenly aware of reticence on the part of Brazilian consumers and farmers, Monsanto, with sales of US$600 million in Brazil during 2002, formulated a counter-attack. Breaking away from its initial policy of disengagement with anti-GM campaigns, Monsanto restructured its public relations department in November 2002. The new department chief reestablished media relations, which had been ignored since 1999, and began to reevaluate its public relations problem with consumer groups and farmers (Mocsanyi 2003). During the first half of December 2002, Monsanto unleashed a marketing effort aimed at convincing farmers of the advantages of biotechnology, especially as regards to cost and crop commercialization; however, Monsanto's marketing director admitted that farmers still have many doubts about GM crops.

Maradona soybeans have gained entry into Brazil's southernmost farms because farmers were curious about a new technology and because the region has similar environmental conditions and latitude as northeastern Argentina's soybean lands. Elsewhere, the spread of GM crops will not be as swift. Adoption of GM crops by farmers in Brazil's central and northern soybean lands will depend on the tropical varieties controlled by Brazilian public institutions and private firms. These transgenic soybeans, which are still under development, will be available only from EMBRAPA-authorized dealers, rather than from individual smugglers. So it is little surprise that Monsanto has targeted these farmers in their recent marketing effort. But evidence from actual farmer groups, such as AIBA and Fundação ABC, suggests that a marketing triumph may be difficult.

Biotechnology Governance under the Lula Administration

A new "Lula consensus" has been evolving since January 2003. The Lula administration circumvented the legal moratorium by offering temporary and limited legal protection to farmers cultivating Maradona soybeans. Lula shifted the broader regulatory issue to the Brazilian congress, which is expected to approve a bill that will legalize GM crops by the 2004 planting season. However, the government also held those individual farmers responsible for damages caused to the environment or third parties. The administration is also trying to centralize decision-making power in the executive branch by reorganizing the biosecurity bureaucracy, principally reducing the power of CTNBio and creating a new mechanism for approval of transgenic experiments and commercial planting. Despite these new policies, biotechnology governance in Brazil is still debated intensely by federal agencies, courts, state governments, consumer groups, and farmer associations.

When the new administration ascended to power, the policy of the PT was to support the moratorium on commercial planting of GM crops. However, the Lula administration's policies indicated a more ambiguous stance on GM crops, one that neither supported nor directly challenged the PT's official position during the 2002 presidential campaign. Once in power, Lula accepted the need for discussion. On a nationally broadcast television program, he acknowledged that there was "a very serious debate within the government, because at some point we will have to say whether we are in favor or opposed. I have been strongly opposed politically; today, scientifically, I have doubts" (Salvador 2003c).

Ambiguity of the Lula administration toward GM crops was reflected in cabinet appointments to key ministries involved in biosafety policy. Lula's agriculture minister, Roberto Rodrigues, is a known advocate of biotechnology. Rodrigues was former president of the Brazilian Agribusiness Association (Associação Brasileira do Agribusiness; ABAG). Rodrigues maintained a cautious position, stating that he will not discuss transgenics until the courts rule on the matter. As president of ABAG, Rodrigues admitted to supporting transgenics, but he later told reporters that "as minister, my position is to wait" (Salvador and Marin 2003). Rodrigues has also been quoted as stating that the "lower costs of production" of GM crops could "reduce the hunger problem" in Brazil (Formenti 2002). After joining the administration, Rodrigues told a reporter that the government was awaiting judicial rulings on GMOs before defining its position: "This is a judicial issue and the executive will not interfere with judicial matters" (Sato 2003b).

Appointments made to head other key federal bureaucracies only raised the uncertainty and ambiguity in biotechnology governance under Lula's leadership. For the environment ministry, Lula selected federal senator Marina Silva, an environmental activist considered to be diametrically opposed to transgenics. Silva urged a "cautious" approach to GM technology that would encourage Brazilian research in transgenic organisms and emphasized that the Lula government did not have an "ideological policy against transgenics" (Formenti 2002). However, Silva demanded that the government withdraw from a key legal case, filed with Monsanto during the Cardoso administration, which would waive the requirement for environmental impact reports on GM crops (Sato 2003a).

Another example of Lula's early ambiguity on GM crops was the appointment of EMBRAPA's new president, Clayton Campanhola, who stated that GM technology research will generate scientific information about the impacts of transgenic organisms on the environment and human health. Campanhola also indicated that the government's position was that transgenics "will only be released when there is sufficient information to guarantee that there is no threat to biosecurity" (Salvador 2003a). Until then, the "precautionary principle" would be adopted. According to another report, Campanhola rejected the idea that "biotechnology will solve everyone's problems" (Escobar 2003a). By contrast, under the Cardoso administration, EMBRAPA's president, at the time Alberto Duque Portugal, forbade staff to oppose GM crops in public. He implicitly prohibited researchers from expressing negative opinions and even issued a leaflet to his staff summarizing what to say in favor of GM crops when questioned (Ripardo and Murakawa 2000).

Two related issues, however, forced the Lula administration to develop a more definitive stance on transgenic crops: Brazil's soybean exports to China and the stark reality that thousands of farmers, who had planted Maradona soybeans illegally, were set to harvest in the first few months of 2003.

Export Protocols for Chinese Markets

In early 2002, China, which is Brazil's largest raw soybean importer (31 percent of Brazil's exports), imposed rules on the importation of transgenic material. The regulations indicated that transgenic soybeans had to be labeled, but the manner by which this was to be done was unclear. These rules, made effective 20 March 2002, stymied the closing of contracts for soybean exports. While not demanding transgenic-free soybean imports, China insisted that exporting countries declare the presence of transgenic material. For Brazil, these rules created an obvious complicating factor. On one hand, the federal government knew that Maradona soybeans were being planted, but did not take preventive action. On the other hand, the legal wrangling, which made commercial GM soybeans illegal, created a judicial problem for the government because federal labeling of transgenic soybean shipments would be tantamount to admitting that it was permitting illegal acts.

China made two concessions to Brazil. From 20 March to 20 December 2002, China permitted third parties, on behalf of the government, to certify the status of Brazil's soybean exports. Upon expiration of this agreement, China and Brazil negotiated certification protocols, which were approved by the Chinese government on 27 January 2003. Until 20 December 2003, China would accept a "provisional certificate," in which Brazil declared that its soybeans are "officially conventional," but "because Brazil borders countries that produce GM soybeans, the presence of GM genes within Brazilian grain is possible." After 20 September, both countries would have three months to resolve the issue again. If the courts had not ruled on the matter by then, Brazil would try to extend the provisional certification system. According to one reporter, Brazil then will be forced either to declare that its soybean exports to China are transgenic, or to prove that they are conventional (Baldi 2003a; Baldi 2003b).

The question of Brazil's soybean exports to China from the 2003–04 harvest was resolved only in March 2004, when China approved a document issued by the CTNBio. The document admits that Brazil's soybean harvest may contain transgenic material, but states that there are no risks to biosecurity from GM soybeans. This opens the way for exporters to obtain necessary documents that meet Chinese import requirements (Salvador 2004). The definitive solution, however, awaits approval of the biosecurity bill the Lula administration recently sent to congress.

Confronting Maradona Soybeans

While the Chinese export protocol issue highlighted potential inconsistencies faced by the government's biosafety policy, a second, more pressing question was at hand. What would be the fate of the "illegal" 2003 harvest in southern Brazil? The government, which was implementing a nationwide project to end hunger in Brazil (Projeto Zero Fome), faced the horrific prospect of burning the crop that had been planted illegally by Rio Grande do Sul's farmers in 2002 (Sato 2003c; Salvador and Nossa

2003). In March 2003 Lula signed a presidential decree (Medida Pro-visória 113), which was subsequently approved by Brazil's congress, to permit the sale of GM soybeans until January 2004. The decree exempted the harvest from regulatory provisions under the original 1995 Biosafety Law. By allowing the sale of GM soybeans to reduce economic suffer-ing, the Lula administration weakened the arguments of the anti-GM lobby. Although this measure prevented Rio Grande do Sul's soybean crop from being incinerated, it did not address the approaching planting season (Brazil 2003a). In September 2003, the Lula administration issued a similar Medida Provisória, permitting farmers to plant "saved seeds" (a euphemism for Maradona soybeans) and commercialize their harvests through December 2004. Individual farmers claimed this exemption and planted Maradona soybeans to sell their harvest only if they signed a document admitting that they violated Brazil's 1995 Biosafety Law by planting transgenic soybeans. Farmers who signed this document also agreed to inform purchasers or consumers that their product may contain transgenic materials and pledged not to sell soybeans as seed. The legal exemption also required that farmers abide by the Biosafety Law for the 2004–05 crop and purchase seed from suppliers authorized by the federal government. By signing the document, farmers also accepted responsi-bility for damage caused by Maradona soybeans to the environment and third parties (Brazil 2003b). By mid-December 2003, 79,000 farmers had signed the document, admitting that they planted GM soybeans and ac-cepting responsibility for damages to the environment or third parties. Ninety-eight percent of these farmers were in Rio Grande do Sul, and only 0.2 percent were from Brazil's top soybean-producing state, Mato Grosso (USDA 2004, 25).

The full agenda of what has become the "Lula consensus" was revealed in October 2003, when the administration sent a Biosecurity Bill (Pro-jeto de Lei 2401/03) to the Brazilian congress. Approved by the Chamber of Deputies in February 2004, the proposed legislation would centralize control of GM policy within the federal executive's political coordination office (Casa Civil). The federal executive would control a fifteen-member National Biotechnology Council (Conselho Nacional da Bio-Segurança, CNBS), which in turn would reduce the power of CTNBio to approving experiments only (Monteiro 2003; Salvador 2003d). Commercial produc-tion of GMOs would require CTNBio's evaluation, a license from en-vironmental and consumer safety agencies, and final approval from the CNBS. However, the bill contains many line items that would be voted on separately, and these may alter the basic text. The Lula administration is committed to congressional approval of the Biosecurity Bill and affirms that it will not issue yet another Medida Provisória for the 2004–05 crop season (Miklasevicius 2004).

In shifting the decision to Brazil's federal legislature, the Lula government has begun to take a clear position on biotechnology governance, particularly in relation to GM crops. The proposed biosecurity bill suggests that the government has decided to treat the issue politically, rather than technically, marking a clear shift from the Cardoso administration's strategy to bureaucratize governance of GM technology. Significantly, the bill drains power from CTNBio, which was hated by NGO activists, and increased the role of IBAMA (Instituto Brasileiro do Meio Ambiente e dos Recursos Naturais), an environmental regulation agency, and the Ministry of the Environment.

While differing from the Cardoso consensus by including environmental and consumer groups in the decision-making process, and significantly weakening CTNBio, the Lula consensus is still highly contested. Legal protection for Maradona soybeans at the federal level has stirred significant opposition at the sub-national level. Just as Rio Grande do Sul contested the Cardoso consensus, Paraná's governor, Roberto Requião, is equally committed to undermining the Lula consensus. The government of Paraná wants to establish certificates of origin for GM-free soybeans that are exported by the state's cooperatives. The certificates, developed by the state's Institute of Technology (Instituto de Tecnologia do Paraná), will allow the cooperatives to pass along to members the premium on conventional soybeans (AS-PTA 2002b). In October 2003, Paraná's legislature approved a biotechnology bill. Not surprisingly, the bill bans the cultivation, processing, and industrialization of transgenic materials until 31 December 2006. Governor Roberto Requião celebrated the state legislature's approval by stating that the state's farmers would be free from "the trap of transgenic crops" (Fadel 2003). More important is the prohibition against exporting transgenic soybeans through the state's Paranaguá port (AS-PTA 2004). In 2003 Paranaguá was Brazil's leading soybean port, shipping nearly one-third of Brazil's total exports to world markets (USDA 2004, 36–37). Greenpeace activists supported Requião's anti-GM policies by preventing ships carrying Argentine GM soybeans from using Paranaguá to top off their holds with Brazilian soybeans (USDA 2004, 38). The Paranaguá conflict is at the center not only of the GM debate, but also of complex local politics and an acrimonious labor dispute involving some 11,000 workers striking for several days, extending waiting times for trucks and ships (Fadel 2004).

Although it is well known that Paraná has considerable area devoted to Maradona soybeans, farmers' organizations hesitate to acknowledge this fact publicly. Since the October 2003 anti-GM bill passed the Paraná legislature, a climate of fear prevails. In May 2003 the MST invaded Monsanto's experimental farm in Ponta Grossa. The courts issued an expulsion order against the alleged invaders of the farm, but the state's military police, under Requião's command, refused to enforce the court order (Maschio 2003). Requião's refusal to enforce the Ponta Grossa court order has sent a clear message to many farmers in Paraná: a pro-GM position will inspire similar invasions of farmland.

Conclusion

The Brazilian case of biotechnology governance is important because it is a leading soybean producer, and soybeans are the leading global transgenic crop. Brazil is an enormous market for commercial GM soybeans, but it is also a large supplier of conventional soybeans. The country is the site of an important battle in global agriculture: will Brazil's farmers become major producers of GM soybeans, or will they, and legal-regulatory apparatus, resist GM crops? In this chapter we have outlined the actors that will determine the outcome of this struggle. Despite the temporary legalization of GM soybeans planted from saved Maradona seeds, it is still uncertain whether Brazil will be open to commercial GM crops. The issue depends on the pending legislation and the outcome of legal challenges to current policies. There are several other issues at play, especially the role of states within Brazil, the evolving position of the judiciary, activism of consumer groups, positions of farmer associations, and recent maneuvers of the Lula administration. Although none of these sectors is sufficient to determine the outcome of the debate, alliances between and among these sectors may prove crucial. State and local governments, and anti-GM farmer and consumer groups, have particular scope for alliances that have yet to be fully explored.

References

Agência Estado. 2004. "Transgênicos: Ibama libera pesquisa de milho da Dow Agrosciences." *Agência Estado*, 20 May, www.aeagro.com.br, accessed 7 June 2004.

AIBA. 2001. "Caderno Especial." *InformaAIBA* (Barreiras, Bahia), 7 January.

Aliski, Ayr. 2002. "Conab estima safra agrícola recorde." *Gazeta Mercantil* (São Paulo), 11 December, p. B-16.

Angelo, Denise. 2002. "Exames confirmam mais 24 áreas de soja transgênica." *Folha de Londrina* (Londrina, Paraná), 9 April, p. 4.

AS-PTA (Assessoria e Serviços a Projetos em Agricultura Alternativa). 2002a. "Laboratório alemão de testes anti-transgênicos abre filial em Itu-SP." *Por um Brasil Livre dos Transgênicos: Boletim 137*, 14 November. Retrieved 14 January 2003. www.idec.org.br.

———. 2002b. "Paraná planeja certificar soja não-transgênica." *Por um Brasil Livre dos Transgênicos: Boletim 140*, 6 December. Retrieved 14 January 2003. www.idec.org.br.

———. 2003. *Por um Brasil Livre dos Transgênicos: Boletim 145*, 31 January. Retrieved 7 February 2003. www.idec.org.br.

———. 2004. *Por um Brasil Livre dos Transgênicos: Boletim 204*, 7 April. Retrieved 31 May 2004. www.aspta.org.br.

Baldi, Neila. 2003a. "Ratificado plano para certificação de soja." *Gazeta Mercantil* (São Paulo), 15 January, p. B-16.

———. 2003b. "A China já liberou a soja brasileira." *Gazeta Mercantil* (São Paulo), 28 January, pp. A-1, B-16.

Brazil, Government of the Federal Republic of. 1995a. Lei No. 8.974. *Diário Oficial da União*, 5 January.

———. 1995b. Decreto No. 1.752. *Diário Oficial da União*, 20 December.

———. 2003a. Lei No. 10.688, 13 June.

———. 2003b. Lei No. 10.814, 15 December.

Campanili, Maura. 2003. "Idec vai recorrer da decisão que liberou os transgênicos." *Agência Estado*, 12 August.

Cardoso, Cíntia. 2002. "Soja transgênica se espalha pelo Brasil." *Folha de São Paulo*, 14 May, B10.

Cardoso, Denis. 2001. "Brasil lidera productividade da soja." *Gazeta Mercantil* (São Paulo), 30 August, p. B-16.

Cardoso, Fátima. 2003. "Na região, soja transgênica já é realidade." *Estado de São Paulo* (São Paulo), 26 January.

Cardoso, Fátima, and Venilson Ferreira. 2003. "Agricultura deve crescer sem dar espetáculo em 2004," Agência Estado, Safra 2003–2004, 31 August.

CTNBio. 2001. "Approved GM Crop Experiments." Retrieved 14 February 2001. www.cntbio.gov.br.

DPV 2000. Brannstrom and Jepson interviews with DPV officials, 18 April 2000, Porto Alegre, Rio Grande do Sul.

EMBRAPA. 2000. "Technical Cooperation Contract with Monsanto." Retrieved 7 February 2001. www.embrapa.br/novidade/palacre/embmon.htm.

Escobar, Herton. 2003a. "Embrapa adota 'precaução' com transgênicos." *Agência Estado* (São Paulo), 22 January.

———. 2003b. "Um grande defensor dos transgênicos." *Estado de São Paulo*, 27 January.

———. 2003c. "Juíza libera comércio de soja transgênica no País." *Estado de São Paulo*, 13 August.

Fadel, Evandro. 2003. "Paraná fica 'livre de transgênicos' até 2006." *Estado de São Paulo*, 15 October.

———. 2004. Paranaguá: Greve dura já 5 dias e mais de 50 navios estão parados. *Estado de São Paulo.*

Fonseca, Marcos. 2000. "Começa colheita de arroz transgênico." *Zero Hora* (Porto Alegre), 18 April, p. 40.

Formenti, Lígia. 2002. "Posição do Brasil sobre transgênicos: Cuatela." *Estado de São Paulo*, 17 December.

Formenti, Lígia, and Mariângela Gallucci. 2003. "Para Ministra, soja transgênica não está liberada." *Agência Estado*, 13 August.

Gallucci, Mariângela. 2002. "Juiz pede vistas e suspende julgamento de soja modificada." *Estado de São Paulo*, 26 February.

———. 2003. "Para procurador, Juíza não poderia liberar transgênico." *Agência Estado*, 12 August.

Gazeta Mercantil. 2002. "Registro-Soja modificada no RS." *Gazeta Mercantil* (São Paulo), 11 November, p. C-5.

Hahn, Sandra. 2001. "Associação alerta para plantio ilegal de transgênicos." *Agência Estado* (São Paulo), 26 December.

IDEC. 2000. "IDEC e Greenpeace encontram transgênicos em testes de alimentos." Retrieved 13 February 2001. www.uol.com.br/idec/not2006012000.

———. 2001. "GM product test results." Retrieved 13 February 2001. http://www.uol.com.br/idec/campanhas/testetran1.htm.

———. 2003. Poster for Brazilian anti-GM campaign "Por um Brasil Livre de Transgênicos." Retrieved 7 February 2003. www.idec.org.br.

ISAAA. 2004. "Double-Digit Growth Continues for Biotech Crops Worldwide." International Service for the Acquisition of Agri-biotech Applications, Press Release, 13 January.

Jepson, W. E. 2002. "Globalization and Brazilian Biosafety: The Politics of Scale over Biotechnology Governance." *Political Geography* 21:905–925.

Keck, Margaret E., and Kathryn Sikkink. 1998. *Activists beyond Borders: Advocacy Networks in International Politics.* Ithaca: Cornell University Press.

Leite, E. C. B. 1997. "Sai a Lei de Proteção de Cultivares." *Agroanalysis*, July, pp. 7–10.

Lopes, J. C. M. 2001, 2002. Personal interviews with Brannstrom. Barreiras, Bahia. 20 July, 29 July.

Luccas, Karina, and Renato Stancato. 2002. "Transgênicos: Plantio deve continuar ilegal na safra de verão." *AgroCast: Safra 2002–2003.* São Paulo: Agência Estado.

Maschio, José. 2003. "Técnicos vão vistoriar unidade invadida no PR." *Folha de São Paulo*, 3 June, B10.

Meurer, Elmar, and Rogério Kiefer. 2002. "Cooperativa Alfa destrói grãos transgênicos em SC." *Gazeta Mercantil* (São Paulo), 9 July, p. B-16.

Miklasevicius, Jane. 2004. "Rodrigues: Lula não editará outra MP para plantio de transgênicos." *Estado de São Paulo*, 27 April.

MMA (Ministério do Meio Ambiente). 2001. *InforMMA: Informativo Diário do Ministério do Meio Ambiente.* No. 221, 13 February.

Monteiro, Tânia. 2003. "Soja transgênica ficará proibida na atual safra." *Estado de São Paulo*, 20 August.

Mocsanyi, L. 2003. Personal interview with Stancato. Sao Paulo, 14 January.

Ogliari, Elder. 2002. "Uma invasão contra o plantio de soja transgênica." *Estado de São Paulo* (São Paulo), 26 February.

O Globo. 2000. "Ministro da Saúde defende moratória para transgênicos." *O Globo* (Rio de Janeiro), 30 June.

Pelaez, V., and W. Schmidt. 2004. "Social Struggles and the Regulation of Transgenic Crops in Brazil" in K. Jansen and S. Vellema, eds., *Agribusiness and Society: Corporate Responses to Environmentalism, Market Opportunities and Public Regulation*, pp. 232–260. London: Zed Books.

Produtor Rural. 1999. "Transgênicos, ter ou não ter? Debate no Fórum Nacional de Agricultura." *Produtor Rural* 80:35–36.

Ripardo, Sérgio, and Fábio Eduardo Murakawa. 2000. "Pesquisa da Embrapa favorece Monsanto." *Folha de São Paulo*, 18 July, F3.

Salvador, Fabíola. 2003a. "Embrapa quer firmar parcerias com o exterior." *Estado de São Paulo* (São Paulo), 24 January.

———. 2003b. "Transgênicos: União quer recurso suspenso." *Estado de São Paulo*, 13 February.

———. 2003c. "Lula vai 'bater martelo' sobre transgênicos." *Estado de São Paulo*, 19 August.

———. 2003d. "Projeto libera transgênicos, mas cria restrições." *Estado de São Paulo*, 21 August.

———. 2003e. "Brasil deve confirmar liderança mundial na exportação da soja." *Estado de São Paulo*, 19 September.

———. 2004. "Soja: China aceita avaliação de transgênicos do Brasil." *Estado de São Paulo*, 24 March.

Salvador, Fabíola, and Denise Chrispim Marin. 2003. "Gasolina voltará a ter 25% de álcool em maio." *Estado de São Paulo* (São Paulo), 25 January.

Salvador, Fabíola, and Leonêncio Nossa. 2003. "Medida Provisória vai liberar soja transgênica." *Estado de São Paulo*, 27 March.

Sato, Sandra. 2003a. "Meio Ambiente quer reestudar transgênicos." *Agência Estado* (São Paulo), 4 February.

———. 2003b. "Ministro da Agricultura aguarda decisão judicial para se manifestar." *Estado de São Paulo* (São Paulo), 7 February.

———. 2003c. "Lula libera soja transgênica para evitar crise." *Estado de São Paulo*, 28 March.

Sato, Sandra, and Demétrio Weber. 2000. "Governo defende produção de transgênicos." *Estado de São Paulo*, 7 July.

Soares, Paulo. 2003. "Briga na Justiça contra transgênicos dura quatro anos." *Gazeta Mercantil* (São Paulo), 20 January, p. B-16.

Soskin, Anthony B. 1988. *Non-Traditional Agriculture and Economic Development: The Brazilian Soybean Expansion, 1964–1982*. New York: Praeger.

Stancato de Souza, Renato J. 2001. "Toward a transnational civil society? The GM debate in Brazil." Master's Dissertation, Institute of Latin American Studies, University of London, London.

USDA. 2001a. "State of Biotechnology in Brazil." GAIN Report #BR1601. Foreign Agricultural Service, Brasília, D.F.

———. 2001b. "Brazil Biotechnology: Update on Biotechnology Issues in Brazil 2001." GAIN Report #BR1623. Foreign Agricultural Service, Brasília, D.F.

———. 2002. "Brazil: Oilseeds and Products Annual 2002." GAIN Report #BR2004. Foreign Agricultural Service, São Paulo.

———. 2003. "Brazil: Oilseeds and Products Annual 2003." GAIN Report #BR3003. Foreign Agricultural Service, São Paulo.

———. 2004. "Brazil: Oilseeds and Products." GAIN Report #BR4611. Brasília: USDA Foreign Agricultural Service.

Valentini, M. 2001, 2002. Personal interviews with Brannstrom. Castro, Paraná. 11 July, 2 April.

Warnken, Philip F. 1999. *The Development and Growth of the Soybean Industry in Brazil*. Ames: Iowa State University Press.

Zero Hora. 2000. "Estado recorre à Justiça." *Zero Hora* (Porto Alegre), 13 April, p. 28.

CHAPTER 10

Brazilian Farmers at a Crossroads: Biotech Industrialization of Agriculture or New Alternatives for Family Farmers?

SHUJI HISANO AND SIMONE ALTOÉ

As the world's second largest soybean producer and exporter, Brazil has emerged as an important battlefield in the global conflict over genetically modified organisms (GMOs) since the late 1990s. Given the fact that the majority of soybean growers in the United States and Argentina have already adopted the associated package of new technologies, European and Asian consumers, looking for non-GMO sources, are curious as to whether or not Brazilian farmers will accept this new package. Its central components are transgenic seeds, genetically modified to resist greater infusions of herbicide, and herbicides that kill most plants, except the target crop. Until recently, growing GM crops in Brazil has been prohibited due to a judicial authority that ruled in favor of the claims made by environmental and consumers' organizations (see Jepson et al. in this volume). While the federal government has not effectively mapped out its policy either against or in favor of GMOs, the state of Rio Grande do Sul (RS), the southernmost state, implemented its "GM-free zone" policy for several years since 1998 (Pelaez and Schmidt 2004). Two other southern states, Paraná and Santa Catarina, have also rejected GMOs.

It is in Rio Grande do Sul, however, where the most contentious problem has occurred: namely the smuggling of GM soybean seeds across the border with Argentina. In spite of the states' policies banning GMO planting, many farmers, ranging from small to large scale, have grown and harvested illegal GM soybeans for years with mixed feelings: the expectation of financial benefits on the one hand; and anxiety about negative environmental and health impacts as well as about breaking the law on the other. This kind of farmers' dilemma is our starting point. But the objective of this chapter is not to discuss GMO politics itself, which is already dealt with in another chapter in this volume (Jepson et al., Chap. 9

in this volume). We focus instead on farmers' responses to GMO issues by contextualizing them in the ongoing biotechnology revolution and the economic policy environment promoted by neoliberal globalism (see Chap. 1 above). We discuss the implications of this context for the capitalist intensification of the agrifood system. The wave of GMOs that hit the market in the mid-1990s is a part of such trends.

Our perspective is also derived from the reality in the southern states, the traditional soybean producing area of Brazil, where many small-family farmers are facing the rapid structural transformation of the soybean sector. This situation has caused socioeconomic difficulties among small-family farmers and led them to believe that adopting Monsanto's Roundup Ready® soybean will solve their problems. We assume that, without addressing this specific socioeconomic reality, any implications for (anti-) GMO politics will not go beyond the level of pro- or anti-ideological discourses. The goal of this chapter is therefore to empirically explore how the joint forces unleashed by the biotechnology revolution and neoliberal globalism are shaping local landscapes through GMOs, and how farmers and civil society respond to them. With this empirically based approach, we hope that our analysis will transcend the simple dichotomies of "to adopt or not to adopt" GMOs.

As argued the sociology of agriculture literature, while the forces unleashed by neoliberal globalism seem to homogenize the reorganization of political and economic systems, the actual forms of reorganization are inevitably diversified by national, regional, and local conditions as well as by the specific social actors involved (Jussaume et al. 2002). Accordingly, an adequate assessment of the diversity and complexity of agrifood systems is required, as well as the analysis of a variety of farmers' responses in such diverse settings. Using field surveys conducted by the authors in August 2000, December 2001, and July 2002 and supplemented by literature surveys, we have selected four different types of local responses led by different key actors:

(1) Managing to survive by integrating to the mainstream technological paradigm and the global market (Agricultural Co-operative of Integrada, northern Paraná);

(2) Seeking to differentiate itself as a niche but still within the export market (Agricultural Co-operative of Cotrimaio, northern RS);

(3) The well-intentioned activities of NGOs and community groups to stay away from the global reality, finding opportunities in the locality (NGOs in Paraná and RS); and

(4) The extension programs of the state extension agency (EMATER/
RS—*Associação Riograndense de Empreendimentos de Assistência Técnica
e Extensão Rural*) in Rio Grande do Sul.

As actor-oriented studies have stressed, it would be useful to focus on the
self-organizing practices of social actors at the local level (Long 2001).
Yet, while there are some notable local movements engaged in or promoting alternative farming in Brazil, most of them have not been very
successful. These movements remain too weak to encompass all the rural farmers across the country. Such a huge vacuum of local-level social
actors in rural Brazil can be best supplemented by institutional actors.
We acknowledge that there are many criticisms and much distrust of the
role of extension programs stemming from the experiences in the United
States (Hightower 1973; Hassanein 1999), as well as those in Southern
countries where extension institutions have been basically used to promote the capitalist-intensification model of agriculture (Shiva 1992;
Caporal 1998). Contrary to some skeptical remarks about public institutions, however, the emphasis on reforming institutional expertise to
serve sustainable agriculture more effectively does not necessarily obscure the role of local actors. This is especially true of the case of the
Associação Riograndense de Empreendimentos de Assistência Técnica e
Extensão Rural in Rio Grande do Sul, EMATER/RS. This state-
promoted institution has worked on the transformation of a locality based
on the combined ideas of agroecology and the participatory and educational approach, thus strengthening local social actors. This institutional
approach contains several promising aspects. These include giving proper
support to family farmers and working together with them, rather than
for them. Such approach is sharply contrasting with hitherto known agricultural extension attempts that ended up by promoting the modern agricultural technological paradigm. On the other hand, it should be said that
EMATER/RS's activities are rather limited by an insufficient budget as
well as a high level of dependence on state and local politics.

This chapter is divided into four sections. We begin by outlining the
reality that surrounds small-family farmers in the southern states in order to underline the relevance of our approach. We then offer an empirical analysis of diversified local responses to that reality in the following two sections. We take into consideration the differences as well
as complementarities between the activities of agricultural cooperatives
and nongovernmental organizations and the institutional activities of the
EMATER/RS. Finally, we conclude by pointing out some limitations of

the bottom-up institutional response, such as political vulnerability and raising the prospect of enhancing the top-down institutional paradigm.

The Contradictory Situation of Farmers in Southern Brazil

Since the 1970s, the soybean sector in Brazil has witnessed rapid growth in modernization, which has launched soybean producers into the world market (Warnken 1999). At the same time, especially since the end of the 1980s, the Brazilian political economy has been confronted by external pressures to open its market (i.e., the introduction of the neoliberal model). Moreover, under the ongoing global restructuring of the agrifood system, transnational corporations such as Monsanto, Novartis (Syngenta), Unilever, Cargill, and ADM have invested in the Brazilian agrifood sector (Hisano 2002; Chiba 1998). These combined processes have caused many changes in rural landscapes. Farmers were under pressure to grow commodities in a competitive market, losing their "freedom for farming" and increasing rural exodus. This process has resulted in the differentiation of commercial-oriented farmers upgraded into huge landowners or corporate farms, on the one hand, from marginalized small-family farmers, on the other.

As mentioned, our research concentrates on the southern part of Brazil—Paraná, Santa Catarina, and Rio Grande do Sul—where many small-family farmers remain key actors in the region. In addition, this southern region is a traditional soybean-producing area. With this crop being at the center of the biotechnology revolution as one of the first transgenic crops to have become available, the region is also facing the rapid restructuring of the Brazilian soybean sector, as shown in Figure 10.1. Restructuring has meant a shift in the heart of soybean production from the southern states to a new producing area called *cerrado*, which includes the Central-west states of Minas Gerais, Mato Grosso, Mato Grosso do Sul, Goias, Tocantins, Distrito Federal, Bahia, and Maranhao. The difference between these two regions is comparable with the difference between small-family farming in the southern region and large corporate farming in the *cerrado* in terms of key social actors (see Table 10.1). In this context, "small-family farmers" are those whose average farm size is around 20–50 hectares, relying mainly on their own labor power (represented by the smallest two size groups in our statistics), while "corporate farmers" are those farming more than several hundred hectares and using capital-intensive technologies (represented by the largest

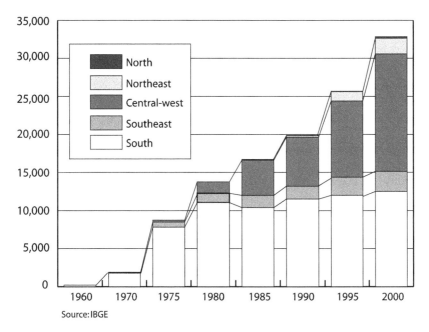

Figure 10.1. Soybean Production by Region in Brazil, 1960–2000, 1,000 tons.

two to three size groups in our statistics). Many in the latter group establish contracts directly and individually with trading companies and/or processing companies.

Although Brazil's position in the international market has been stable for many years, it is reported that the productivity of the world's first and third largest soybean producers—the United States and Argentina—is increasing due to improvements in technology. These competitors' advantages do not come from technology alone, but also from public support. Unlike these countries, the Brazilian federal government is so far neither investing nor providing enough incentives to its farmers (except those in the *cerrado*), leading many of them to question their future competitiveness in the world market. Required to maintain their high levels of production and concerned about losing market opportunities, Brazilian farmers tend to seek out and implement the latest technologies to improve production and reduce costs. In this mainstream of modern agriculture, the latest technology and products are transgenic crops or GMOs.

Designed to improve soybean production on industrialized farms, GMOs are being commercialized as an indispensable product to increase yields. This marketing strategy is pushing small-family farmers, too, to

Table 10.1. Comparisons between Brazil (total), Central-west (*cerrado*), and South (traditional), by Size Groups in 1995/96

a. Number of Farms Producing Soybeans by Size Groups

Size Groups of Total Area (ha)	Brazil		Central-west*		South*	
	number	%	number	%	number	%
Less than 10	57,203	**23.5**	567	5.3	55,771	**25.1**
From 10 to less than 100	157,148	**64.7**	2,574	24.2	149,921	**67.5**
From 100 to less than 1000	24,713	10.2	5,298	**49.8**	15,670	7.1
From 1,000 to less than 10,000	3,774	1.6	2,099	**19.7**	895	0.4
More than 10,000	153	0.1	96	0.9	4	0.0
TOTAL	242,999	100.0	10,634	100.0	222,265	100.0

b. Quantity of Soybean Production by Size Groups

Size Groups of Total Area (ha)	Brazil		Central-west		South	
	1,000t	%	1,000t	%	1,000t	%
Less than 10	357	1.7	10	0.1	338	3.1
From 10 to less than 100	5,060	**23.4**	186	2.3	4,585	**42.7**
From 100 to less than 1000	8,602	**39.8**	2,695	**32.7**	4,761	**44.4**
From 1,000 to less than 10,000	6,657	**30.8**	4,626	**56.1**	1,035	9.6
More than 10,000	912	4.2	729	8.8	14	0.1
TOTAL	21,588	100.0	8,246	100.0	10,733	100.0

c. Production Value of Soybeans by Size Groups

Size Groups of Total Area (ha)	Brazil		Central-west		South	
	1,000 R$	%	1,000 R$	%	1,000 R$	%
Less than 10	75,336	1.8	1,869	0.1	71,391	3.2
From 10 to less than 100	1,042,391	**24.6**	34,309	2.4	949,008	**42.3**
From 100 to less than 1000	1,728,286	**40.7**	483,472	**33.2**	998,958	**44.6**
From 1,000 to less than 10,000	1,229,123	**29.0**	807,854	**55.4**	218,966	9.8
More than 10,000	166,952	3.9	130,068	8.9	3,177	0.1
TOTAL	4,242,124	100.0	1,457,571	100.0	2,241,501	100.0

(continued)

Table 10.1. (continued)

d. Percentages of quantity and value in total, by Size Groups

Size Groups of Total Area (ha)	Brazil		Central-west		South	
	quantity	value	quantity	value	quantity	value
Less than 10	1.7	1.8	0.0	0.0	1.6	1.7
From 10 to less than 100	23.4	24.6	0.9	0.8	21.2	22.4
From 100 to less than 1000	39.8	40.7	12.5	11.4	22.1	23.5
From 1,000 to less than 10,000	30.8	29.0	21.4	19.0	4.8	5.2
More than 10,000	4.2	3.9	3.4	3.1	0.1	0.1
TOTAL	100.0	100.0	38.2	34.4	49.7	52.8

Note: Central-west consists of Mato Grosso, Mato Grosso do Sul, and Goias (as well as these states, Minas Gerais, Tocatins, Distrito Federal, Bahia, and Maranhao are usually included in the *cerrado*). South consists of Rio Grande do Sul, Santa Catarina, and Paraná (as well as these states, Sao Paulo is usually included in the traditional area).
Source: IBGE, Censo Agropecuario, available on the website (www.sidra.ibge.gov.br) accessed on October 30, 2002.

question whether or not to adopt the new technological package. Scientists, government officials, and representatives of transnational corporations often speak out about how this new technology will perform in the best interests of farmers. Yet policies and biotechnologies directed toward small-family farmers are rarely implemented, and actual risks and benefits to them remain uncertain. The strategies of transnational corporations (TNCs) vary according to which sector they are primarily dealing with, and certainly not as a function of small-family farmers' interests. For crop-feed/food complex companies like Cargill, dissemination of GMOs in Brazil would enable them to source a cheaper and stable amount of raw materials. For agrochemical companies like Monsanto, biotech-products adoption would expand their market both for seeds and agrochemicals outside their saturated market in the North, contrary to their propaganda. To sum up, the biotechnology revolution combined with neoliberal globalism is promoting the intensification of capitalist agriculture.

It is estimated that for Brazilian small-family farmers to maintain their market position, they will have to overcome at least four main barriers, namely: transportation costs, imposition of a complicated taxation system, inequality of resource allocation, and lack of adequate credit facilities (Warnken 1999). The government has so far addressed none of these factors seriously, thus keeping small farmers at a considerable disadvantage. Reality shows that purchased inputs are very expensive, farmers are subjected to the volatile and competitive export market, and the revenue they get from their farmland is not enough to make a living. Both internal and external pressures to reduce the cost of production, so as to remain competitive, have made more and more small-family farmers abandon their farmlands and migrate to urban slums or *favelas* in search of jobs, or to sell their labor power to large producers. In many cases, bankrupt farmers end up in situations of poverty and misery. As long as such disadvantageous circumstances remain unchallenged, even "successful" adoption of the latest technology is unlikely to improve the socioeconomic condition of small-family farmers.

As mentioned above, the majority of small-family farmers are concentrated in South Brazil, especially in the state of Rio Grande do Sul (RS). Partly because of this concentration in RS, the state government has stood against the adoption of GMOs. It considers that the new biotechnology has been developed primarily for large-scale capitalist agriculture and is therefore not a feasible option for small-family farmers in the region. Also aware of the fact that a great number of consumers in Europe and East Asia are not willing to buy GM food products, the state government believed that blocking GM crops, rather than promoting them, would be a wise trade policy (Bell 1999). Contrary to this line of thought, however, farmers are still curious to try GM technology. Believing that they will pay cheaper prices for GM seeds (because they do not actually pay royalties) and use fewer chemicals on the farmland, farmers in the state are smuggling GM seeds from Argentina and planting them illegally. Indeed, it is reported that 70 percent of soybeans planted in 2002 had originated from smuggled seeds. The "GM-free zone" policy in RS was no longer working and was eventually repealed by national legislation in 2004. Paraná and Santa Catarina were roughly in the same situation.

Understanding such a condition, EMATER/RS developed a variety of programs to stimulate small-family farmers to refocus on a locally based, agroecological farming model. This model contrasts with the notion in which small farmers had previously been absorbed, the modernization model—that input-intensive and industrialized farming is the viable solution to provide high-yield and high-quality products for the mainstream market. The perception of EMATER/RS reminds us that the question of whether small farmers should adopt GM technology or not overlaps with another question: whether they should stay in export-oriented commodity farming or turn to local-oriented agroecological farming. The latter question is important because the mainstream path in terms of both technology and market is unfeasible for small-family farmers. While it is becoming increasingly difficult to remain unaffected by the processes of the biotechnology revolution and neoliberal globalism, local actors cannot help but respond in some way to this overwhelming reality.

A Variety of Local Responses

Local responses vary according to individual actors and their diverse socioeconomic settings. We will look more closely at some local responses in this and the following section.

Linking to the Mainstream Market: The Agricultural Co-op of Integrada

Even in the southern states of Brazil, there are increasing numbers of middle- and large-sized farmers who have been aggressively adopting new technologies to intensify their farming. There are also some areas where not only agricultural cooperatives but also subsidiaries of multinationals are doing business by linking localities directly with the world market. A typical example is the case of the Agricultural Co-operative of Integrada (Cooperativa Agropecuária de Produção Integrada do Paraná). This agricultural co-op is located in the northern part of Paraná, where a relatively large number of middle- to large-scale farms exists. The co-op members' average farm size is around 100–200 hectares. This agricultural co-op is one of the successors of the Agricultural Co-operative of Cotia (Cooperativa Agrícola de Cotia), which was founded by Japanese immigrants in the 1920s and had grown to become one of the largest agricultural co-operatives in Brazil until the head office was dissolved in 1994. As a successor, Integrada has become the largest agricultural co-op in northern Paraná. Initially, the purpose of establishing Cotia was to protect the member farms from the whims of the market and to improve their livelihood. As is often the case with agricultural co-operatives during periods of economic growth, however, its original aims receded into the background as the business grew. The more successful it became, the more the pursuit of business became profit-oriented. Although the deterioration of the Brazilian economy accompanied by the drastic change in politics is considered the direct cause of Cotia's failure, it is more likely that its dissolution occurred due to the huge amount of debt produced by its expansionist business style.

According to Tanaka (2002), Integrada's trajectory mirrors the later stage of Cotia rather than embracing a legacy of early dates, that is, decentralization and bottom-up producers' activities. Integrada's main business is the marketing of crops, including soybeans, wheat, and corn. Among them, soybeans' share of total sales is around 60 percent. The volume of soybean marketing has been constantly increasing during the past five years (see Table 10.2). About 90 percent of the main crops are sold to transnational grain trading companies such as Cargill, through the Chicago Board of Trade, and about 30 percent of soybeans are traded at the futures market via these companies (Tanaka 2002). As for the purchasing business, it is worth noting that sales of agrochemicals and fertilizer have increased more than fourfold during the past five years. This indicates that the co-op is still following the mainstream path of modern agriculture (see Chap. 1 above).

In regard to GMOs, the co-op had no particular vision at the time we conducted an interview with its executive staff members in Londrina, Paraná, in August 2000. They stated that the production of GM soybeans in the region would be more than half of the total cropping area in a few years. Whether farmers would adopt GMOs or not would depend on the extent to which this technology reduced their production costs. The staff understandably expressed their concern about external market trends; however, they did not show us any strategy for coping with this issue. This was despite the suggestion of market consultants connected to the co-op, who said that it would be possible to segregate non-GMOs from conventional soybeans (including GMOs because they are not officially handled as such) due to the close relationship between the co-op and member farmers (ACP 2000). Rather, it seemed to us that they took the dissemination of GMOs for granted as an inevitable path for the intensification of agriculture. Subordinating their business to the world market order is for them a rational and acceptable option. So long as TNCs keep purchasing their products, the agricultural co-op and its member farmers do not need to concern themselves with GMO issues. We can say that this is still one probable response for local social actors to survive the harsh international competition under neoliberal globalism—that is, to accept what happens at the global level.

Seeking a Niche Market: The Agricultural Co-op of Cotrimaio

In contrast to the above example, organic soybean production was one of the effective options against GMOs. Organic soybean production began in Brazil as an isolated initiative in some regions, chiefly in the southern states—Paraná, Santa Catarina, and Rio Grande do Sul (Fonseca and Feliconio 2000). Although commercial production is still limited, the growth rate of organic farming remains in the double digits in recent years. The

Table 10.2. Five-year Summary of Integrada's Marketing Business

Year	1996				1997				1998				1999				2000			
Crops	Volume	I96	Sales	%	Volume	I96	Sales	%	Volume	I96	Sales	%	Volume	I96	Sales	%	Volume	I96	Sales	%
Soybean	125,523	100	31,627	64	175,782	140	52,037	57	184,797	147	43,646	47	214,260	171	60,961	49	232,137	185	69,826	62
Wheat	74,353	100	4,237	9	103,659	139	20,053	22	133,156	179	24,242	26	101,235	136	23,397	19	37,775	51	16,109	14
Corn	55,648	100	5,827	12	88,942	160	10,003	11	102,947	185	14,298	15	136,943	246	23,420	19	85,795	154	17,587	15
Cotton	17,387	100	4,335	9	8,700	50	1,498	2	13,016	75	4,241	5	11,239	65	5,643	5	13,046	75	5,019	4
Others	4,991	100	3,690	7	7,886	158	8,410	9	8,702	174	5,894	6	7,532	151	10,185	8	4,878	98	4,973	4
Total			49,716	100			92,001	100			92,321	100			123,606	100			113,514	100

Source: Integrada, Relatorio Annual de Atividades 1997–2000.

BNDES (Brazilian Development Bank) survey in 2001 recorded 12,590 organic farms cultivating 203,000 hectares of certified and in-conversion farmland (Neves et al. 2001). In addition, the number of supermarket chains interested in commercializing organic products has increased. In April 1999, representatives of a consortium of leading European supermarkets visited the state of RS, including Sainsbury (UK) and Carrefour (France), which have been committed to eliminating GM ingredients from their own-brand products (USDA-FAS 2000). There are already market opportunities for selling organic foods in Brazil, but the demand for these products is much higher for vegetables, fruits, and refrigerated processed foods. It is expected that organic soybean will be grown mainly for the export market in Europe and Japan. According to Ormond et al. (2002), the number of organic soybean farmers is 593 (8.40 percent of all organic farmers in Brazil), covering an area of 12,516 hectares (4.64 percent of the total organic area in Brazil).

An organic boom has emerged because of growing consumers' concerns over food safety and environmental problems. In addition, according to the farmers interviewed, organic farming requires harder work than conventional farming, but productivity seems to be about the same. Because of the extreme high cost of conventional agricultural inputs in Brazil, farmers benefit from the reduced production costs with organic products, and the attractive premium prices tend to compensate for other disadvantages.

There are already some agricultural co-operatives dealing with organic products. Among them, we focus on the Agricultural Co-op of Cotrimaio (Cooperativa Agropecuária Alto Uruguai), since it represents a remarkable example of agricultural co-operatives, contrasting with the case of Integrada. Data on the Cotrimaio were collected during interviews and a field survey conducted by the authors in December 2001. Cotrimaio is also a big co-op, consisting of about 6,200 farmers with more than R$120 million sales in 2001. A remarkable characteristic of Cotrimaio's operation is its diversified business, ranging from animal products and the processing of member farmers' produce to retailing. This can be contrasted with Integrada, which concentrates on marketing grains. This co-op, located in the city of Três de Maio, RS, is aware that there is an increasing demand in Europe for organic and GM-free soybeans with a premium price. Negotiating with consumer co-operatives in France and the Netherlands since 1998, Cotrimaio has been providing member farmers with the idea of growing organic and/or non-GM soybeans for the European market.

Table 10.3. Soybean Handled by Cotrimaio (tons)

Type of Product	2000/01	2001/02 (estimated)
GM-free	74,000	108,000
Conventional	24,000	18,000
Organic (in-conversion)	500	500
Organic (certified)	250	600

Source: Interview with Cotrimaio conducted by authors in December 2001.

As shown in Table 10.3, in the year 2000/01, Cotrimaio produced 74,000 tons of non-GMO soybeans and 750 tons of organic soybeans (including 500 tons of "transitional" organic) for export, in contrast to 24,000 tons of conventional soybeans that may contain GMOs. In the following year, the co-op was planning to produce 108,000 tons of non-GMO soybeans, 1,100 tons of organic and transitional organic soybeans, and reduce to 18,000 tons of conventional soybeans. To be certified as "organic," soybeans have to be grown in a field that has been cultivated organically for more than two years. Soybeans in the second year of this in-conversion process are called "transitional organic" here.

This co-op is working closely with farmers, sending technical consultants to help them understand how profitable it can be to become organic and to sell their crops in the international market (Cotrimaio 2001). The co-op is paying a 40–60 percent premium for organic edible soybeans, a 10 percent premium for transitional organic, and a 3.5 percent premium for non-GMO soybeans. In the latter case, the market price for non-GMOs is just 2–3 percent more than for conventional soybeans, meaning that Cotrimaio cannot receive much financial compensation for dealing with merely non-GMOs. In the interview, an executive staff member from the co-op complained that an 8 percent or more premium price in the consumer market is required to fully cover the cost of handling (e.g., DNA screening and segregated distribution) and to ensure proper benefit for its business.

Thus, the response of this agricultural co-operative is a clear example of pursuing an alternative or niche market, that is, non-GMO and organic soybeans for export. By doing so, local farmers can avoid severe competition in the conventional market, and hence protect themselves from the pressures of agricultural intensification and marginalization. Its limitation, however, comes from the extent to which it depends on the ex-

ternal market. Considering the price the co-op receives, this kind of relationship between consumer co-ops or supermarket chains in the North
and agricultural co-ops or producers in the South may not be sustainable.
When it comes to farmers' preference, some of them told us that the reason they started to grow organic soybeans was the high premium price
promised by the co-op staff for them. There are also some farmers who
prefer to grow organic crops because of environmental concerns, and because they are against GMOs, understanding that GM technology is not
compatible with agroecological farming. Those perceptions of concerned
farmers, however, cannot be expected to continue without proper support
from the government through policies and extension services (i.e., EMA-
TER/RS as described below).

Dependency on organic markets has another limitation for small-
family farmers. The rapid growth of organic farming has prompted the
Brazilian government to regulate the sector. In October 1998, the Ministry
of Agriculture and Food Supply published Directive 505 with the purpose
of establishing national standards for the certification of organic products.
The Directive was legislated in April 1999. Such legislation of standards is
controversial. So long as the establishment of nationwide standards helps
legitimize organic production in the eyes of consumers and makes it become easier for small farmers to sell products to a broad consumer base,
it should be welcomed. As is often the case in some developed countries,
however, it would also increase the chance for large processors/shippers as
well as export-oriented large-scale growers to enter this growing market
and compete directly against small farmers (Jussaume 2000).

Well-Intentioned Activities of an NGO and Community Groups in a Locality

In the state of Paraná and in several other states, an NGO named AS-
PTA (Assesoria e Serviços a Projetos em Agricultura Alternativa) has
been helping small-family farmers to succeed in using agroecology. The
approach implemented by AS-PTA is to identify constraints in their
farming through participatory methods and to develop solutions through
research involving the farmers. This participatory process is exemplified
in the program in the central/southern part of Paraná (Weid and Tardin
2001). AS-PTA's staff members offer agroecological alternatives and let
farmers choose technologies and ideas which are thought to be adaptable to the local conditions. In order to structure its work, AS-PTA put
the proposed alternatives into several divisions: genetic resources, eco-

logical soil management, agro-forestry, and family gardens. Seed production on the farm has also increased with the aim of financial savings as well as diversification and yield increase (Weid 2001). The program now covers twenty-two municipalities with a population of 250,000, including roughly 55,000 family farmers, among whom around 10,000 family farmers are directly involved in the intensive experimentation with agroecological practices.

Such experimental activity using sustainable farming methods has been also initiated by a church group through the Diocese de Santa Maria in the state of RS (Dill and Buske 2001). For the purpose of educating people to become better entrepreneurs and to produce in an ecologically and economically viable way, a project named Projeto Esperança involving eighty-two families was introduced in the region of Santa Maria in 1982. Since then, farmers started rethinking the conventional farming style, opening their minds to agroecological alternatives such as the multiple farming of organic rice, fish, and ducks. The wet rice fields can be rotated with fish farming and the ducks are expected to control weeds and insects in the rice fields (Dill and Buske 2001). Farmers have also been working on sharing experiences and trying other feasible alternatives suggested in their farm-level trial and error. Another notable achievement is the marketing of their products at local markets based on the concepts of "alternative cooperativism" and "popular/solidarity economy." A commerce center called Cooesperança was established in 1989 to distribute their products directly, and to foster fair business and solidarity between producers and local consumers.

To sum up, these two examples, AS-PTA and Projeto Esperança, can be seen as cases of local responses to the biotechnology revolution and neoliberal globalism in favor of local solidarity and an agroecological approach: They have promoted the diversification of farm styles and encouraged farmers' participation. Although the impressive results of these local activities are undeniable, we need to evaluate these experiences carefully. In the first example, Weid (2001), an executive staff member of AS-PTA, points out several obstacles and limitations: the farmers' lack of access to capital and adequate credit systems slows down the process; their access to seeds for green manure is also restricted due to cost and lack of availability; the regional market is controlled by a handful of intermediaries, and because of this, prices have been lowered and production has been discouraged; and the funding available for AS-PTA to support farmers is very limited despite the fact that public authorities are providing loans for the use of "technological packages." In Santa Catarina, for instance,

AS-PTA is collaborating with EPAGRI, a state extension service, to stimulate the use of green manure and cover crop technology among all state regions, involving more than 100,000 farmers (de Freitas 1995).

On the other hand, *Projeto Esperança* has also expanded to involve 123 groups and has reached more than 10,000 consumers. The reason for such success can be attributed to its focus on creating alternative markets. Another reason is its close partnership with public institutions, that is, the state government (through several relevant departments), local municipalities, the regional EMATER/RS, and the Federal University of Santa Maria. Indeed, many noteworthy examples of alternative, agroecological agriculture can be observed across the state of RS. In many cases, the EMATER/RS plays the vital role of an "institutional host" (Houtzager 2001) to facilitate and mediate among small-family farmers. In this partnership, farmers are organizing themselves collectively toward adopting agroecological alternatives, some of which are initiated locally by NGOs and community groups, as discussed above. We now turn to a close examination of the EMATER/RS's activities.

Institutional Alternatives: Agroecological Extension in Rio Grande do Sul

According to Caporal (2002a), we can divide the EMATER's history into four phases, each of which is characterized by the philosophy and emphasis of extension programs. The first period is from 1948 to 1960; the extension focused on supporting poor families and communities by using "rural credit" as a tool to help transform and modernize agriculture. The second period is from 1961 to 1980; modernization of agriculture was promoted further, and rural extension services prioritized the development and diffusion of modern technologies. The education process in this period was based on motivating farmers to adopt new practices and Green-Revolution-type technologies (see Chap. 2, above). When the national extension agency (EMBRATER—Empresa Brasileira de Assistência Técnica e Extensão Rural) was established under the military government in 1976 and imposed limits on the state-level extension services, some states tried to establish their own rural extension agencies, including EMATER/RS (1977) in the state of RS. The third period from 1980 to 1990 is characterized by "Reflexive Criticism"; the extension agencies were becoming concerned about the environment and the consequences of modern agriculture, while the priorities were still to improve production and productivity. Influenced by Paulo Freire, the famous Brazilian educator, changes were suggested for the method of rural extension. The fourth period is from 1990 until now under an "Environmental Transition" to a more environmentally sound agricultural style. After the dissolution of EMBRATER in 1990, some EMATERs have enforced their environment-oriented programs.

EMATER/RS, linked to the state government of Rio Grande do Sul, is famous for its strong and well-organized extension system, as well as for adopting an environmentally sound policy. About 80 percent of its budget comes from the state government, supplemented by 8 percent from the counties (Caporal 2002a). Since 1997, EMATER/RS's priority has been given to small-family farmers, the majority and prevailing rural actor in the state. The institution promotes rural development, using technical assistance and educational processes to strengthen family-based farming and stimulate farmers to improve the quality of their lives. Farmers usually welcome the extension services, and also join in projects provided by central and regional offices of EMATER/RS in collaboration with many other institutions, including rural workers' associations, women's associations, state or local schools, municipalities, and churches.

The idea adopted by EMATER/RS is "agroecology," which has become one of its most important projects, drawing attention to the need to preserve ecological and cultural biodiversity, act locally, and make use of diversified practices (Caporal and Costabeber 2001). A chief staff member of EMATER/RS's regional office interviewed by the authors argued that the Brazilian national agricultural research institution (EMBRAPA—Empresa Brasileira de Pesquisa Agropecuária) still adheres to the "old-fashioned idea" of the Green Revolution. "Agroecology is a more innovative way of thinking," he added. Another important aspect of EMATER/RS's agroecology program is to integrate farmers' skills with academic knowledge. This program reflects on past experiences, in which the knowledge of agricultural experts prevailed in agricultural matters, sometimes ignoring the wisdom of farmers acquired during many years of experience in their local settings. On the other hand, although some farmers' experience-based knowledge may lead to the development of a better agricultural model, most farmers in the state are immigrants from European countries and have lost even their "transplanted" knowledge on sustainable agriculture during the Green Revolution. In such a situation, a critical integration of their rediscovered knowledge with scientific expertise is an inevitable attempt to create an innovative way of thinking to convert the conventional farm to a new agroecological one (Altieri 2001). With this view, EMATER/RS launched a new program entitled ATER (Ações de Assistência Técnica e Extensão Rural) for working together with farmers to make them aware of their experience-based expertise, by drafting and conducting a Rural Participatory Diagnosis (DRP— Diagnóstico Rápido Participativo) for each local community (Caporal 2002b). The sole purpose of the DRP is to mobilize rural communities to identify their main problems and opportunities. To prepare specific plans to address them, potential technologies are evaluated based on environmental, economic, and social aspects expressed by local people. Participatory measures are employed throughout the process in order to involve the entire community.

The goal of EMATER/RS is not only to change the way of farming but also to create local marketing systems that are economically affordable, ecologically balanced, socially fair, and culturally acceptable. In 2001, the number of agroecological groups totaled 160, with 2,436 members. Their agroecological products are sold at farmers' outlets or in local/regional markets with the support of public institutions, NGOs, and farmers' associations as well as some Catholic-church organizations. According to Caporal (2002a), there are already 107 weekly markets selling agroecological products with more than 800 farmers participating. Additionally, EMATER/RS has initiated a number of collaborative relationships with agricultural co-operatives working to produce local-oriented agroecological and organic products (see Table 10.4).

The role of EMATER/RS in agroecology is not much different from those of NGOs and community groups, which can also create feasible programs to develop a sustainable agricultural model. It can be said, however, that EMATER/RS has the most important and pivotal role as a facilitator or "institutional host" given the institution's broad range in terms of areas and beneficiaries. Indeed, out of 623,000 families (as estimated in 1999) in the state's rural area, it gave assistance to more than 350,000 families in 2001, and among these nearly 285,000 families received continual assistance. EMATER/RS has more than 2,300 staff located in one central office, ten regional offices, and 470 county offices (95 percent of all 497 counties in the state) (Caporal 2002a). There is no other service available to so many farmers and rural areas.

Since 2000, EMATER/RS has also organized international seminars on agroecology, in partnership with the state government, with the aim of disseminating these alternative-agriculture ideas (Felippi 2000; Caporal 2002a). The first was held in Porto Alegre and had around a thousand participants. The second was held in 2001, again in Porto Alegre, and had some 2,300 participants from countries throughout the world. The lecturers were from the United States, Europe, and Latin America,

Table 10.4. Co-operatives Working with EMATER/RS for Agroecological Production

Area/Region	Products	Cooperatives
Serra do Sudeste and RS	Vegetables, Chicken	Cooperativa da Coolmeia
Ipe, Antonio Prado and Sarandi	Grapes, Wine	Cooperativa da Coolmeia
Northeast RS and Alto Uruguai	Subtropical fruits	Cooperativa da Coolmeia
Vale do Cai	Citrus	Cooperativa dos Citricultores do Vale do Cai
Constantina	Wheat	Cooperativa de Pequenos Agricultores (Coopac)
Barra do Rio Azul (Alto Uruguai)	Brown Sugar	Associacao de Agricultores Familiares Agroecologicos de Campo Alegre
Barra do Rio Azul (Alto Uruguai)	Brown Sugar, jam, sweets, pickled fruits, juice, milk, bread, juice, milk, bread, cheese	Centro de Apoio ao Pequeno Agricultor (CAPA); Cooperativa Central Alto Uruguai (COCEL); Sindicato dos Trabalhadores Rurais de Aratiba; Cooperativa de Produção Agropecuária Aratiba (COPAAL) and Movimento das Mulheres Trabalhadoras Rurais
Centro-Serra	Vegetables, Fruits	Cooperativa Ecologica (COAGRICEL)

Source: EMATER/RS, 2002.

including Professors Miguel Altieri (University of California, Berkeley) and E. S. Guzman (University of Cordoba). The number of participants in the third seminar, held in September 2002 at the same venue, increased to 3,087. The fourth seminar was held in November 2003, also in Porto Alegre.

These international seminars are not the only events organized by the institution. In 2001, EMATER/RS organized 2,945 local events across the state, involving 141,649 farmers. In the region of Santa Rosa, for example, 26 events were held in April, 37 in May, and 25 in June, 2002. These events include field trips, technical demonstrations, technical speeches, campaigns, and meetings for farmers, especially women and young farmers

Table 10.5. Farmers' Opinions in Santa Rosa Regarding Agroecology and GMOs

	Age	Size (ha)	Growing Organic	Agroecology Positive	Not Sure	Negative	GMOs Positive	Not Sure	Negative
A	41	9	X	X					X
B	60	12		X					X
C	28	14	X	X					X
D	39	21		X					X
E	58	28	X	X					X
F	63	31		x (if profitable)			x (if legal)		
G	54	36	X		X			X	
H	25	40		X					X
I	51	80			X			X	
J	76	103				X		X	
K	25	120			X		X		

Source: Authors' Survey in Santa Rosa, RS, Brazil, in December 2001 (A, C–J) and July 2002 (B, K).

among others. All the events are free of charge and open to anyone who is interested. Sometimes other enterprises or public institutions interested in rural services sponsor these events. Furthermore, EMATER/RS tries to reach farmers across the state through radio and TV programs.

The key point in this discussion is that the promotion of agroecological ideas serves to divert small-family farmers away from GMOs and external competitive pressures, nationally or internationally. In Santa Rosa, the northern part of the state that shares a border with Argentina, despite a lot of Roundup Ready® soybean seeds having been smuggled into the country, small soybean farmers interviewed by the authors in December 2001 and July 2002 showed a strong interest in changing their farming model from conventional to agroecological rather than to biotech intensification (see Table 10.5). In addition, many of them said that they relied on EMATER/RS for such a transformation and appreciated its close relationship with local farmers. Even in this region, however, some agricultural co-operatives still pursue the intensification of agriculture, promoting Roundup® herbicide together with smuggled GM soybean seeds among member farmers. Given that small-family farmers generally trust agricultural co-operatives as well, it is not surprising that the EMATER/RS's local activities are sometimes constrained in such cases. Moreover, the fact that there are some extension agents and agronomists backing GMOs as a "new tool for sustainability" should be acknowledged.

Conclusion

Under the advancement of neoliberal globalism and the biotechnology revolution, small-family farmers in the southern states of Brazil are becoming entangled in a relentless survival game in the competitive market. While soybean, one of their main crops, has been high up on the list of the country's most promising export goods, the role of small-family farmers in soybean production is diminishing rapidly. The federal government and TNCs have given priority to the high-inputs, intensive middle- and large-scale capitalist farmers located mainly in the *cerrado* region. Small-family farmers in the south have been faced with an apparently single choice: they either enter fully into the neoliberal market in a bid to improve their socioeconomic status, or face the imminent prospect of leaving farming altogether to migrate to urban areas. There is no reason to presuppose, however, that this is the only choice for farmers. As we have discussed, some farmers opted to produce non-transgenic soybeans, also for the export market. But given that this alternative is also firmly rooted in the modern agricultural paradigm and subject to the perils of neoliberal globalism, it has been a nonstarter for most farmers. Yet they also have the alternative of transcending the modern agriculture paradigm and the export market; they can move toward an agroecological approach focused on the local market. This is the first major conclusion of our chapter.

The second conclusion concerns the role of extension services in helping small-family farmers to continue farming. Saving and revitalizing rural Brazil requires abundant financial and human resources. Even successful experiences led by NGOs and community groups cannot progress without intensive and continuous support from public institutions. Especially in the case of the state of Rio Grande do Sul, EMATER/RS's activities based on agroecology play a key role in enabling farmers to access alternative technologies and markets, and diverting them away from GMOs without addressing a political ideology or emotional perception.

EMATER/RS is not free from limitations, however. First, EMATER/RS's budget is insufficient to fulfill its needs, pay its workforce, and ensure statewide activities, in spite of favorable support from the state government. Second, the institution's dependence on state politics and on local governments makes it both unstable and vulnerable. Although its initial idea of promoting agroecology and an alternative style of extension is not directly related to the state government's policy, its activities were heavily supported by the state government as well as the local left wing party (i.e., the PT or Workers' Party). Because of this, no one can predict how much the right-wing state government elected in 2004 will affect EMATER/RS's activities. Furthermore, it should be noted that the federal government, headed by the Workers' Party since 2002, decided to provisionally legalize the planting and sale of GM soybeans as of October 2003. As such, the political instability concerning GMO issues is expected to continue and hinder the extension activities and other local actors for some time. Finally, even in southern Brazil, which is known for being well organized as well as for having the most democratically and equitably distributed locality in the country, we cannot deny the fact that, compared with those in "Northern" countries, grassroots movements are relatively weak and small-family farmers remain rather passive actors. This fact could inhibit the extension from carrying out successfully the participatory and educational approach required. Hence an agroecological awareness program must be part and parcel of an educational effort geared toward an alternative, sustainable agriculture.

References

ACP (Agribusiness, Consultoria, e Planejamento S/C Ltda). 2000. Interview conducted by the authors in August 2000 in Londrina, Paraná, Brazil.

Altieri, M. 2001. "Agroecologia: A Dinamica Produtiva da Agricultura Sustentavel." UFRGS, Editora da Universidade, Porto Alegre, Brazil.

Bell, J. 1999. "Brazil's Transgene-Free Zone." *Seedling* 16(3): 2–10.

Caporal, F. R. 1998. "La Extension agraria del sector público ante los desafíos del desarrollo sostenible: El caso de Rio Grande do Sul, Brasil." Unpublished doctoral thesis submitted to the University of Cordoba, Spain.

———. 2002a. "La Extension Rural en Rio Grande do Sul: De la Doctrina 'Made in USA' hacia el Paradigma Agroecologico." Paper presented at the VI Maestría en Agroecologia y Desarollo Rural Sostenible en America Latina y Espana, International University of Andalucia, Spain, from 1 April to 12 June.

———. 2002b. Interview conducted by the author in July 2002.

Caporal, F. R., and J. A. Costabeber. 2001. "Agroecologia e Desenvolvimento Rural Sustentavel: Perspectivas para uma Nova Extensao Rural." EMATER/RS.

Chiba, T. 1998. "Agribusiness and Environmental Problems in Latin America" in I. Nakano, ed., *Agribusiness-ron.* Tokyo: Yuhikaku. Written in Japanese.

Cotrimaio (Cooperativa Agro-pecuária Alto Uruguai). 2001. Interview conducted by the authors in Tres de Maio, Rio Grande do Sul, Brazil, 6 December.

de Freitas, V. H. 1995. "Green Manures, a New Chance for Small Farmers." *ILEIA Newsletter* 11(3): 16.

Dill, L., and Buske, S. D. 2001. "Cooperation, Management and Sympathetic Popular Economy: An Experience of Development from the Projeto Esperanca." Paper presented at the II Seminário Internacional Sobre Agroecologia, Porto Alegre: Brazil, 26–28 November.

Felippi, A. 2000. "Reportagem: Extensao Rural inicia transicao agroecological." *Agroecologia e Desenvolvimento Rural Sustentavel*, Revista numero 01, EMATER/RS, Porto Alegre-RS, Brazil.

Fonseca, M. F., and A. E. Feliconio. 2000. "The Production and Commercial Network of Organic Food in Natural in Brazil: A Way for Citizen-ship." Paper presented at the X World Congress of Rural Sociology, Rio de Janeiro, Brazil, 30 July to 5 August.

Hassanein, N. 1999. *Changing the Way America Farms: Knowledge and Community in the Sustainable Agriculture Movement*. University of Nebraska Press.

Hightower, J. 1973. *Hard Tomatoes, Hard Times*. Mass.: Schenkman Publication.

Hisano, S. 2002. *Agribusiness and GMOs: A Political Economy Approach*. Tokyo: Nihonkeizaihyoron-sha. Written in Japanese.

Houtzager, P. P. 2001. "Collective Action and Political Authority: Rural Workers, Church, and State in Brazil." *Theory and Society* 30(1): 1–45.

Jussaume, R. A. Jr. 2000. "Building Trust with Consumers." *Farming West of the Cascades*, EB1889, Washington State University.

Jussaume, R. A. Jr., S. Hisano, C.-K. Kim, P. McMichael, S. Otsuka, Y. Taniguchi, and L. Zhibin. 2002. "Local Dimensions of East Asian Agri-Food Restructuring." Paper in process.

Long, N. 2001. *Development Sociology: Actor Perspectives*. London: Routledge.

Neves, M. C. P., P. F. Filho, and J. G. P. Ormond. 2001. "Organic Agriculture in Brazil: Panorama and Perspectives." *Organic Standard*, Issue 8, pp. 3–7.

Ormond, J. G. P., S. R. de Paula, P. F. Filho, and L. T. da Rocha. 2002. "Agricultura organica: Quando o passado e futuro." *BNDES Setorial*, Rio de Janeiro, No. 15.

Pelaez, V., and W. Schmidt. 2004. "Social Struggles and the Regulation of Transgenic Crops in Brazil." In K. Jansen and S. Vellema, eds., *Agribusiness and Society: Corporate Responses to Environmentalism, Market Opportunities and Public Regulation*, pp. 232–260. London: Zed Books.

Shiva, V. 1992. *The Violence of the Green Revolution*. London: Zed Books.

Tanaka, N. 2002. "Directions of Agricultural Cooperatives in Brazil: A Case Study on the Agricultural Co-op Created after the Dissolution of the Agricultural Co-op of Cotia." Paper presented at the Annual Conference of the Agricultural Economics Society of Japan, Ibaraki University, Japan, 31 March.

USDA-FAS (Foreign Agricultural Service). 2000. "Annual Planting Seed Report, Brazil 2000." *USDA GAIN Report*.

Warnken, P. F. 1999. *The Development and Growth of the Soybean Industry in Brazil*. Iowa: Iowa State University Press.

Weid, J. M. 2001. "Scaling Up, and Further Scaling Up Participatory Development," *LEISA Magazine*, Vol. 17, No. 3, pp. 23–26.

Weid, J. M., and J. M. Tardin. 2001. "Genetically Modified Soybeans: Blessing or Curse for Brazilian Agriculture?" *LEISA Magazine*, Vol. 17, No. 4, pp. 19–20.

CHAPTER 11

Social Movements and Techno-Democracy: Reclaiming the Genetic Commons

MANUEL POITRAS

The commercialization of genetically engineered products from the mid-1990s on was accompanied by the steady rise of activism and social movements against their use. Genetic engineering is by now a common issue in mass gatherings of the movement contesting the neoliberal character of globalization, and has motivated the creation of permanent campaigns on this issue in environmental organizations such as Greenpeace and Friends of the Earth. Social movements and activism around issues of genetic engineering have also emerged in parts of Latin America, with a timing, and with political and organizational characteristics, specific to the region in which they emerged and thus distinct from those of their counterparts in North America and Europe.

In this chapter I will examine the movements that have emerged in Mexico, and offer an interpretation of the meaning of these movements in the context of contemporary Mexico and Latin America. My goal is to provide an explanation for the rise of these movements that goes beyond the usual assumption that resistance arose because the movements involved are essentially anti-technology and anti-progress, or alternatively because the technology is essentially bad and thus must be rejected. The analytical framework used to interpret these movements draws from a specific understanding of the relationship between technology and politics, whereby the establishment of technological hegemonies is seen as an essential part of broader hegemonic politics. These concepts will be explained in the next section.

The interpretation proposed here views the resistance against commercial genetic engineering in agriculture as an essential part of the broader counter-hegemonic movement against neoliberal globalism. More specifically, I argue that the resistance to genetically modified organisms

(GMOs) undermines one central aspect of the bid for neoliberal hegemonic control, the technological one, by displacing the scientific and allegedly neutral discourse about GMOs, and thus by uncovering the power
relations at play in the use of this technology. Resistance to GMOs, therefore, seeks to overcome the fetishism of technology prevalent in modern
development discourse. The first section presents the conceptual framework on technology and politics used in this essay. I will then situate
GMOs in the context of Mexican politics. The next section offers a brief
analysis of the groups and movements that have recently taken up the issue of GMOs in Mexico. I then discuss the activism performed by these
groups on one crucial front: the contamination by genetically engineered
maize imported from the United States of central and southern agricultural fields in Mexico, the world's center of origin and mega-biodiversity
of maize. In the conclusion I offer an interpretation of what resistance to
GMOs in Mexico signifies in the wider context of technological hegemonies. My main conclusion is that the parties promoting genetic engineering and GMOs have failed to attain hegemonic consensus for reasons that
are specific to the region and to the political-economic context of the
time.

The Politics of Technology:
Genetic Engineering as Passive Revolution

Technology is at the core of the production of goods. Through the production process it also becomes embedded in the commodities that we
consume, and occupies an ever larger place in our daily lives. Technology
therefore occupies a central place in the dynamics of societies. It can thus
seem surprising that technology has only seldom been a topic of political
debate or struggle. Part of the explanation lies in the fact that the en-
lightenment thinking on social progress still common in Western societies views technology as a gift of modernity and as necessarily beneficial,
inevitably leading to the greater benefit of all. Such view of technology
has been transmitted to Latin American state formation projects from
the nineteenth century on, notably at first through the dissemination of
French positivism. According to this modernist view, technology does not
need explanation or debate. It is neutral socially, and there are no alternatives to it if society is to progress. It is apolitical, the realm of specialists
and scientists, not of politicians or social movements. This is the view
that underlies the position of many governments and regulatory institu-

tions on genetic engineering. It is also most notably the position adopted by the U.S. government on the international scene, with its constant insistence that GMOs should be regulated on the basis of "science" only, not ideology or politics. Another consequence of this view of technology is that resistance to new technologies is considered necessarily regressive and against the progress of human societies.

The mirror view of this perspective on technology sees it as the root of all problems in modern society, and calls for a return to traditional ways. While this view may be held by some in the opposition to GMOs, a more sophisticated understanding of technology is usually behind the social movements described in this essay.

In this third perspective, technology is considered "underdetermined," as there are a number of alternative developments that can arise from one technological advance; branching out is one of technology's intrinsic properties (Feenberg 1999). Therefore, "the choice between alternatives ultimately depends neither on technical nor economic efficiency, but on the fit between devices and the interests and beliefs of the various social groups that influence the design process. What singles out an artifact is its relationship to the social environment, not some intrinsic property" (Feenberg 1999, 79).

At a more abstract level, one could argue that to gloss over the technology embedded in commodities amounts to a certain form of fetishism, similar to the fetishism of commodities discussed by Marx. In other words, technology is part and parcel of the social relations of production that the fetishism of commodities hides. To debate the technologies embedded in commodities, such as the genetic engineering embedded in commercial transgenic seeds, is thus to remove the veil of fetishism and to question the production process and the social relations underlying them. In this conception, technology should be regarded as political, layered with ideological as well as cultural components. A version of this perspective, prevalent among techno-critics today, views technocracy, or the association of powerful technology with dominant power relations, as the problem. It is the way technology is developed and used and not necessarily technology itself that is at stake. Many contemporary social movements thus view technology as something that must be reappropriated by society and not left in the sole hands of governments or corporations.

For the purposes of this paper, the relationship between technology and politics is best understood through the concept of *technological hegemonies*, which, based on the Gramscian understanding of hegemony (cf. Gill 1993; Cox 1993; Gramsci 1971), refers to the emergence of a certain form of social consensus on the use of a predominant set of technologies. In a context of technological hegemony, dominant technologies are thus seen either as neutral or as necessarily progressive, and questioning their use becomes an act of transgression, of questioning the social order itself. Since the beginning of the commercialization of GMOs in the mid-1990s, genetic engineering has been the focus of an attempt by agrochemical corporations, government officials, and a number of scientists to form a hegemonic consensus on its use and benefits. The topic is still hotly debated, however. Quite unlike the Green Revolution technologies and other technologies that form part of our everyday life, genetic engineering has not reached hegemonic consensus but has rather unleashed multiple waves of protests around the world against its use. In the Gramscian lexicon, the kind of top-down restructuring attempted by the supporters of genetic engineering is called a passive revolution. Because of the top-down nature of the changes proposed, without the active support of the users and consumers of the technology (the farmers and the consumers of the GMOs), hegemonic consent, if reached, is not organic; it is thus more fragile and subject to contestation. The hegemonic debate regarding GMOs will be discussed here using the case of Mexico as my empirical referent.

The Non-Hegemonic Politics of GMOs

A number of elements have prevented the formation of a hegemonic consensus on GMOs in Mexico. First of all, GMOs entered Mexico at the end of the 1980s, at the height of the neoliberal restructuring of state-society relations, which Susanne Soederberg (2001) identifies as a form of passive revolution. One of the crucial aspects of this neoliberal passive revolution that is of particular import here is the shift in the legitimating rhetoric of the state: it shifted from a socially inclusive revolutionary nationalism, which was sounding increasingly hollow in a context of deepening economic crisis from the mid-1960s on, to the disciplining imperative of the world market. As Soederberg (2001) put it, the state repositioned itself from a developmentalist state to a national competitive one. This shift was most clearly felt by the peasantry, who had until then drawn considerable political strength from its role in the 1910s revolution, and from its strategic role of providing cheap food and raw materials to the industrial sector, at least until the agricultural crisis of the mid-1960s. While in fact the peasant sector had become increasingly marginalized in the postwar era, the state kept appealing to the revolutionary legacy to sustain its hegemonic claim (Hellman 1983; Otero 1999). The neoliberal passive revolution and its "compete or perish" logic only confirmed this marginalization, but also added numbers to the ranks of the marginalized farming population as the state also abandoned once-privileged farmers (Rubio 1998). The loss of political legitimacy for the ruling Institutional Revolutionary Party (PRI) was substantial. The Zapatista upheaval is only the most spectacular and well-known illustration of this (Harvey 1998; Gilbreth and Otero 2001; Cockcroft 1998). But another consequence is that GMOs enter the picture at a time when the great majority of Mexican rural producers are being explicitly set aside from the national development project, making them increasingly redundant as food producers and as the source of labor power (Bartra 2004).

The other most relevant change introduced by the neoliberal passive revolution was the removal of the pretence of "Mexicanization" of the economy and of technology. It is well established that the developmentalist project had given priority to capital-intensive technology for industrialization toward the development of a Mexican industrial bourgeoisie, rather than full employment policies and labor-intensive technological development (Cockcroft 1998; Hellman 1983). Nonetheless, technology was defined as a public good to be appropriated and developed for the benefit of the Mexican people, used by the state-supported and protected national bourgeoisie toward the development of a domestic industry that would provide employment for all. A considerable public research and scientific infrastructure was built to support that aim. For instance, the Green Revolution technologies that spread throughout the world in the postwar era originated in the Office of Special Studies of the Mexican Secretariat of Agriculture in the 1940s. In addition, so-called Mexicanization policies were set up to ensure that all investment had to be controlled by Mexicans, thus also providing for technology transfer to the domestic economy, though they ended up being ineffective in accomplishing that aim because of loopholes and lax enforcement (Hellman 1983; Cockcroft 1998).

Given the reliance on foreign investment to finance import substitution industrialization, however, most technology was imported at great cost from abroad. The neoliberal passive revolution removed this nationalist tension and its attendant contradiction between reliance on foreign import and technology, and the goals of endogenous technology development. By the same token, it confirmed the ascendancy of transnational and foreign technology over the national one. Two important legislative

changes in this regard were the liberalization of foreign investment regulations in 1989, allowing for full foreign ownership of Mexican companies (Otero 1996), and the 1987 and 1991 reforms of the intellectual
property law, making patent protection more stringent, and eliminating
nationalist and protectionist measures such as the exclusion of patents on
strategic products (including chemical, pharmaceutical, agrochemical,
and food products and processes) (Solleiro and Coutiño 1998; Aboites
and Martínez 1995). Both measures struck a final blow to the Mexicanization policy.

Genetic engineering could thus not be covered with the mantle of the
nationalist and developmentalist technological hegemony that was the political foundation of the Green Revolution. For a while, at the beginning
of the 1990s, Mexico's public research system was poised to become quite
active in genetic engineering research. Many governmental and academic
documents identified it as a promising avenue for Mexican agriculture,
and saw adequate human and basic scientific resources available in Mexico (UNIDO 1991; Mexico 1991; Quintero 1996). An intergovernmental
program, supported by the UNDP, to promote a comprehensive biotechnology development plan in Mexico was also designed during that period
(Mexico 1991). While this plan would have allegedly been directed to the
needs of Mexican agricultural producers, it was soon abandoned (Quintero
1999). Instead, a number of cuts were imposed on the public agricultural
research system, with no exception made of the fledging genetic engineering research agenda. Probably as a result of the cuts endured, INIFAP (the
National Institute for Forestry, Agricultural and Livestock Research),
the leading Mexican agricultural research center and institutional heir
of the Green Revolution, has allegedly been very slow and ineffective in
introducing the new biotechnologies in its research program (Pedraza
et al. 1998). This is not to say, however, that there is no public genetic engineering research in the country. In the last fifteen years, there has been
an important growth of genetic engineering research groups in public institutions, making Mexico one of the leading Latin American countries
in this regard (Quintero 1999). But most of the research being done at
this point does not involve genetic engineering, but rather micropropagation and tissue culture—lower technologies within the biorevolution.
Only two research centers reportedly have an ongoing genetic engineering
program, and only one of the two has had concrete results so far (UNIDO
1991; Pedraza et al. 1998; Quintero 1996 and 1999).

As for private corporate research, only one corporation of Mexican capital, the transnational Pulsar, uses genetic engineering, although only in

the laboratories of its subsidiaries in North America and Europe (Briceño 1999). The International Centre for Agricultural Research and Training (CIICA), the main agricultural Pulsar subsidiary located in Mexico, used to perform genetic engineering in its Chiapas laboratories, but the genetic engineering programs have now been transferred to a subsidiary, DNAP, located in California. Plants genetically modified in California are then sent to Mexico for micro-propagation and field testing. Micro-propagation is labor intensive, and this explains why it is performed in Mexico, where labor is cheaper. The failure to situate genetic engineering within a hegemonic framework, therefore, also comes from the fact that the technology is clearly controlled by transnational corporations rather than by the domestic bourgeoisie. The abandonment of revolutionary nationalism during and after the 1980s sapped one major source of legitimation for new technologies.

An attendant element that explains the resistance to GMOs and the failure of the latter to become hegemonic is the fact that the state has patently failed to protect domestic genetic resources and the peasantry against the interests of transnational capital. In fact, as we will see, much of the strength of the social movements against corporate genetic engineering comes from their successful demonstration of this lacuna.

Social Actors' Resistance against the Neoliberal Technological Hegemony

Mexico's political history greatly affected the way in which technology dissent could emerge. The political arena was closed for most of the developmentalist years, circumscribed by the corporatist system of affiliating workers, peasants, and the popular sector as separate organizations into the ruling PRI. That system started to crack from the 1960s on, however, with peasant guerrilla movements, student protests, and independent unions emerging, and with the formation of new opposition parties and the legalization of older ones. The cracks grew much wider from the 1980s on, with a renewal of party politics accompanying the electoral victories of opposition parties, and the emergence of new social and political actors. Among these, the independent peasant, indigenous, and environmental movements have become three core social actors of contestation of the new agricultural biotechnologies.

Rural social forces had been integrated since 1938 into a strict corporatist structure under the mantle of the Confederación Nacional Campesina (CNC). The CNC came to assume the role of political control of peasants by regulating access to land and containing discontent. It was the exclusive intermediary between the *campesinos* and the state, and pervasive clientelist relationships were created over time (Mackinlay 1996; Mackinlay and Otero 2004). The control of *campesino* politics by the CNC started to break down at the end of the 1950s, with the formation of independent peasant and rural workers' unions and a number of armed groups, both directed at the counter-agrarian policies of most postwar governments. In the 1970s, a large number of commodity- and service-based organizations were created, at first under the control of the CNC, but soon reaching beyond that control (Mackinlay 1996). Demands for autonomy from state domination, and from the control of the *caciques* it implied, soon became a core issue among peasant groups, ultimately leading to the formation of the Union Nacional de Organizaciones Regionales Campesinas Autónomas (UNORCA) in 1985 (Harvey 1996; Mackinlay 1996). Other autonomist peasant groups have emerged in the wake of the neoliberal withdrawal of the state from the countryside and of the signing of the North American Free Trade Agreement (NAFTA) in 1993. UNORCA has been the peasant organization most involved in the debate on GMOs in agriculture. This orientation has been bolstered by the integration in 1999 of UNORCA into Via Campesina, an international consortium of peasant organizations from around the world, set up to defend peasant lifestyle and livelihoods in international forums. Via Campesina stands for farmers' rights and argues against stronger forms of intellectual property (such as patents) on seeds and on plant varieties, for the recognition of the contribution of peasant communities to the preservation of biodiversity, and for the precautionary principle in dealing with the introduction of GMOs. UNORCA integrated these issues in its political agenda from 1999 on (Via Campesina 1999; Ladrón de Guevara 1999a; UNORCA 1999).

At the end of 2002, just before a 1 January 2003 NAFTA deregulation deadline for most agricultural products (excepting only corn, beans, and powdered milk), a large number of groups including UNORCA banded together under the banner of *El campo no aguanta más* ("The countryside can take no more"). The movement was aimed most forcefully at demanding that the government renegotiate the agricultural chapters of NAFTA, but also protests the fact that the government has constantly failed to enforce quotas and tariffs on maize imports since the coming into force of NAFTA in 1994 (*La Jornada* 1999; Enciso 2003; Acuña Rodarte 2003, 134). According to some commentators, the wave of mobilization sparked by the creation of *El campo no aguanta más* was such that it represented the rebirth of the *campesino* movement in Mexico (Hernández Navarro 2003). This rebirth was, however, short-lived, the group being disbanded in July 2004.

The GMOs debate was part of the negotiations that led to the signing of a new *Acuerdo para el Campo* (Agreement for the countryside) in April 2003 between the government and a number of peasant organizations, including *El Campo no aguanta más*. The *Acuerdo* stipulated that a new Law on Biosafety and Genetically Engineered Organisms would be promulgated. The manner in which the demands for autonomy and for the preservation of peasant livelihoods and biodiversity were to be promoted by this law, however, was not clear. What was clearly stated, however, was that the purpose of the law was to ensure that "biotechnology has better chances to contribute to the development of agriculture" and "to design and implement a national policy in relation to the experimentation, production and commercialization of products of biotechnology and genetic engineering" (*Acuerdo* 2003, art. 225, author's translation). Such clause seems to contradict UNORCA's and other peasant organizations' earlier rejection of transgenic agriculture as a tool for rural development (Ribeiro 2003). It thus remains to be seen whether this agreement between independent peasant organizations and the state will lead to their cooptation, as happened with the policy of *concertación* at the beginning of the 1990s (Harvey 1996; Mackinlay 1996), or if it really means, as Armando Bartra (2003) recently suggested, a victory in a gradualist scenario of reform toward a more inclusive agricultural policy and toward continued peasant mobilization.

Another set of agents of techno-resistance in Mexico is the indigenous autonomist movement, which also emerged from the breakdown of the corporatist structure. As the indigenous communities had been formally integrated into the CNC in 1965 (Mackinlay 1996), the indigenous movement took shape within the autonomist peasant movement, sharing much of its agenda against the CNC and in favor of autonomy from state clientelism and paternalism. It gained strength and independence with the support of international human rights groups, and with the emergence of indigenous issues in high-profile forums such as the UN Conference on Development and the Environment in Rio in 1992 (Sarmiento 1996). During the 1980s and 1990s, indigenous groups started to differentiate their political programs from the *campesinista* (peasant, land-reform-oriented) claims to land, toward the claim for territory and self-determination as a people. These demands were met repeatedly with rebuttal. Consistently failing attempts at legislative reform were a key motivation behind the Chiapas uprising of 1994, and the accompanying unilateral proclamation of autonomy of a number of indigenous communities (Otero and Jugenitz 2003).

The new Indian Rights Law voted by Congress in April 2001 is not likely to resolve the conflict. It is, in fact, a mutilated version of the San Andres Accord, which had been signed between government representatives and the Zapatistas in 1996, finally presented to Congress by President Fox at the beginning of his administration in 2000 and then modified by the Senate before ratification. Most indigenous groups, including the EZLN, were quick to condemn the new law. Among other issues, the central concepts for both territory and self-determination that were part of the San Andres Accord were gutted from federal recognition, and it was left for each state legislation to decide on specific forms of recognition. The new law does not prevent indigenous communities from owning natural resources, but also allows for their private appropriation by outsiders (Ross 2001a; Thompson 2001; Associated Press 2001). Moreover, faithful to the longstanding legacy of *indigenismo*, indigenous peoples are still regarded by the law as objects of public assistance, rather than as subjects of the rights that they have been demanding (Otero 2004b). Despite these setbacks, the indigenous movement has gained much political prominence over the last decade, a prominence that has been put to use in regard to biotechnology. Their main struggle has targeted a number of bioprospecting projects, whereby traditional indigenous knowledge is sought by pharmaceutical companies in their search for useful biological compounds. Groups such as the Chiapas Council of Traditional Indigenous Midwives and Healers (COMPITCH), founded in 1994 (the year of the Zapatista uprising), have been very active against such projects, which they have come to view as attempts at biopiracy (RAFI 1999; Raghavan 1999; Weinberg 2001; ETC 2001; Ross 2001b; Belejack 2002; Barreda 2003).

Through their claims for autonomy and for control over natural and productive resources, peasant and indigenous groups have implicitly, and at times explicitly, voiced an environmentalist discourse. This is especially true of the indigenous and small-scale farmers' groups, with traditional practices reputed to conserve biodiversity (such as traditional agriculture based on a diversity of crops rather than on monoculture; traditional botany; shade coffee growing that helps to preserve the rainforest; and so on). This contribution to biodiversity has drawn considerable additional support for their cause. While "traditional" practices were once viewed as backward and a feature to be done away with, they are now becoming viewed by some as ecologically necessary (Bartra 2004). In fact, many of the environmental claims of these groups echo or are informed by agroecology, an ascending agricultural productive paradigm promoting environmentally and culturally balanced productive practices, often drawing substantially from traditional practices and knowledge as well as from Western science (Altieri 1995).

A group that has environmental issues as an exclusive target is the Mexican chapter of Greenpeace. In fact, Greenpeace-Mexico has been the most vocal nongovernmental organization mobilizing against genetic engineering. It started to seriously tackle the issue at the end of 1998, focusing on what is most likely to resonate with a broad sector of Mexican society—genetically engineered maize. Greenpeace-Mexico is quite different from the other Mexican civil-society organizations involved with genetic engineering, as it corresponds more to the model of social movements found in industrialized countries and is not linked organically with the political history of the country. Greenpeace-Mexico has no broad popular constituency or large membership—hence it should perhaps be characterized as an NGO rather than a civil-society organization. Nor does it get much financing from local sources, relying instead largely on Greenpeace International (Covantes 1999). It also extensively uses the media, focuses on nontraditional issues, and relies on a transnational network of support. In contrast, while the peasant and indigenous movements that emerged out of the decline of the agrarian corporatist structure also espouse some of these mobilization characteristics (use of the media, transnational linkages, and nontraditional issues such as biodiversity and cultural identity), their mass membership and the prominence of productive and class issues keep them closer to a traditional model of political participation and collective action.

Counter-Hegemonic Struggles to Reclaim the Genetic and Technology Commons

While the introduction of GMOs in Mexico dates back to the early 1990s with the deregulation of the productivity enhancing dairy hormone rBST, a broad political debate over the issue is only now surfacing. Of all the issues and products of genetic engineering up for debate, transgenic maize is the one that has most galvanized the Mexican public and a number of social movements.

Maize is Mexico's staple food, the most widely cultivated crop, which started to be domesticated in this region about 10,000 years ago. It holds great symbolic importance, particularly in indigenous communities, where people themselves are regarded as being made of corn. British anthropologist Eric Thompson wrote in 1966: "Love of the soil is found among peasants the world over, but I doubt that there is a more strongly mystical attitude toward its produce than in Middle America. . . . Even today, after four centuries of Christian influence, [maize] is still spoken of with reverence and addressed ritualistically as 'Your Grace'" (Thompson 2002, 86). Southern Mexico is also the center of origin and of megabiodiversity for this crop, with more than 50 distinct races of maize and some 10,000 varieties kept in the maize germplasm collections of INIFAP as well as of the International Maize and Wheat Improvement Centre (CIMMYT). In Mexico one can also find in the wild the two closest parents of maize: teosinte and tripsacum (Serratos 1998). It also carries tremendous economic importance: about 70 percent of the cultivated surface is planted with local landraces (5.5 out of 8 million hectares), and 68 percent of the maize cultivated is for domestic human consumption (Mellen 2003; see Chap. 3 by McAfee and Chap. 6 by Fitting, this volume). Among the potential environmental problems identified with the eventual introduction of *Bt* maize in Mexico is the loss of biodiversity that transgenic maize could cause. *Bt* maize is a variety that has been genetically engineered to produce its own insecticide, the bacterium *Bacillus thuringiensis*. Concerns are greatest especially if maize is modified to include the Terminator gene (a gene that makes the seeds of a plant sterile [Jefferson et al. 1999]); but also because of the rapid development of resistance by pests to *Bt* (an otherwise very useful biological insecticide widely used in organic agriculture); and the potential development of new pests or the intensification of already existing ones. The latter could become a problem since the pest that *Bt* maize was sold for is not present in Mexico, and close relatives of that pest could become resistant to the *Bt* toxin if exposed to *Bt* maize (Serratos 1998; Gálvez and González 1998). These issues prompted some debate among specialists and academics when *Bt* maize started to be commercialized in the United States in the mid-1990s. The Mexican government finally banned the cultivation of transgenic maize in 1998, and field tests in 1999.

But a ban on cultivation has not been enough to bar *Bt* maize from entering Mexico. Even as early as 1996, some specialists were predicting that transgenic maize would enter the country in the short term: "The introduction of transgenic maize in Mexico is not only a possibility; we can foresee that it will happen in the short term, legally or not" (Serratos 1998, 80; author's translation). The reason for this is that Mexico imports increasing amounts of maize from the United States, some 5 million tons a year (almost twice as much as the amount planned under NAFTA), accounting for 24 percent of total corn consumption in Mexico and 11 percent of U.S. maize exports (Mellen 2003). Transgenic maize in the United States accounts for an increasing proportion of the maize crop. While these imports were supposed to be dedicated exclusively for animal feed and the production of high fructose corn syrup, its distribution through public food agencies in many parts of the country ensured that it would eventually be used as seed for planting. In the debate that ensued, farmers admitted that they had planted maize distributed through the food agency (Serratos 1998; Covantes 1999; Mellen 2003).

Beyond the environmental issues already explained, other issues that have been raised include the liability for the intellectual property embedded in transgenic seeds in cases of contamination. Would farmers who inadvertently planted GM seeds be made to pay royalties to the transnational corporation owning patents on the technology? In Canada, a farmer named Percy Schmeiser lost a legal battle against Monsanto, which accused him of using its technology without paying the royalties. Schmeiser claimed he is the victim of genetic contamination and never willingly planted Monsanto's GM seeds. A final issue for farmers raised by the contamination of landraces is the potential loss of non-GMO grain markets.

Greenpeace has drawn considerably on the symbolic power of maize in Mexico to make its case, noting not only the dangers for this millennial crop, but also the hardship suffered by national producers confronted with increasing amounts of allegedly "dumped" maize from the United States. Moreover, through its numerous actions around the issue, Greenpeace taps into a still very resonant nationalist and anti-imperialist chord among the Mexican population. For instance, during 1999, both the *Angel de la Independencia*, a monument to the heroes of the Independence in Mexico City, and a historic fortress in Veracruz, also symbolizing Mexican resistance to imperialism, have been occupied by Greenpeace activists, who announced the "Mexican declaration of genetic independence." Moreover, a huge banner was displayed denouncing U.S. imperialism underpinning the import of transgenic maize in Mexico.

Greenpeace has also largely taken advantage of the institutional failures and loopholes in the regulation of genetic engineering in Mexico, and the fact that the import of transgenic maize in such quantities demonstrates that the government is not respecting its own regulations regarding NAFTA quotas and tariffs or the protection of Mexican biodiversity. During other demonstrations, Greenpeace activists have attached themselves to rails in front of trains which contained maize imported from the United States, or demonstrated in the port of Veracruz in front of shipments of U.S. maize, asking to see the permits for the import of transgenic grain (Pérez 1999). In order to prove its allegations that transgenic maize was being imported into Mexico—something which the government was denying—it contracted a laboratory from Austria to test samples of U.S. maize, and then went to the Ministry of Agriculture with the evidence. Agriculture officials then sent Greenpeace to the Ministry of Health, arguing that since the claim is that transgenic maize is being consumed by humans, the issue pertains to the latter administration. In turn, the Ministry of Health declined to deal with the issue, saying that it was the Ministry of Agriculture's responsibility. Greenpeace thus successfully exposed the contradictions and gaps in the regulation of the introduction of transgenic crops in the country, and pressed the government to take action. Greenpeace has also participated in forums on genetic engineering, produced a video on genetically engineered maize and its consequences for Mexico, and argued for the opening up of GMO regulation and monitoring institutions to civil society organizations.[1]

Other important social forces mobilizing against transgenic maize are UNORCA and its network of regional peasant associations. UNORCA has held forums dealing with this issue, and has managed to gather farmers' support against the idea of using transgenic crops by showing the dangers of this technology for small-scale peasant agriculture (UNORCA 1999; Ladrón de Guevara 1999b). Activism against transgenic crops also spread in the agricultural regions of Mexico, notably in Chiapas, where more than 100 Chiapas Indian communities' representatives held a Maize Meeting in San Cristóbal de las Casas in April 2001. A vow was made not to use transgenic corn in Chiapas (Weinberg 2001). The seminar "In the Defense of Maize" was organized in Mexico City in January of 2002, and attended by a consortium of 138 Mexican and international organizations. One of the outcomes was an international declaration calling for an immediate moratorium on the imports of transgenic corn into Mexico (Notimex 2002; UPI 2002; Biodiversidad en América Latina 2003; see Chap. 3 by McAfee and Chap. 6 by Fitting, this volume). An international day of action against the contamination of the center of origin of maize also took place in April 2002, with actions in a number of countries across the Americas. Food companies also started to react to public concern on the issue. Although Mexico has no consumer organizations or consumer activism to speak of, Maseca, the largest corn processor in the country, changed its mind during 2000 about publicly supporting GMOs and using transgenic corn in its products, and instead pledged not to use it (Massieu Trigo and González 2000).

When an article in the British scientific journal *Nature*, published at the end of 2001, claimed to have proof of the transgenic contamination of maize in the southern states of Mexico, the issue became international, and provided new material to bolster the global and domestic anti-GMO movement. The study was at first denounced by Ministry of Agriculture officials and scientists associated with the genetic engineering industry and even disavowed by the editors of *Nature*, an unprecedented move (Freeman 2002; Kaufman 2002; Elias 2002; Mellen 2003; *Milenio* 2001). But the Environment Ministry eventually confirmed the findings with its own and other studies. Ministerial declarations since then have shown that the contamination might be even more widespread than initially thought (Kaufman 2002; Ho 2002; Brown 2002). These studies and the mounting public and international pressure sparked an intense debate within the Mexican Congress, which voted in December 2001 to advise the Department of Agriculture to unilaterally halt imports of transgenic corn from the United States, in addition to undertaking a comprehensive study of the scope and magnitude of the contamination and to formulate a remediation plan. A new proposal for a Biosecurity Law is blocked in Congress. In the meantime, the administration does not seem intent on stopping or even slowing down the import of U.S. transgenic maize.

The battle was also waged in the courts. Greenpeace, UNORCA, a number of other peasant organizations, academic research groups, and prominent individuals began a formal complaint process against the Mexican government at the Federal Tribunal for the Protection of the Environment (PROFEPA) to seek an immediate ban on the import of transgenic corn from the United States. After the repeated calls for a moratorium on imports of transgenic maize were not heeded by the Mexican government, an appeal to the Commission for Environmental Cooperation of NAFTA was made in April 2002. The CEC is designed as a formal complaint process for citizens of the NAFTA region against any of the three governments when these fail to enforce their own environmental laws. The decisions of the CEC are nonbinding, but are considered a significant recourse for civil society organizations to push a government into action. In the appeal to the CEC, Greenpeace was joined by the Unión de Grupos Ambientalistas, as well as by Oaxaca-based indigenous and peasant groups such as the Unión de Comunidades Zapotecas y Chinantecas and the Comité de Recursos Naturales de la Sierra Norte de Oaxaca (Notimex 2002; UPI 2002). At the CEC Secretariat, the file was dealt with under Chapter 15 rather than Chapter 14 of the North American Agreement on Environmental Cooperation, thus giving the issue priority status, more resources, more leeway in the design process, and a higher profile (Carpentier 2003). The final report was issued in September 2004, recommending that the Mexican government suspend imports of genetically modified corn, but was met with the skepticism of the U.S. and Canadian governments (*SourceMex* 2004), which could otherwise have been important levers in pressuring the Mexican government to act on this matter.

Overall, despite the numerous attempts by supporters of genetic engineering and by industry representatives or front groups to cover up the contamination issue and discredit the studies, the anti-GMO and environmental movements have managed to position this issue as a new symbol of the incapacity of the current regulatory environment and of industry to deal with GMOs in a way that would be safe for the environment and for humans,[2] thus undermining any hegemonic appeal to this technology as a new basis for the development of the countryside.

Steps toward Techno-Democracy?

As demonstrated in this essay, there is currently a substantial amount of mobilization around the issue of genetic engineering by a diversity of social actors. What brings these groups together around this issue is not the technology as such, but more crucially the power relations that genetic engineering currently implies, namely the top-down, undemocratic, and arbitrary nature of its introduction into and use in the Mexican countryside. Binding these groups (within Mexico and internationally) are also certain conceptions of power and democracy that inform their rejection of current genetic engineering. At stake is a rejection of the capitalist state, and of the liberal and neoliberal electoral democracy which adapts to capital accumulation and its technological needs rather than to the needs of food consumers and indigenous or peasant communities.

The end of the authoritarian developmentalist era brought about the potential for indigenous and peasant communities to reclaim their environmental, biodiversity, genetic, cultural, and knowledge commons from the state. But the transition to neoliberalism has introduced a new threat, that of privatizing these commons. Because genetic engineering is controlled by giant transnational corporations and because it is emerging in a context of intensification of strict intellectual property rights regimes, it has become a potent example of that threat. The policy response to the privatization of the biodiversity commons has been weak, with a regulatory regime full of loopholes and contradictions, apparently bent on protecting private accumulation interests rather than consumers and indigenous and peasant communities. The struggle of the Mexican peasant and indigenous movements has thus become one that centers on autonomy from centralized, top-down state power as well as one against the commodification and depletion of their commons. In doing so, these movements are taking crucial steps toward the end of the technological fetishism behind neoliberal globalism, toward self-management and the democratic control over their natural resources as well as technology itself. At the same time, they are becoming political actors that cannot be ignored when considering the interaction between technological change and more participatory and sustainable forms of development. Thus, while some of their goals and strategies are immediate, their "success" and importance should not be evaluated solely on the basis of immediate legislative change. Rather, their significance is broader and should be viewed in the long term. Their success will be measured ultimately by changes in the way technology is understood and incorporated in our daily lives, in food production systems, and in local, national, and global technology governance.

Notes

1. The sources for this section on Greenpeace include Mexican press coverage of their actions, an interview with Liza Covantes, head of the Genetic Engineering Campaign at Greenpeace-Mexico in Mexico City (Covantes 1999), and various press releases and documents produced by Greenpeace.

2. An earlier major issue drawing similar fire was in 2001, when StarLink® corn, a transgenic variety of maize approved by the FDA only for animal consumption, was found in the U.S. human food chain.

References

Aboites, Gilberto, and Francisco Martínez. 1995. "Situación de la Legislación Mexicana en Materia de Biotecnología y Recursos Genéticos." In *El Campo Mexicano en el Umbral del Siglo XXI*, ed. A. E. Rodríguez. Mexico D.F.: Espasa Calpe.

Acuerdo Nacional Para el Campo: Por el Desarrollo de la Sociedad Rural, la soberanía y seguridad alimentarias. 2003. www.economia.gob.mx/pics/p/p2/Acu_Nac_Campo.pdf. Accessed June 2003.

Acuña Rodarte, Olivia. 2003. "Toward an Equitable, Inclusive, and Sustainable Agriculture: Mexico's Basic Grains Producers Unite." In *Confronting Globalization: Economic Integration and Popular Resistance in Mexico*, ed. Timothy A. Wise et al. Bloomfield, Conn.: Kumarian.

Altieri, Miguel. 1995. *Agroecology: The Science of Sustainable Agriculture*, 2nd ed. Boulder, Colo.: Westview Press.

Associated Press. 2001. "Mexico Rebels Reject Indian Bill." *New York Times*, 1 May.

Barreda, Andrés. 2003. "Biopiracy, Bioprospecting and Resistance: Four Cases in Mexico." In *Confronting Globalization: Economic Integration and Popular Resistance in Mexico*, ed. Timothy Wise et al. Bloomfield, Conn.: Kumarian.

Bartra, Armando. 2003. "El Campo no Aguanta Más, a Báscula." *La Jornada*, 23 April.

———. 2004. "Rebellious Cornfields: Toward Food and Labour Self-Sufficiency." pp. 18–36 in Otero 2004a.

Belejack, Barbara. 2002. "Bio 'Gold' Rush in Chiapas on Hold." *NACLA Report on the Americas* 35:5 (April/March): 23–28.

Biodiversidad en América Latina. 2003. "Conclusiones del Seminario 'En Defensa del Maíz.'" Biodiversidad en América Latina. http://biodiversidadla.org/article/articleprint/983/-1/10/. Accessed June 2003.

Briceño, Guillermo. 1999. Personal interview at CIICA, Tapachula, Chiapas. 1 July.

Brown, Paul. 2002. "Mexico's Vital Gene Reservoir Polluted by Modified Maize." *Guardian*, 19 April.

Carpentier, Chantal Line. 2003. Personal interview at CEC's Secretariat, Montreal. 9 June.

Cockcroft, James D. 1998. *Mexico's Hope: An Encounter with Politics and History*. New York: Monthly Review Press.

Covantes, Lisa. 1999. Personal interview at Greenpeace Mexico headquarters. Mexico City, 13 April.

Cox, Robert. 1993. "Gramsci, Hegemony and International Relations: an Essay in Method." In Gramsci, *Historical Materialism and International Relations*, ed. Stephen Gill. Cambridge: Cambridge University Press.

Elias, Paul. 2002. "Corn Study Spurs Debate over Corporate Meddling in Academia." Associated Press, 21 April.

Enciso, Angelica. 2003. "El Acuerdo para el Agro, en Riesgo de Venirse Abajo." *La Jornada*, 25 April 2003.

ETC. 2001. "US Government's $2.5 Million Biopiracy Project in Mexico Cancelled." ETC Group News Release (9 November), www.etcgroup.org. Accessed December 2001.

Feenberg, Andrew. 1999. *Questioning Technology*. London: Routledge.

Freeman, James. 2002. "Scientist Claims Vendetta over GM Research; Biotech Industry 'Fighting Back against Maize Attack.'" *Herald* (Glasgow), 5 April.

Gálvez M., Amanda, and Rosa Luz González A. 1998. *Armonización de Reglementaciones en Bioseguridad.* Mexico City: UNAM.

Gilbreth, Chris, and Gerardo Otero. 2001. "Democratization in Mexico: The Zapatista Uprising and Civil Society." *Latin American Perspectives.* Issue 119, 28(4): 7–29.

Gill, Stephen. 1993. "Epistemology, Ontology and the 'Italian School.'" In Gramsci, *Historical Materialism and International Relations,* ed. Stephen Gill. Cambridge: Cambridge University Press.

Gramsci, Antonio. 1971. *Selections from the Prison Notebooks,* ed. and trans. Quintin Hoare and Geoffrey N. Smith. New York: International Publishers.

Harvey, Neil. 1996. "Nuevas Formas de Representación en el Campo Mexicano: La Unión Nacional de Organisaciones Regionales Campesinas Autónomas (UNORCA), 1985–1993." In *Neoliberalismo y Organización Social en el Campo Mexicano,* ed. Hubert Carton de Grammont. Mexico D.F.: Plaza y Valdés/UNAM.

———. 1998. *The Chiapas Rebellion: The Struggle for Land and Democracy.* Durham, N.C.: Duke University Press.

Hellman, Judith Adler. 1983. *Mexico in Crisis,* 2nd ed. New York: Holmes and Meier.

Hernández Navarro, Luis. 2003. "Renacimiento del Movimiento Campesino." *Ojarasca* 69 (January).

Ho, Mae-Wan. 2002. "'Worst Ever' Contamination of Mexican Landraces." ISIS Report, 29 April. http://www.i-sis.org.uk, accessed April 2002.

Jefferson, Richard A., Don Byth, Carlos Correa, Gerardo Otero, and Calvin Qualset. 1999. "Genetic Use Restriction Technologies: Technical Assessment of the Set of New Technologies which Sterilize or Reduce the Agronomic Value of Second Generation Seed, as Exemplified by U.S. Patent 5,723,765, and WO 94/03619." Prepared for the United Nations Convention on Biological Diversity. Available in English at http://www.biodiv.org/doc/meetings/sbstta/sbstta-04/official/sbstta-04-09-rev1-en.pdf.

Kaufman, Marc. 2002. "The Biotech Corn Debate Grows Hot in Mexico." *Washington Post.* 25 March, p. A09.

Ladrón de Guevara, Ernesto. 1999a. "Contaminación Genética, Alimentos Transgénicos." Unpublished internal document. Mexico City: UNORCA.

———. 1999b. Personal interview. Mexico City, 30 July.

La Jornada. 1999. "Cayó la Inversión Pública en el Campo y Obligó al Abandono del Cultivo." *La Jornada,* 29 March, p. 19.

Mackinlay, Horacio. 1996. "La CNC y el 'Nuevo Movimiento Campesino.'" In *Neoliberalismo y Organización Social en el Campo Mexicano,* ed. Hubert Carton de Grammont. Mexico D.F.: Plaza y Valdés/UNAM.

Mackinlay, Horacio, and Gerardo Otero. 2004. "State Corporatism and Peasant Organizations: Toward New Institutional Arrangements." Pp. 72–88 in Otero 2004a.

Massieu Trigo, Yolanda, and Rosa Luz González. 2000. "Ingeniería Genética y Biodiversidad: La Necesidad de la Regulación." Presented at the 22nd Congress of the Latin American Studies Association. Miami, 16 March.

Mellen, Matt. 2003. "Interview with David Quist," *Seedling*, April 2003. www .grain.org/seedling/seed-interview-d-en.cfm. Accessed June 2003.

Mexico. Secretaría de Relaciones Exteriores. 1991. *Preparación del Programa Nacional de Cooperación en Biotecnología*. Mexico City: Secretaria de Relaciones Exteriores.

Milenio. 2001. "No Hay Maíz Transgénico en el Estado de Oaxaca." *Milenio*, 12 April.

Notimex. 2002. "Pedirán Comisión Ambiental del TLC Analizar Contaminación del Maíz en Oaxaca." *Noticias de Oaxaca*, 25 April.

Otero, Gerardo. 1996. "Neoliberal Reform and Politics in Mexico." pp. 1–25 in *Neoliberalism Revisited: Economic Restructuring and Mexico's Political Future*. ed. Gerardo Otero. Boulder, Colo.: Westview Press.

———. 1999. *Farewell to the Peasantry: Political Class Formation in Rural Mexico*. Boulder, CO: Westview.

———, ed. 2004a. *Mexico in Transition: Neoliberal Globalism, the State and Civil Society*. London: Zed Books.

———. 2004b. "Global Economy, Local Politics: Indigenous Struggles, Civil Society and Democracy." *Canadian Journal of Political Science* 37(2) 325–346.

Otero, Gerardo, and Heidi Jugenitz. 2003. "Challenging National Borders from Within: The Political Class Formation of Indigenous Peasants in Latin America." *Canadian Review of Sociology and Anthropology* 40(5): 503–524. Special Issue on Borders and Identities.

Pedraza, Lorena, et al. 1998. "La Biotecnología en México: Una Reflexion Retrospectiva 1982–1997." *Biotecnología: Nueva Era* 3:3 (Sept.–Dec.).

Peréz U., Matilde. 1999. "Míembros de Greenpeace, Encadenados a Vías de Tren con Maíz Transgenico." *La Jornada*, 8 December, p. 53.

Quintero, Rodolfo. 1996. "Biotecnología para la Agricultura." In *Posibilidades para el desarrollo tecnológico del campo Mexicano*, Tomo 1. Ed. José Luis Solleiro, María del Carmen del Valle, and Ernesto Moreno. Mexico City: UNAM.

———. 1999. Personal interview. Mexico City, 10 December.

RAFI. 1999. "Biopiracy Project in Chiapas, Mexico, Denounced by Mayan Indigenous Groups." RAFI News Release, 1 December. www.etcgroup.org. Accessed December 2001.

———. 2000. "Indigenous Peoples' Organizations from Chiapas Demand Immediate Moratorium." RAFI Geno-Types, 23 October. www.etcgroup.org. Accessed December 2001.

Raghavan, Chakravarthi. 1999. "Mexico: It's Bio-Plunder, Not Benefit-Sharing." SUNS, 3 October.

Ribeiro, Silvia. 2003. "Los Transgénicos y el Acuerdo para el Campo." *La Jornada*, 28 April.

Ross, John. 2001a. "'Indian Rights' Law: Last-Ditch Stand by Mexico's Indians to Right White Wrongs." *Mexico Bárbaro* 273, 11 August.

———. 2001b. "Indian Doctors Battle the Devil to Save Their Native Plants from Transnational Bio-Pirates." *Mexico Bárbaro*, 271, 28 July.

Rubio, Blanca, 1998. "El Dominio 'Desarticulado' de la Industria Sobre la Agricultura: La Fase Agroexportadora Excluyente." Presented at the Fifth Latin American Congress of Rural Sociology. Chapingo, Mexico.

Sarmiento, Sergio. 1996. "Movimiento Indio, Autonomía y Agenda Nacional." In *Neoliberalismo y Organización Social en el Campo Mexicano*, ed. Hubert Carton de Grammont. Mexico D.F.: Plaza y Valdés/UNAM.

Serratos H., J. Antonio. 1998. "Evaluación de Variedades Novedosas de Cultivos Agrícolas en su Centro de Origen y Diversidad: El Caso del Maíz en México." In *Transformación de las Prioridades en Programas Viables: Actas del Seminario de Política Biotecnología para Latina América*, ed. John Komen et al. The Hague: Intermediary Biotechnology Service.

Soederberg, Susanne. 2001. "From Neoliberalism to Social Liberalism: Situating the National Solidarity Program within Mexico's Passive Revolutions." *Latin American Perspectives*, Issue 118, 28:3 (May), 104–123.

Solleiro, J. L., and Beatriz Coutiño. 1998. *Políticas de Biotecnología y Biodiversidad: Estrategias de Gestión de la Propiedad Intelectual para la Industria de Semillas*. Mexico City: UNAM.

SourceMex. 2004. "Nafta Environmental Commission Recommends That Mexico Suspend Imports of Genetically Modified Corn." *SourceMex* 15:35 (22 September).

Thompson, Eric. 2002. "The Meaning of Maize for the Maya." In *The Mexico Reader: History, Culture, Politics*, ed. Gilbert M. Joseph and Timothy J. Henderson. Durham, N.C.: Duke University Press.

Thompson, Ginger. 2001. "Mexico Congress Approves Altered Rights Bill." *New York Times*, 30 April.

UNIDO. 1991. *Biotechnology Policies and Programmes in Developing Countries: Survey and Analysis*. New York: United Nations Industrial Development Organization.

UNORCA. 1999. "Reunión Sobre Biodiversidad, Ingenería Genética y Manejo de Recursos Naturales." Mexico City: UNORCA.

UPI. 2002. "Activists Want Mexican Crackdown on Genetic Corn." United Press International, 26 April.

Via Campesina. 1999. "Biodiversidad, Bioseguridad y Recursos Geneticos." Unpublished internal document. Mexico City: Via Campesina.

Weinberg, Bill. 2001. "Bio-Piracy in Chiapas." *Nation*, 20 August.

CHAPTER 12

Conclusion: Food for the Few?

GERARDO OTERO AND GABRIELA PECHLANER

The purpose of this concluding chapter is to offer a conceptual wrap-up of the foregoing discussions, address potential alternatives, and propose a research agenda. What does it all mean for Latin American countries? First, looking at the general overview chapters of Latin America and the case studies from three of its largest countries makes it very clear that the export of the U.S. model of modern agriculture has some very particular effects for this region. In the introductory chapter, Gerardo Otero outlined some potential concerns over the export of this model to developing countries. As we indicated in Chapter 2, a number of social issues with this model already became apparent with the introduction of Green Revolution technologies in the postwar period. Kathy McAfee highlighted the "geographies of difference" and how these make a direct transfer of genetic engineering products, designed for U.S. conditions, highly problematic. Both socioeconomic profiles of most Latin American producers, and the specific ecological conditions of the region, make it imperative to move toward a locally based, bottom-up approach to plant breeding.

The case studies presented here substantiate many of these concerns and draw further attention to three issue areas of particular salience to the Latin American agricultural biotechnology experience: social polarization at the national level, international equity, and environmental impacts. These will be discussed in more detail below. The case for the development of an alternative to the technological paradigm offered by biotechnology as the continuation of modern agriculture is weak, but imagining an outline is possible. Finally, we will offer some areas for future research on how the biotechnology revolution is affecting agrarian social structures, biodiversity, and the environment.

Latin America and Agricultural Biotechnology

This collection of studies provides a conceptual and empirical overview of Latin American agricultural biotechnology: from adoption to impact, there are definite differences between advanced industrial countries and developing ones with respect to their relationship with agricultural biotechnologies. Some of these differences can be seen in the negotiation and implementation of regulatory frameworks around biotechnology, such as McAfee (Chap. 3) indicates, as do Jansen and Roquas (Chap. 4), while others can be seen in the more general patterns around particular issues of hunger or biodiversity, such as discussed by Poitras (Chap. 5) and Fitting (Chap. 6). One thing that binds all these experiences is the seemingly irrevocable linking between neoliberal economic reforms, modern agriculture, and genetic engineering technologies. The technological package of modern agriculture, now extended by the addition of biotechnology, is inevitably associated with economic (free-market) policies. While this is not necessary for the technology itself, it has been the specific policy context for its dissemination.

Perhaps no chapter more explicitly draws the link between biotechnology and neoliberal economic restructuring than Miguel Teubal's chapter on the adoption of genetically modified soya in Argentina (Chap. 8). Argentina is one country that has fully embraced the biotechnology/free-market agricultural program. The country has undertaken the most radical structural adjustments toward neoliberal reformation, including the rapid deregulation and dismantling of agricultural boards and supports that could otherwise mitigate impacts on food supplies and on medium and peasant farmers (Chap. 8). In about a decade, Argentina's agriculture has been reoriented away from production for the domestic market and toward the production of GM soy for the export market. Other countries responding to the call of the export market and pressures to open their markets undergo similar reorientations, though few to the extent of Argentina (see, for example, Fitting on Mexican restructuring, Chap. 6).

Not all connections to neoliberal restructuring and biotechnologies occur at the national or even the officially sanctioned level, however. In some cases the connection is made through the cobbled-together responses of subsistence producers under stress. Hisano and Altoé (Chap. 10), for example, argue that despite an official national prohibition against GM crops, many small soybean farmers in Brazil turned to the use of genetically modified soybeans when faced with rapid economic restructuring in the hopes that these crops would circumvent their economic imperilment. And yet, in spite of such attempts at economic sur-vival, these case studies vividly illustrate how the combination of neolib-eral policies and modern agriculture negatively affects subsistence and other small farmers in Latin America. The reaction of these farmers dif-fers by country and by circumstance, but some basic strategies are com-mon to all. Some producers rush to grab a place in the global economy, maximizing their capital investment, as they try to "flee toward the fu-ture," as Teubal put it in his chapter. While successful for some, eco-nomic devastation is the result for most. For many subsistence farmers, even the attempt is not an option and they are forced to other strategies for survival. Social polarization is the inevitable result.

While broad patterns can be found, there are also distinctions among Latin American countries that signal the potential for an emerging re-gional stratification based on differential incorporation and adoption of biotechnology and modern agriculture. The neoliberal paradigm, spe-cifically manifested here in deregulated agriculture and a transition to GM technologies, is prompting different regional incorporation into the global economy. Argentina's profound neoliberal restructuring and full-scale transition to GM soya production for export has transformed it into a major player in the global soybeans market, for example (Chap. 8). The international role to be played by many other Latin American countries is not yet as clear as Argentina's. The role of Brazil, another major soy-beans producer, will highly depend on the resolution of massive inter-nal struggles over the regulation of biotechnology (Jepson, Brannstrom, and Stancato de Souza, Chap. 9). Similarly, Mexico currently addresses its GM adoption somewhat inconsistently, on a case-by-case assessment (Poitras, Chap. 5). Mexico's relationship with GM crops takes place in a context of a high level of peasant and environmental opposition, espe-cially with respect to corn (Poitras, Chap. 11).

The outcome of many of these national dynamics remains to be seen. Our current data are still insufficient to predict the outcome of specific regional stratification patterns. Yet the cases presented here do provide enough evidence to flag a clear shifting of the international division of labor in agricultural production.

Emerging Patterns of Impact: Food for the Few

In our assessment of this collection of case studies we find overwhelm-ing evidence that the adoption of the technological paradigm of modern agriculture in general, and the addition of biotechnology to that model in particular, are having profound effects on agrarian social organization in developing countries. These changes are evident in the increasing social polarization, the growing dissimilarity and differential viability between commercial and subsistence production, the consequent uprooting of peasants, the replacement of traditional staples production with production for export, and the increased dependency on food imports.

Some would argue that social reorganization is a necessary component of technological change to production, and to avoid any change that has negative impacts would be to exclude any potential for good as well. In this situation, the rate of change may be the key to mitigating negative impacts, if the technology is deemed to have social utility. The changes brought about by agricultural technologies in an industrialized country, however, where only 2 to 3 percent of the economically active population engages in agriculture, are going to be drastically different than the social experience of that change in a developing country, where 25 to 60 percent of the economically active population engages in agricultural production. The difference in these percentages needs to be placed at the forefront when considering the "necessary" proletarianization of subsistence peasants who are no longer viable in the transition to commercial agriculture and neoliberal economic restructuring. In the developing-country experience represented here, poverty and hunger appear to be the most consistent companions of such impacts on social organization.

As Teubal (Chap. 8) clearly illustrates with respect to Argentina, hunger, in many situations, is less about food-production capacity than about food entitlement. Argentina's dramatic transition from one of the few developing countries self-sufficient in food to a country in crisis—where half the population is below the official poverty line and one quarter is so impoverished as to be unable to cover basic food needs—occurred in the context of increasing agricultural production. In sum, the question of hunger and whether biotechnology will, in fact, become the savior of the poor appears to be answered in these chapters with a resounding "No," at least in the current social and political context. While the technology is purportedly pro-poor, there is ample evidence that numerous factors preclude the poor from accessing its intended benefits. As we outline in our investigation of the Green Revolution (Chap. 2), increased food production does not necessarily lead to decreased hunger. Despite the Green Revolution's theoretical "scale neutrality," and despite its phenomenal success with respect to increasing cereal yields, hunger and poverty actually increased in many regions as a result of it. The social and economic context of the technology, not just its ability to increase food production, was revealed to be central to food security.

Manuel Poitras (Chap. 5) addressed the question of biotechnology's proposed scale neutrality directly, through an investigation of its potential benefits to small producers. Poitras's investigation of the modification of patent laws, the style of regulation in Mexico, and the impacts of shifts away from publicly funded research supports the conclusion that the overwhelming direction of biotechnology is not likely to positively affect small holders. The problems of technology dissemination are therefore clearly not the only factors contributing to increased social polarization, as in many cases these technologies are inapplicable to the needs of small producers. In the case of rBST use in Mexico, for example (Otero, Poitras, and Pechlaner, Chap. 7), no amount of the technology's dissemination would be likely to improve the milk productivity of cows on the smaller dual-purpose or seasonal farms, where proper nutrition is often a significant barrier to production.

The prospects for those agricultural producers outside of the increasingly globally integrated food system are consistently bleak. This may be due to partial technology diffusion, technology inapplicability, or international inequality (as will be discussed). Repeatedly, these case studies tell a story of declining subsistence and small-commodity producers with limited or no alternative means of support. In this respect, the addition of biotechnology to the technological package of modern agriculture has not changed the anti-poor technological dynamics of modern agricultural development in any way, other than to exacerbate its concomitant negative effects of capital concentration and centralization that the technology facilitates.

Biotechnology in the Latin American Environment

As noted in a number of these chapters, the environmental concerns raised by agricultural biotechnologies are diverse, ranging from food safety concerns to "superweeds" and other unexpected repercussions, with everything in between. Many of the environmental concerns raised about the technology lack a sufficient degree of scientific knowledge and certainty; but they are nonetheless serious in their potential. Two very significant environmental problems that are supported by a fair amount of scientific knowledge are biotechnology-specific exacerbations of problems initiated by the importation of the Green Revolution: petrochemical reliance and threats to biodiversity. As noted in Chapters 2 and 3, the position of herbicide tolerance as the most prevalent genetically modified trait has an indisputable impact on the further entrenchment of petrochemical reliance. This petrochemical dependency is arguably already unsustainable in the American agricultural system. The threat to biodiversity is much more multicausal—and graver—in Latin America, a region of biological origin of many plant species with great importance to world food production.

Genetic contamination, in itself, greatly expands on the threats that the Green Revolution's industrial monocropping introduced to biological diversity. Weaknesses in many Latin American state structures can mean that even when regulations are put in place—such as those prohibiting the cultivation of GM corn in Mexico—negative environmental repercussions (i.e., the contamination of landraces) still occur. The social control over biodiversity does not begin and end with regulation, however. These case studies of Latin America also reveal a proliferation of connections between the social and the environmental (e.g., McAfee, Chap. 3). At a most basic level, agricultural producers pushed to the edge of economic viability are going to engage in their practice in the most profitable way; environmental considerations are necessarily pushed to a second or third order of priorities. Sometimes the very adoption of the technology, with all its attendant environmental issues, is directly geared just to stay economically afloat in the face of restructuring, as exemplified by Brazilian farmers. Hisano and Altoé (Chap. 10) call this a "farmer's dilemma." Many of the links between social and environmental effects, however, are not so direct. Elizabeth Fitting (Chap. 6) demonstrates how even those actions seemingly not directly affecting a negative environmental response can result in long-term compromises to something as significant as in situ maize biodiversity in Mexico. Specifically, Fitting outlines how the adaptation strategies of rural households struggling under neoliberal reforms (i.e., out-migration and increased corn production) occur with accompanying risks to long-term corn production, and hence maintenance of the local landraces. Fitting calls these adaptation strategies the social aspect of biodiversity.

In all of these cases, the social context of biotechnology's diffusion can be seen to exacerbate its undesirable environmental impacts. At the same time, the negative social effects of its associated agricultural restructuring are decreasing the capacity for social debate over these impacts.

Global Governance and International Equity

Regionally our case studies hint at the internal differentiation that Latin America is undergoing with respect to different countries' responses to biotechnology and subsequent adoption into the global food market. Yet there still remain numerous commonalities among developing countries negotiating at the global policy table. Jansen and Roquas (Chap. 4) provide insight into the different ways that Latin American countries confronted the need for biotechnology regulations. While some countries were stimulated to respond by their early involvement with the industry, others were only prompted by becoming signatories to the Cartagena Protocol on Biosafety. The Cartagena Protocol aims to harmonize biosafety regulations; however, there is room for country-specific approaches to creating these regulations, and each country was found to arrive at its particular regulatory response through a number of different internal and external factors. The high cost and necessary expertise required for developing such a system, however, led many developing countries to rely on international—"absent expertise"—models and standards, rather than engaging in internal processes of lawmaking adapted to local conditions.

There are a number of consequences to this international modeling, which Jansen and Roquas argue tends to "forget" the weak state character of many Latin American governments. The "inadequacy of state performance" and "inability to enforce laws" that the authors outline is given full illustration with GMO prohibition in Brazil (Jepson et al., Chap. 9) and on the contamination of landraces in Mexico (Fitting, Chap. 6), where regulations and the enforcement of regulations are demonstrated to be two very separate things. Two other consequences of this international support speak more directly to the global power imbalance around such technologies: an emphasis on "sound science," which supplants the precautionary approach, and a silencing of public debate over the technology as a result of this emphasis (McAfee, Chap. 3). The disadvantage of developing countries in addressing their societal needs is exacerbated by a reliance on expert systems detached from local concerns and contexts (Jansen and Roquas, Chap. 4).

The "developing country disadvantage" regarding regulation manifests itself in various ways. First, as mentioned, there may be a considerable mismatch between regulation and enforcement. Second, the selection of regulations with social criteria in mind is usually absent, as it is the most powerful lobby groups that push for regulations that favor their economic interests. Third, the economic disadvantage of these countries also affects their ability to conduct the necessary assessments on the extent to which technologies are in their best interests. In some cases, the profitability and national benefit of the technology is itself in question (see Chaps. 3 and 7). In such cases, the economic benefits of the technology appear to be transferred almost wholly to its producers, while negative social and environmental effects imbalance the cost-benefit ratio. Stronger state involvement in assessment and regulation would be required to prevent these imbalances.

In a context of international support, these disadvantages could find eventual solutions. More ominously for the question of international equity, however, a significant part of international negotiations involves the betterment of the interests of industrialized countries. The neoliberal restructuring embraced in some of the developing countries discussed here is dubiously matched by a similar ideology in industrialized countries. Trade liberalization for developing countries in the context of hefty U.S., Japanese, and EU agricultural subsidies—whether direct subsidies or indirectly through non-agricultural subsidies, such as on environmental grounds—subjects developing countries to cheap imports against which local producers cannot compete. Similarly, changes in the international legal context of biotechnology, specifically the intellectual property rights component of the WTO, are leading to numerous inequalities and opening the door to biopiracy (Poitras, Chap. 5). In sum, even the limited treatment provided here seems to provide strong indication that the international legal context emerging around biotechnology is skewed against the benefit of developing countries.

Potential for Alternatives?

Returning to the introductory discussion of biotechnology's revolutionary potential for a moment, a number of conclusions can be drawn from our case studies. First, while agriculture may be a declining sector, particularly in developed countries, the importance of changes to this sector in developing countries is clearly demonstrated here to be profound. The increasingly universal model of a globally marketed, industrial "farming without farmers," as Teubal put it (Chap. 8), is not without deep social consequences in countries where this renders vast masses of people without an adequate livelihood in agriculture, or alternatives for gainful employment in other sectors of the national economies. Therefore, while biotechnology is indeed "substitutionist" in its current agricultural application, this has not minimized its polarizing socioeconomic impact on these countries. The increasing concentration of control over the food supply by a handful of TNCs, the imbalanced international context for trade and for intellectual property rights, the emphasis on production for export in a context where new biotechnologies could dramatically change the international division of labor—these all speak to the relative national weakness of developing countries. Locally, our case studies provide full testament to the technology's negative social and environmental impacts.

Even where the experience of biotechnology may not be direct, the integration of technology and policy in the context of neoliberal restructuring ensures that no producers are immune from its effects. Hisano and Altoé assess a number of responses to the agricultural restructuring in Brazil—attempts to integrate into the mainstream technological paradigm; differentiating into niche market production; reemphasizing production for local markets; and extension programs—to conclude that small family farmers are engaged in a "relentless survival game" (Chap. 10), which requires significant financial and human resources to mitigate.

Given the repository of negative consequences, it seems fair to question whether agricultural biotechnologies should be allowed at all. It should be emphasized, however, that the majority of authors here do not make claims for the inherent good or evil of biotechnology as a technology. They merely document the impacts of the technology in its current social, political, and economic context. In this context, the dissemination of the technology in Latin American countries has been highly problematic. Therefore, drawing on our collection of case studies, we now briefly turn our attention to the question of alternatives. Does there appear to be any potential for an alternative to the negative direction of modern agriculture in Latin America? Could a socially and environmentally sustainable agriculture, for example, still develop in those regions where the Green Revolution and biotechnology have not yet permeated? Or could a democratization of control over technology development and food policy direct biotechnology toward social and environmental aims?

In the global agricultural structure, with its emphasis on production for export, the voices of consumers in developed countries attentive to environmental and social sustainability could have an impact on the form of agricultural production in developing countries. While a positive prospect, this is weak with respect to country-specific evolution. A social debate regarding the proper integration of the new biotechnology policy would be necessary to establish its best integration into the social life of any particular country. Where these impacts of agricultural restructuring have been social polarization, increased poverty, and significant social dislocation—as the majority of these case studies on Latin America indicate—then the potential for the development of an alternative is weakened by the vulnerability of the actors involved.

One key issue remains unresolved, however, with respect to the future of genetic engineering (GE) technology, and that is its hegemonic status. As a number of these case studies have pointed out, there is high contestation around GE technology, and a broad resistance to it both for its environmental and its social implications. McAfee highlighted in Chapter 3 that there are increasing questions about the main productivity and profitability claims of new biotechnologies made by industry proponents. While yield increases had been expected to be dramatic, actual performance has been mediocre at best. The same result was found and discussed for rBST in Chap. 7. As Poitras (Chap. 11) conveys, GMOs entered the scene in Mexico just as rural producers felt they were being "set aside" for neoliberal competitiveness. In consequence, resistance to the technology was incorporated into the broader struggles of peasant social movements. In this light, the biggest impetus for biotechnology's adoption—sweeping neoliberal economic reforms and deregulation—may also be the biggest factor behind its inability to gain hegemonic status, as the number and diversity of groups negatively affected breeds significant resistance. While the potentials for a reversal or for the emergence of an alternative are far from assured, at the very least it is clear that GE technology is not undergoing a seamless adoption into the neoliberal project. Alternatives such as agroecology are not currently positioned as strong contenders to biotechnology; however, the high level of contestation around the implementation of GE technology and neoliberal restructuring adds a wild card to considerations of future possibilities. Progressive alternatives may evolve out of inroads to democratize the regulatory process and devise new roles for state intervention, putting the needs of small farmers on center stage.

Hisano and Altoé discussed the critical role that state expansion agencies can play in promoting education toward an alternative, agroecological approach to farming. In the case of Rio Grande do Sul, in southern Brazil, however, such intervention took place when a left-of-center, environmentally inclined political party was in government. The extent to which parties with different ideological leanings will continue to promote agroecology remains to be seen. It should be clear, though, that only bottom-up pressure and mobilization from peasant, consumer, and environmen-

talist groups will enhance the possibility that governments become responsive to promoting an equitable and sustainable agricultural model.

Agenda for Future Research

Last, we need to consider what areas of future research can help us assess
the socioeconomic impact of biotechnology in developing countries. We
have four suggestions for this. First, drawing on the first hints of regional
differentiation demonstrated here, we should further clarify the new
stratification of countries in the international division of labor, and determine which forms of integration to the world economy are most promising for the majority of the people in terms of the distribution of benefits from development. This would also involve the study of "structural
processes of technological innovation," which should combine structural
analysis with that of the protagonist actors in the economic dynamic of
developing countries. We suggest at least the following actors for close
scrutiny: transnational corporations, governments, local entrepreneurs,
and international agencies. What does the information era involve for the
changing relationships among these actors, and how will they distribute
the benefits of development? Is there any chance that at least intermediary countries, with substantial natural resources, can modify the neoliberal globalist paradigm currently guiding development?

Second, we should evaluate the potential of the systems of science and
technology in developing countries and their existing links with industry.
What is the extent of technology transfer, and how could this be further
promoted? Where scientific capacity or such links are non-existent, technological dependency or marginalization are inevitable. Where technology development and transfer can be contemplated, however, the question
also emerges as to how such development will be guided: by the needs of
the local people, to be defined according to some bottom-up democratic
process; or by the profitability needs of domestic and/or transnational
corporations?

Third, what is the character of legal structures in regard to intellectual
and industrial property? Do they promote or hinder the development of
a local biotechnology industry? To what extent have legal frameworks
become conformed to the requirements of neoliberal globalization? Assuming that such development has been profound, favoring the com-
moditization or privatization of formerly public commons, what are the
possibilities for returning some areas of legislation to the public realm?

Fourth, we need an analysis of the various industries which have been affected by products of biotechnology, and of the industries or institutions which have been charged with their dissemination. In contrast to the Green Revolution, which was promoted mostly by public and semi-public institutions, private industry has been the main promoter of biotechnology. Given its unequivocal interest in profit-maximization, it is likely that the impact of this new institutional framework will be even more socially polarizing than with the case of the Green Revolution. What are the conditions for developing public institutions and policies that might regulate and direct some of the research conducted by private firms, so that it fulfills a more public and socially oriented function? What are the conditions required to develop a bottom-up linkages approach in technology development, such that farmers and the consuming public may participate in decision making?

Moreover, it is also important to study the structure of industries that have adopted biotech's products. Their effects will vary depending on whether the industry is homogeneous (prevalence of large, medium, or small producers) or heterogeneous. The more heterogeneous the structure, the more polarizing biotech is likely to be if most of its products are scale-biased toward large-scale operations. Ideally, such studies should be done with a "dynamic equilibrium" approach, taking into account simultaneous changes in various sectors of the economy. Otherwise, our analysis will be incomplete and may lose sight of important effects of the new technology.

About the Contributors

SIMONE ALTOÉ obtained her Ph.D. from Hokkaido University in Japan and is now an independent consultant in Amsterdam. Her Ph.D. thesis is "Sustainable Rural Development and Institutional Support in Brazil: The Case of Small Soybean Farmers in the State of Rio Grande do Sul" (2004).

CHRISTIAN BRANNSTROM is Associate Professor of Geography at Texas A&M University. His research focuses on environmental governance and land-use and land-cover change in the Brazilian cerrado. He edited *Territories, Commodities and Knowledges*, a book on Latin American environmental history, and his recent research articles have appeared in *Geoforum, Geocarto International, Environment and History*, and *Bulletin of Latin American Research*.

ELIZABETH FITTING is an Assistant Professor in the Department of Sociology and Social Anthropology at Dalhousie University in Canada. Her work examines changing rural livelihoods and cultural practices under state reform and economic globalization. She has researched and published articles on the Mexican debates over GM corn as a focal point for competing narratives about neoliberal policy and globalization, and contextualized these debates in relation to the ways small-scale maize producers and migrants engage and contest such policies and processes in southeastern Puebla.

SHUJI HISANO is an Associate Professor in the Graduate School of Economics at Kyoto University, Japan. He works on the international political economy of food and agriculture, especially focusing on global biotechnology regulation. He has published many articles and book chapters about the issue as well as a book titled *Agribusiness and GMOs: Political Economy Approach* (2002, written in Japanese).

KEES JANSEN lectures at the Technology and Agrarian Development Group, Wageningen University, the Netherlands. Among his publications are *Agribusiness and Society: Corporate Responses to Environmentalism, Market Opportunities and Public Regulation* (co-editor; Zed, 2004) and *Political Ecology, Mountain Agriculture, and Knowledge in Honduras* (Thela, 1998), and journal articles in *Development and Change, The Journal of Peasant Studies,* and *World Development*. His research interests include the governance of high-risk, agricultural technologies, the sociology of new standards in global commodity chains (particularly environmental and fair trade labels), and the political ecology of agriculture and development. E-mail: Kees.Jansen@wur.nl.

WENDY JEPSON is an Assistant Professor of Geography at Texas A&M University, College Station. She received her Ph.D. (2003) in Geography from the University of California – Los Angeles. She has published on topics including the history of geography, globalization, biotechnology, and land-use and land-cover change in the Brazilian *cerrado* for journals such as *Political Geography, Environment and Planning C: Government and Policy, The Geographical Journal, Economic Geography,* and *Journal of Historical Geography*. Her research has been supported by the National Science Foundation (United States), Organization of American States, and Society for Women Geographers.

KATHLEEN MCAFEE is an economic geographer on the faculty of San Francisco State University's International Relations Department. She teaches about political economy and the global politics of agriculture, food, and the environment. She received her doctorate from the University of California at Berkeley after a career in international development. She has been a consultant to multilateral agencies and NGOs on biocultural diversity, biotechnology, and food sovereignty. Her current research in Mexico and other regions concerns "selling nature to save it"—the treatment of living things, genes, knowledge, and environmental services as marketable commodities. E-mail: kmcafee @sfsu.edu.

GERARDO OTERO is Professor of Sociology and Latin American Studies at Simon Fraser University. He is the author of *Farewell to the Peasantry? Political Class Formation in Rural Mexico* (Westview 1999; revised and expanded version published in Spanish by M. A. Porrúa in Mexico in 2004). He is editor of *Neoliberalism Revisited: Economic Restructuring and Mexico's Political Future* (Westview 1996) and *Mexico in Transition: Neoliberal Globalism, the State and Civil Society* (Zed Books, 2004; Spanish edition with M. A. Porrúa, 2006). He is completing research about health and safety conditions for Mexican migrant and Indo-Canadian immigrant agricultural workers in British Columbia, Canada, and his other project concerns the neoliberal food regime in North America: state policies, biotechnology, and the new division of labor. E-mail: otero@sfu.ca.

GABRIELA PECHLANER obtained her Ph.D. in sociology at Simon Fraser University in 2007, researching the social changes brought to agricultural production by the advent of biotechnology. Her dissertation focuses on the technology's proprietary aspects, particularly on the lawsuits emerging as a result of the use of transgenic crops in Canada and the United States. E-mail: gabepech@telus.net.

MANUEL POITRAS obtained his Ph.D. from York University in Toronto. His dissertation, *Engineering Genomes, Engineering Societies? Genetic Imperialism and the Politics of Biotechnology in Mexico*, won the 2003 York University's Dissertation Prize.

ESTHER ROQUAS is a research fellow at Rural Development Sociology Group, Wageningen University, the Netherlands. She is interested in policy and legal reforms in weak states, as well as the role played by expertise and knowledge claims. She has published various articles and a book, *Stacked Law: Land, Property and Conflict in Honduras* (2002). E-mail: Esther.Roquas@wur.nl.

RENATO STANCATO DE SOUZA worked for several years as a journalist reporting on agricultural markets for Agência Estado. He took a sabbatical in 2000–2001 to do a master's dissertation at the Institute of Latin American Studies (presently the Institute for the Study of the Americas), London, where he researched the anti-GM movement. After returning to Brazil to work as a journalist, he earned a position in the

Brazilian Foreign Relations Ministry, and is currently being trained at the Instituto Rio Branco in Brasília.

MIGUEL TEUBAL is a Professor of Economics at the University of Buenos Aires, Argentina, and a researcher with the National Council for Scientific and Technological Research (CONICET). He has published widely on the political economy of food, agriculture, and poverty in Latin America and Argentina. Some recent articles and books include "Expansión del modelo sojero en la Argentina: de la producción de alimentos a los commodities" (*Realidad Económica* 220, 2006); "Tierra y reforma agraria en América Latina" (*Realidad Económica* 220, 2003); *Agro y alimentos en la globalización: una perspectiva crítica* (with Javier Rodríguez; La Colmena, 2002); and *Globalización y expansión agroindustrial: ¿superación de la pobreza en América Latina?* (Ediciones El Corregidor, 1995).

Index

absentee expertise, 104–106, 109–110, 295
Aceitera General Deheza, 204
Action Aid Brasil, 230
Ad Hoc Intergovernmental Task Force on Foods Derived from Biotechnology, 93
ADM company, 55, 246
Africa: biosafety in, 70; and donations of U.S. transgenic maize, 66; Green Revolution in, 33; maize production in, 75; opposition of, to globalized intellectual property rights (IPR), 68; rBST approval in, 162
AGD company, 205
Agency for International Development, U.S., 67, 70
Agrevo, 128
agricultural biotechnology. *See* biotechnology
Agricultural Co-op of Cotrimaio, 254–256
Agricultural Co-operative of Integrada, 251–253
agriculture: alternative agriculture, 31–32, 34, 82–83, 245, 254–264; contract farming, 205, 208; disappearance of medium and small farmers and peasants in Argentina, 192, 206–210; and geographies of

difference distinguishing U.S. and developing countries, 35, 62, 74–81, 289; and irrigation, 33, 35, 43, 144, 145, 173–174; labor force participation in, 18; modern agriculture in U.S., 7–18, 32, 35; and mono-cropping, 196, 213, 294; organic agriculture, 31–32, 54, 254–256; productivity rate of, in U.S., 18–19; in Tehuacán Valley, Mexico, 144–152; in United States before 1860s, 7–8. *See also* biotechnology; Green Revolution; specific countries and specific crops
agro-biodiversity. *See* biodiversity
agrochemicals, 2, 34, 41–42
agroecology, 259–262
Agromod, 129
Ahmed, Iftikhar, 6
AIBA, 231–232
AID (Agency for International Development), 67, 70
ALIDES (Central American Alliance for Sustainable Development), 102
Almeida, Selene Maria de, 222–223
alternative agriculture, 31–32, 34, 82–83, 245, 254–264
Altieri, Miguel, 261
Altoé, Simone, 23–24, 243–264, 290–291, 294, 297, 298–299, 301
American Cyanamid, 164